Structural Geology Algorithms

Vectors and Tensors

State-of-the-art analysis of geological structures has become increasingly quantitative, but traditionally, graphical methods are used in teaching and in textbooks. Now, this innovative lab book provides a unified methodology for problem solving in structural geology using linear algebra and computation. Assuming only limited mathematical training, the book builds from the basics, providing the fundamental background mathematics, and demonstrating the application of geometry and kinematics in geoscience without requiring students to take a supplementary mathematics course.

Starting with classic orientation problems that are easily grasped, the authors then progress to more fundamental topics of stress, strain, and error propagation. They introduce linear algebra methods as the foundation for understanding vectors and tensors. Connections with earlier material are emphasized to allow students to develop an intuitive understanding of the underlying mathematics before introducing more advanced concepts. All algorithms are fully illustrated with a comprehensive suite of online MATLAB® functions, which build on and incorporate earlier functions, and which also allow users to modify the code to solve their own structural problems.

Containing 20 worked examples and over 60 exercises, this is the ideal lab book for advanced undergraduates or beginning graduate students. It will also provide professional structural geologists with a valuable reference and refresher for calculations.

RICHARD W. ALLMENDINGER is a structural geologist and a professor in the Earth and Atmospheric Sciences Department at Cornell University. He is widely known for his work on thrust tectonics and earthquake geology in South America, where much of his work over the past three decades has been based, as part of the Cornell Andes Project. Professor Allmendinger is the author of more than 100 publications and numerous widely used structural geology programs for Macs and PCs.

NESTOR CARDOZO is a structural geologist and an associate professor at the University of Stavanger, Norway, where he teaches undergraduate and graduate courses on structural geology and its application to petroleum geosciences. He has been involved in several multidisciplinary research projects to realistically include faults and their associated deformation in reservoir models. He is the author of several widely used structural geology and basin analysis programs for Macs.

DONALD M. FISHER is a structural geologist and professor at Penn State University, where he leads a structural geology and tectonics research group. His research on active structures, strain histories, and deformation along convergent plate boundaries has taken him to field areas in Central America, Kodiak Alaska, northern Japan, Taiwan, and offshore Sumatra. He has been teaching structural geology to undergraduate and graduate students for more than 20 years.

STRUCTURAL GEOLOGY ALGORITHMS

VECTORS AND TENSORS

RICHARD W. ALLMENDINGER
Cornell University, USA
NESTOR CARDOZO
University of Stavanger, Norway
DONALD M. FISHER
Pennsylvania State University, USA

CAMBRIDGE UNIVERSITY PRESS
Cambridge, New York, Melbourne, Madrid, Cape Town,
Singapore, São Paulo, Delhi, Tokyo, Mexico City

Cambridge University Press
The Edinburgh Building, Cambridge CB2 8RU, UK

Published in the United States of America by Cambridge University Press, New York

www.cambridge.org
Information on this title: www.cambridge.org/9781107012004

© Richard W. Allmendinger, Nestor Cardozo and Donald M. Fisher 2012

This publication is in copyright. Subject to statutory exception
and to the provisions of relevant collective licensing agreements,
no reproduction of any part may take place without the written
permission of Cambridge University Press.

First published 2012

Printed in the United Kingdom at the University Press, Cambridge

Internal book layout follows a design by G. K. Vallis

A catalog record for this publication is available from the British Library

Library of Congress Cataloging in Publication data
Allmendinger, Richard Waldron.
Structural geology algorithms : vectors and tensors / Richard W. Allmendinger,
Nestor Cardozo, Donald M. Fisher.
p. cm.
ISBN 978-1-107-01200-4 (hardback) – ISBN 978-1-107-40138-9 (pbk.)
1. Geology, Structural – Mathematics.   2. Rock deformation – Mathematical models.
I. Cardozo, Nestor.   II. Fisher, Donald M.   III. Title.
QE601.3.M38A45   2011
551.801′5181–dc23
2011030685

ISBN 978-1-107-01200-4 Hardback
ISBN 978-1-107-40138-9 Paperback

Additional resources for this publication at www.cambridge.org/allmendinger

Cambridge University Press has no responsibility for the persistence or
accuracy of URLs for external or third-party internet websites referred to
in this publication, and does not guarantee that any content on such
websites is, or will remain, accurate or appropriate.

Contents

Preface *page* ix

1 Problem solving in structural geology   1
   1.1 Objectives of structural analysis   1
   1.2 Orthographic projection and plane trigonometry   3
   1.3 Solving problems by computation   6
   1.4 Spherical projections   8
   1.5 Map projections   18

2 Coordinate systems, scalars, and vectors   23
   2.1 Coordinate systems   23
   2.2 Scalars   25
   2.3 Vectors   25
   2.4 Examples of structure problems using vector operations   34
   2.5 Exercises   43

3 Transformations of coordinate axes and vectors   44
   3.1 What are transformations and why are they important?   44
   3.2 Transformation of axes   45
   3.3 Transformation of vectors   48
   3.4 Examples of transformations in structural geology   50
   3.5 Exercises   65

4 Matrix operations and indicial notation   66
   4.1 Introduction   66
   4.2 Indicial notation   66
   4.3 Matrix notation and operations   69
   4.4 Transformations of coordinates and vectors revisited   77
   4.5 Exercises   79

5 Tensors — 81
- 5.1 What are tensors? — 81
- 5.2 Tensor notation and the summation convention — 82
- 5.3 Tensor transformations — 85
- 5.4 Principal axes and rotation axis of a tensor — 88
- 5.5 Example of eigenvalues and eigenvectors in structural geology — 91
- 5.6 Exercises — 97

6 Stress — 98
- 6.1 Stress "vectors" and stress tensors — 98
- 6.2 Cauchy's Law — 99
- 6.3 Basic characteristics of stress — 104
- 6.4 The deviatoric stress tensor — 112
- 6.5 A problem involving stress — 113
- 6.6 Exercises — 119

7 Introduction to deformation — 120
- 7.1 Introduction — 120
- 7.2 Deformation and displacement gradients — 121
- 7.3 Displacement and deformation gradients in three dimensions — 125
- 7.4 Geological application: GPS transects — 128
- 7.5 Exercises — 132

8 Infinitesimal strain — 135
- 8.1 Smaller is simpler — 135
- 8.2 Infinitesimal strain in three dimensions — 138
- 8.3 Tensor shear strain vs. engineering shear strain — 140
- 8.4 Strain invariants — 141
- 8.5 Strain quadric and strain ellipsoid — 142
- 8.6 Mohr circle for infinitesimal strain — 143
- 8.7 Example of calculations — 144
- 8.8 Geological applications of infinitesimal strain — 147
- 8.9 Exercises — 164

9 Finite strain — 165
- 9.1 Introduction — 165
- 9.2 Derivation of the Lagrangian strain tensor — 166
- 9.3 Eulerian finite strain tensor — 167
- 9.4 Derivation of the Green deformation tensor — 167
- 9.5 Relations between the finite strain and deformation tensors — 168
- 9.6 Relations to the deformation gradient, **F** — 169
- 9.7 Practical measures of strain — 170
- 9.8 The rotation and stretch tensors — 173
- 9.9 Multiple deformations — 176
- 9.10 Mohr circle for finite strain — 176
- 9.11 Compatibility equations — 178
- 9.12 Exercises — 180

10	**Progressive strain histories and kinematics**	**183**
10.1	Finite versus incremental strain	183
10.2	Determination of a strain history	199
10.3	Exercises	213

11	**Velocity description of deformation**	**217**
11.1	Introduction	217
11.2	The continuity equation	218
11.3	Pure and simple shear in terms of velocities	219
11.4	Geological application: Fault-related folding	220
11.5	Exercises	252

12	**Error analysis**	**254**
12.1	Introduction	254
12.2	Error propagation	255
12.3	Geological application: Cross-section balancing	256
12.4	Uncertainties in structural data and their representation	266
12.5	Geological application: Trishear inverse modeling	270
12.6	Exercises	279

References	**281**
Index	**286**

Preface

Structural geology has been taught, largely unchanged, for the last 50 years or more. The lecture part of most courses introduces students to concepts such as stress and strain, as well as more descriptive material like fault and fold terminology. The lab part of the course usually focuses on practical problem solving, mostly traditional methods for describing quantitatively the geometry of structures. While the lecture may introduce advanced concepts such as tensors, the lab commonly trains the student to use a combination of graphical methods, such as orthographic or spherical projection, and a variety of plane trigonometry solutions to various problems. This leads to a disconnect between lecture concepts that require a very precise understanding of coordinate systems (e.g., tensors) and lab methods that appear to have no common spatial or mathematical foundation. Students have no chance to understand that, for example, seemingly unconnected constructions such as down-plunge projections and Mohr circles share a common mathematical heritage: They are both graphical representations of coordinate transformations. In fact, it is literally impossible to understand the concept of tensors without understanding coordinate transformations. And yet, we try to teach students about tensors without teaching them about the most basic operations that they need to know to understand them.

The basic math behind all of these seemingly diverse topics consists of linear algebra and vector operations. Many geology students learn something about vectors in their first two semesters of college math, but are seldom given the opportunity to apply those concepts in their chosen major. Fewer students have learned linear algebra, as that topic is often reserved for the third or fourth semester math. Nonetheless, these basic concepts needed for an introductory structural geology course can easily be mastered without a formal course; we assume no prior knowledge of either. On one level, then, this book teaches a consistent approach to a subset of structural geology problems using linear algebra and vector operations. This subset of structural geology problems coincides with those that are usually treated in the lab portion of a structural geology course.

The linear algebra approach is ideally suited to computation. Thirty years after the widespread deployment of personal computers, most labs in structural geology teach students increasingly arcane graphical methods to solve problems. Students are taught the operations needed to solve orientation problems on a stereonet, but that does not teach them the mathematics of rotation. Thus, a stereonet, either digital or analog version, is nothing more than a graphical black box. When the time comes for the student to solve a more involved problem – say, the rotation of principal stresses into a fault plane coordinate system – how will they know how to proceed? Thus, on another level, one can look at this book as a structural geology lab manual for the twenty-first century, one that teaches students how to solve problems by computation rather than by graphical manipulation.

The concept of a twenty-first century lab manual is important because this book is not a general structural geology text. We make no attempt to provide an understanding of deformation, rather we focus on how to describe and analyze structures quantitatively. Nonetheless, the background and approach is common to that of modern continuum mechanics treatments. As such, the book would make a fine accompaniment to recent structural texts such as Pollard & Fletcher (2005) or Fossen (2010).

Chapter 1 provides an overview of problem solving in structural geology and presents some classical orientation problems commonly found in the lab portion of a structural geology course. Throughout the chapter (and the book) we make only a brief attempt to explain why a student might want to carry out a particular calculation; instead we focus on how to solve it. Chapters 2 and 3 focus on the critically important topic of coordinate systems and coordinate transformations. These topics are essential to the understanding of vectors and tensors. Chapter 4 presents a review (for some students, at least) of basic matrix operations and indicial notation, shorthand that makes it easy to see the essence of an operation without getting bogged down in the details. Then, in Chapter 5, we address head on the topic of what, exactly, is a tensor as well as essential operations for analyzing tensors. With this background, we venture on to stress in Chapter 6 and deformation in Chapters 7 to 11. In the final chapter, we address a topic that all people solving problems quantitatively should know how to do: error analysis. All chapters are accompanied by well-known examples from structural geology, as well as exercises that will help students grasp these operations. Allmendinger was the principal author of chapters 1–9, Cardozo of chapters 11–12, and Fisher of chapter 10. All authors contributed algorithms, which were implemented in MATLAB® by Cardozo. Any bug reports should be sent to him.

Many of the exercises involve computation, which is the ideal way to learn the linear algebra approach. Some of the exercises in the earlier chapters can be solved using a spreadsheet program, but, as the exercises get more complicated and the programs more complex, we clearly need a more functional approach. Throughout the book, we provide code snippets that follow the syntax of MATLAB® functions. MATLAB is a popular scientific computing platform that is specifically oriented towards linear algebra operations. MATLAB is an interpreted language (i.e., no compilation needed) that is easy to program, and from which results are easily obtained in numerical and graphical form. Teaching the basic syntax of MATLAB is beyond the scope of this book, but the basic concepts should be familiar to anyone who is conversant with any programming language. The first author programs in FORTRAN and the second in Objective C, however, neither has trouble reading the MATLAB code. Additionally, the code snippets are richly commented to help even the novice reader capture the basic approach. Many of these code snippets come directly from programs by the first two authors, which are widely used by structural geologists. Thus, on a third level, this book can be viewed as a sort of "Numerical Recipes" (Press *et al.*, 1986) for structural geology.

Many colleagues and students have helped us to learn these methods and have influenced our own teaching of these topics. Foremost among them is Win Means, whose own little book, *Stress and Strain* (Means, 1976), unfortunately now out-of-print, was the first introduction that many of our generation had to this approach. Win was kind enough to read an earlier copy of this manuscript. Allmendinger was first introduced to these methods through a class that used Nye's excellent and concise treatment (Nye, 1985). Classes, and many discussions, with Ray

Fletcher, Arvid Johnson, and David Pollard about structural geology were fundamental to forming his worldview. We thank generations of our students and colleagues who have learned these topics from us and have, through painful experience, exposed the errors in our problem sets and computer code. Allmendinger would especially like to thank Ben Brooks, Trent Cladouhos, Ernesto Cristallini, Stuart Hardy, Phoebe Judge, Jack Loveless, Randy Marrett, and Alan Zehnder for sharing many programming adventures. He is particularly grateful to the US National Science Foundation for supporting his research over the years, much of which led to the methods described here. Cardozo would like to thank Alvar Braathen, Haakon Fossen, and Jan Tveranger for their interest in the description and modeling of structures, and Sigurd Aanonsen for introducing inverse problems. Our families have suffered, mostly silently, with our long hours spent programming, not to mention in preparation of this manuscript.

CHAPTER

ONE

Problem solving in structural geology

1.1 OBJECTIVES OF STRUCTURAL ANALYSIS

In structural analysis, a fundamental objective is to describe as accurately as possible the geological structures in which we are interested. Commonly, we want to quantify three types of observations.

Orientations are the angles that describe how a line or plane is positioned in space. We commonly use either *strike* and *true dip* or true dip and *dip direction* to define planes, and *trend* and *plunge* for the orientations of lines (Fig. 1.1). The trend of the true dip is always at 90° to the strike, but the true dip is not the only angle that we can measure between the plane and the horizontal. An *apparent dip* is any angle between the plane and the horizontal that is not measured perpendicular to strike. For example, the angle labeled "plunge" in Figure 1.1 is also an apparent dip because line A lies in the gray plane. Strike, dip direction, and trend are all horizontal azimuths, usually measured with respect to the geographic north pole of the Earth. Dip and plunge are vertical angles measured downwards from the horizontal. Where a line lies in an inclined plane, we also use a measure known as the *rake* or the *pitch*, which is the angle between the strike direction and the line. There are few things more fundamental to structural geology than the accurate description of these quantities.

Whereas orientations are described using angles only, *magnitudes* describe how big, or small, the quantity of interest is. Magnitudes are, essentially, dimensions and thus have units of length, area, or volume. Some examples of magnitudes include the amplitude of a fold, the thickness of a bed, the length of a stretched cobble in a deformed conglomerate, the area of rupture during an earthquake, or the width of a vein. With magnitudes, size matters, whereas with orientations it does not.

The third type of observation compares both orientation and magnitude of something at two different times. The difference between an initial and a final state is known as *deformation*. Determining deformation involves measuring the feature in the final state and making

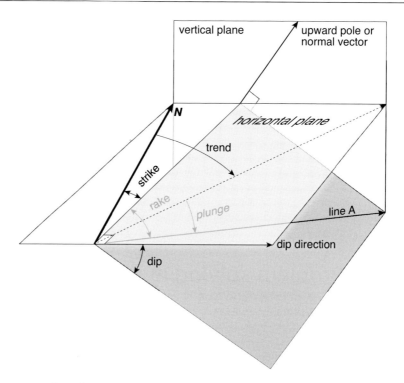

Figure 1.1 Three-dimensional perspective diagram showing the definition of typical structural geology terms. Strike and dip give the orientation of the gray plane with respect to geographic north (N) and the horizontal. Trend and plunge describe the orientation of line A. Because line A lies within the gray plane, we can specify the rake, the angle that the line makes with respect to the strike of the plane. The pole or normal vector is perpendicular to the plane. Note that because dip and plunge are measured from the horizontal, there is an implicit sign convention that down is positive and up negative.

inferences about its size, position, and orientation in the initial state. Deformation is commonly broken down into translation, rotation, and strain (or distortion) and each can be analyzed separately (Fig. 1.2), although when strains are large the sequence in which those effects are analyzed is important.

To determine orientations, magnitudes, or deformations, we need to make measurements. All measurements have some degree of *uncertainty*: is the length of that deformed cobble 10.0 or 10.3 cm? Is the strike of bedding on the limb of a fold 047° or 052°? In structural geology, the measurements that we make of natural, inherently irregular objects usually have a high degree of uncertainty. Typically, uncertainties, or *errors*, are estimated by making multiple measurements and averaging the result. However, we often want to calculate a quantity based on measurements of different quantities. *Error propagation* allows us to attach meaningful uncertainties to calculated quantities; this important operation is the subject of Chapter 12.

A complete structural analysis, of course, involves much more than just orientations, magnitudes, and deformations. These quantities tell us the "what" but not the "why." They may tell us that the rocks surrounding pyrite grains and curved pressure shadows suffered a rotation of 37° and a stretch of 2, but they tell us nothing about why the deformation occurred nor, for example, why the rocks surrounding the pyrite changed shape continuously whereas the pyrite itself did not deform at all. Nor does the fact that a thrust belt was shortened

1.2 Orthographic projection and plane trigonometry

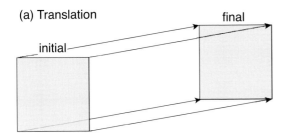

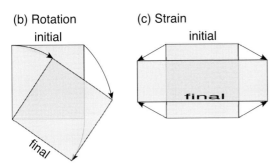

Figure 1.2 The three components of deformation – (a) translation, (b) rotation, and (c) strain – all require the comparison of an initial and a final state.

horizontally by 50% tell us anything about why the thrust belt formed in the first place. This complete understanding of structures is beyond the scope of this book, but the reader should never lose sight of the fact that accurate description based on measurements and their errors is just one aspect of a modern structural analysis.

1.2 ORTHOGRAPHIC PROJECTION AND PLANE TRIGONOMETRY

The methods we use to describe structures serve another purpose besides just providing an answer to a problem: They help us visualize complex, three-dimensional forms, thereby giving us a better intuitive understanding. Thus, many structural methods are graphical in nature, or are simple plane trigonometry solutions that have been derived from graphical constructions. Maps and cross sections constitute some of our most basic ways of graphically representing structural data and interpretations. Simpler graphical constructions using folding lines, front, side, and top views, etc. help us to visualize structures in three dimensions (Fig. 1.3). Until the 1980s, most structural geologists did not have knowledge of, or access to, the computing power needed to analyze complicated structural problems in any way except via graphical methods. Graphical methods, including spherical projection, were necessary to reduce complex three-dimensional geometries to two-dimensional sheets of paper.

Beginning structural geology students typically learn two types of graphical constructions: *orthographic* and *spherical* projections. In orthographic projection, one views the simple three-dimensional geometries as if they formed the sides of a box. Because one can only measure true angles with a protractor when looking perpendicularly down on the surface in which they occur, the sides of the box have to be unfolded before one can measure the angles of interest.

Consider the problem depicted in Figure 1.3: The gray plane has a strike, a true dip, δ, measured in a direction perpendicular to the strike, and an apparent dip, α, in a different direction. If one knows two out of the three quantities – the strike, true dip, and apparent dip – one can determine the third quantity. In orthographic projection, the true dip direction and the apparent dip direction are used as *folding lines*; they are literally like the creases on an unfolded

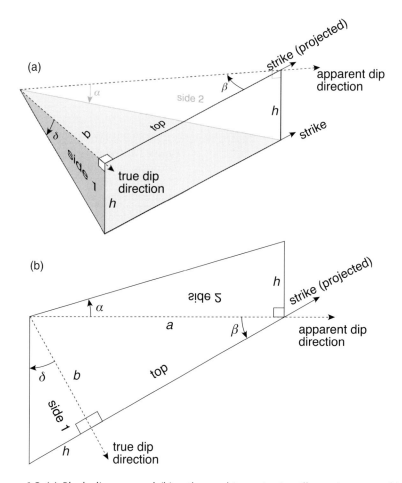

Figure 1.3 (a) Block diagram and (b) orthographic projection illustrating a graphical approach to the apparent dip problem. The dashed lines corresponding to the true and apparent dip directions are folding lines along which sides 1 and 2 have been folded up to lie in the same plane as the top of the block. h is the height of the block, which is the same everywhere along the strike line. δ, α, and β are the true dip, apparent dip, and angle between the strike and apparent dip directions, respectively.

cardboard box. By folding up the sides so that the top and the two sides all lie in the same plane, one can simply measure with a protractor whichever angle is needed.

The orthographic projection also provides the geometry necessary for deriving a simple trigonometric relationship that allows us to solve for the angle of interest by introducing a new angle from the top of the block (Fig. 1.3b): the angle between the strike and the apparent dip direction, β. Edge b of the top of the block is equal to

$$b = \frac{h}{\tan \delta}$$

The edge between the top and side 2, a is

$$a = \frac{b}{\sin \beta} = \frac{h}{\tan \delta \sin \beta}$$

And, from side 2 we get

$$a = \frac{h}{\tan \alpha}$$

Thus, using *plane trigonometry*, we can write the equation for the apparent dip:

$$\tan \delta \sin \beta = \tan \alpha \qquad (1.1)$$

where δ is the true dip, β the angle between the strike and the apparent dip direction, and α the apparent dip. Plane trigonometry works very well for simple problems but is more cumbersome, or more likely impossible, for more complex problems.

A different approach, which has the flexibility to handle more difficult computations, is spherical trigonometry. To visualize this situation, imagine that the plane in which we are interested intersects the lower half of a sphere (Fig. 1.4) rather than a box. In general, with power comes complexity, and spherical trigonometry is no exception. To calculate the apparent

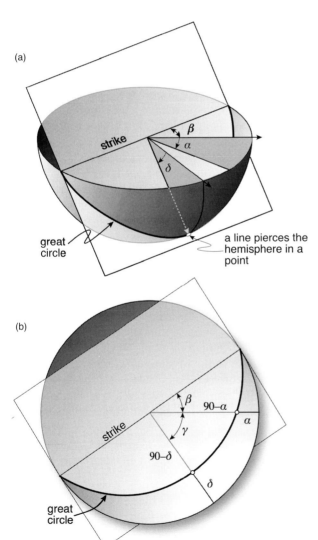

Figure 1.4 (a) Perspective view of a plane intersecting the lower half of a sphere. The angular relations are the same as those shown in Figure 1.3. The intersection of a sphere with any plane that goes through its center is a great circle. (b) Same geometry as in (a) but viewed from directly overhead as if one were looking down into the bowl of the lower hemisphere. View (b) was constructed using a stereographic projection. γ is the angle between true and apparent dip directions and other symbols are as in Figure 1.3.

dip, one must realize that, for the right spherical triangle shown (Fig. 1.4b), we know two angles (γ, which is the difference between the true and apparent dip directions, and the angle 90 because it is a true dip) and the included side (90 – the true dip δ). Thus, we can calculate the other side of the triangle (90 – the apparent dip α) from the following equation:

$$\cos \gamma = \tan(90 - \delta) * \cot(90 - \alpha) \tag{1.2}$$

A problem with both trigonometric methods is that one must guard against a multitude of special cases such as taking the tangent of 90°, the sign changes associated with sine and cosine functions, etc. On a more basic level, they give one little insight into the physical nature of what it is we are trying to determine. For most people, they are merely formulas associated with a complex geometric construction. And, the mathematical solution to this problem bears no obvious relation to other, more complicated problems we might wish to solve in structural geology.

1.3 SOLVING PROBLEMS BY COMPUTATION

One of the primary purposes of this book is to show you how to solve problems in structural geology by computation. There are many reasons for this emphasis: As a practicing geologist, you will use computer programs written by other people most of your professional life, so you should know how those programs work. Furthermore, computation is an important skill for any modern research scientist and allows you to solve problems that others cannot. Most importantly, the language of computation is linear algebra, and linear algebra is fundamental to developing a complete understanding of structures and continuum mechanics.

There are lots of different choices of computer platform and language that one could make. Perhaps simplest would be the humble spreadsheet program. In fact, many of the calculations that we ask you to do early in the book can easily be done in a spreadsheet program without even using its programming language (Visual Basic in the case of the popular program Excel). However, when you get to more complicated programs, spreadsheets are inadequate. Most commercial software these days is written in C, C++, or a variety of other platforms. In those programs, implementing the interface – that is, the windows, menus, drawing, dialog boxes, and so on – commonly takes up 95% or more of the lines of code. In this book, however, we want you to focus on the scientific algorithms rather than the interface.

Thus, we have chosen to illustrate this approach using the commercial software package, MATLAB®. Many universities now teach computer science and scientific computing using MATLAB, and many research geologists use MATLAB as their computing platform of choice. Because MATLAB is an interpreted language, it removes much of the fussiness of traditional compiled languages such as FORTRAN, Pascal, and C among a myriad of others. MATLAB also allows you to get results conveniently without worrying about the interface. You will be introduced to MATLAB in the next section, so we wanted to say a few general words about programming and syntax here.

First, programming languages, including spreadsheets and MATLAB, do trigonometric calculations in radians, not degrees. The relationship between radians and degrees is

$$\begin{aligned} 1 \text{ radian} &= \frac{180°}{\pi} = 57.295\,779\,513\,1° \\ 1° &= \frac{\pi}{180} = 0.017\,453\,292\,5 \text{ radians} \end{aligned} \tag{1.3}$$

1.3 Solving problems by computation

The four points of the compass – N, E, S, and W – can be defined in radians quite easily:

$$\begin{aligned} \textit{North } 0° &= 0 \text{ radians} = 360° = 2\pi \text{ radians} \\ \textit{East} &= 90° = \frac{\pi}{2} \text{ radians} \\ \textit{South} &= 180° = \pi \text{ radians} \\ \textit{West} &= 270° = \frac{3\pi}{2} \text{ radians} \end{aligned} \qquad (1.4)$$

Second, any good computer code should have explanatory comments that tell the reader what the program is doing and why. Comments are for humans and are totally ignored by the computer. In all computer languages, a special character precedes comments; in MATLAB, that character is %, the percent character. We have tried to use comments liberally in this book to help you understand what is going on in the functions we provide.

In all computer programs, the things to be calculated are held in *variables*. Variables can hold a single number, but they can also hold more complicated groups of numbers called *arrays*. The best way to think about arrays is that they represent a list of related data (in one dimension) or a table of related data in two dimensions. Mathematically, arrays are matrices. When one has their data in an array, repetitive calculations can be made very easily via what are known as *loops*. Let's say we need to add together 25 random numbers. We could write

```
x1 + x2 + x3 + x4 + x5 + x6 + x7 + ... + x22 + x23 + x24 + x25
```

Alternatively, one can do this calculation using an array and a loop:

```
x = randn(1,25); %x is an array of 25 random numbers
Sum = 0; %Initialize a variable to hold the sum of the array elements
for i=1:25 %Start of the loop. i starts at 1 and ends at 25
    Sum = Sum + x(i); %Add the current value x(i) to Sum
end %End of the loop
```

We will see later on in the book that the arrays and loops are what make the marriage of computing and linear algebra so seamless. Though the above example is trivial, arrays and loops will really help when we get to something like a tensor transformation that involves nine equations with nine terms each!

In computer programs, we can also select at run-time which operations or block of code are executed. We do this through the if control statement. Suppose we want to add the even but subtract the odd elements of array **x**. We can do this by modifying the loop above as follows:

```
for i=1:25 %Start of the loop. i starts at 1 and ends at 25
    if rem(i,2) == 0 %Start if statement. If remainder i/2=zero (i.e., even)
        Sum = Sum + x(i); %Add even element to Sum
    else %Else if odd element
        Sum = Sum - x(i); %Subtract odd element from Sum
    end %End of if statement
end %End of the loop
```

Finally (for now), many multi-step calculations are repeatedly used in a variety of contexts. Just as the tangent is used in both Equations 1.1 and 1.2, you can imagine more complicated calculations being used multiple times with different values. All programming languages have a variety of built-in functions, including trigonometric functions. The above code snippets use two such built-in functions: **randn**, which assigns random numbers to the array **x**, and **rem**, which determines the remainder of a division by an integer. Programming makes it easy to write your code in modular snippets that can be reused. You will see multiple examples in this book where one chunk of code, called a *function* in MATLAB and a function or *subroutine* in other languages, calls another chunk of code. Table 1.1 lists all of the MATLAB functions written especially for this book and shows which functions call, or are called by, other functions. All the functions follow the MATLAB help syntax. To get information about one of the functions, for example function **StCoordLine**, just type in MATLAB: `help StCoordLine`

1.4 SPHERICAL PROJECTIONS

The image in Figure 1.4b is known as a *spherical projection*, which is an elegant way of representing angular relationships on a sphere on a two-dimensional piece of paper. It should not be surprising that spherical projections are closely related to map projections, with the exception that in structural geology we use the lower hemisphere, as shown in Figure 1.4, whereas map projections use the upper hemisphere. Spherical projections are one of the most published types of plots in structural geology. They are used to carry out angular calculations such as rotations, apparent dip problems, and so on, as well as to present orientation data in papers and reports. Visualizing "stereonets," as they are commonly called, is one of the most important tasks a structural geology student can learn.

1.4.1 Data formats in spherical coordinates

Before diving in to stereonets, however, we need to examine briefly how orientations are generally specified in spherical coordinates (Fig. 1.5). In North America, planes are commonly recorded using their strike and dip. But, the strike can correspond to either of two

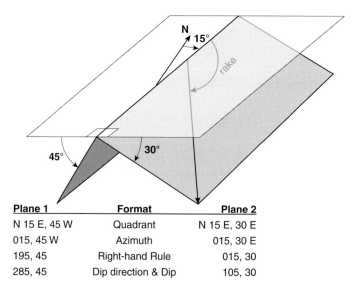

Figure 1.5 Common data formats for two planes that share the same strike but dip in opposite directions. Plane 1 is dark gray and plane 2 light gray. We do not recommend the quadrant format!

Chapter	Function	Description	Called by	Calls Function(s)
1	StCoordLine	Coordinates of a line in an equal angle or equal area stereonet of unit radius	GreatCircle, SmallCircle, Bingham, PTAxes	ZeroTwoPi
1	ZeroTwoPi	Constrains azimuth to lie between 0 and 2 radians	StCoordLine, CartToSph, Pole, SmallCircle, GeogrToView, Bingham, InfStrain	
2	SphToCart	Converts from spherical to Cartesian coordinates	CalcMV, Angles, Pole, Rotate, GeogrToView, Bingham, Cauchy, DirCosAxes, PTAxes	
2	CartToSph	Converts from Cartesian to spherical coordinates	CalcMV, Angles, Pole, Rotate, GeogrToView, Bingham, PrincipalStress, ShearOnPlane, InfStrain, PTAxes, FinStrain	ZeroTwoPi
2	CalcMV	Calculates the mean vector for a given series of lines		SphToCart, CartToSph
2	Angles	Calculates the angles between two lines, between two planes, etc.		SphToCart, CartToSph, Pole
2	Pole	Returns the pole to a plane or the plane that correspond to a pole	Angles, GreatCircle, Stereonet	ZeroTwoPi, SphToCart, CartToSph
3	DownPlunge	Constructs the down plunge projection of a bed		
3	Rotate	Rotates a line by performing a coordinate transformation	GreatCircle, SmallCircle	SphToCart, CartToSph
3	GreatCircle	Computes the great circle path of a plane in an equal angle or equal area stereonet of unit radius	Stereonet, Bingham, PTAxes	StCoordLine, Pole, Rotate
3	SmallCircle	Computes the paths of a small circle defined by its axis and cone angle, for an equal angle or equal area stereonet of unit radius	Stereonet	ZeroTwoPi, StCoordLine, Rotate
3	GeogrToView	Transforms a line from NED to view direction	Stereonet	ZeroTwoPi, SphToCart, CartToSph
3	Stereonet	Plots an equal angle or equal area stereonet of unit radius in any view direction	Bingham, PTAxes	Pole, GeogrToView, SmallCircle, GreatCircle
4	MultMatrix	Multiplies two conformable matrices		
4	Transpose	Calculates the transpose of a matrix		
4	CalcCofac	Calculates all of the cofactor elements for a 3×3 matrix	Determinant	
4	Determinant	Calculates the determinant and cofactors for a 3×3 matrix	Invert	CalcCofac

(cont.)

Chapter	Function	Description	Called by	Calls Function(s)
4	Invert	Calculates the inverse of a 3 × 3 matrix		Determinant
5	Bingham	Calculates and plots a cylindrical best fit to a pole distribution		ZeroTwoPi, SphToCart, CartToSph, Stereonet, StCoordLine, GreatCircle
6	Cauchy	Computes the tractions on an arbitrarily oriented plane	ShearOnPlane	DirCosAxes, SphToCart
6	DirCosAxes	Calculates the direction cosines of a right-handed, Cartesian coordinate system of any orientation	Cauchy, PrincipalStress, TransformStress	SphToCart
6	TransformStress	Transforms a stress tensor from old to new coordinates		DirCosAxes
6	PrincipalStress	Calculates the principal stresses and their orientations	ShearOnPlane	DirCosAxes, CartToSph
6	ShearOnPlane	Calculates the direction and magnitudes of the normal and shear tractions on an arbitrarily oriented plane		PrincipalStress, Cauchy, CartToSph
8	InfStrain	Computes infinitesimal strain from an input displacement gradient tensor	GridStrain	CartToSph, ZeroTwoPi
8	PTAxes	Computes the P and T axes from the orientation of fault planes and their slip vectors		SphToCart, CartToSph, Stereonet, GreatCircle, StCoordLine
8	GridStrain	Computes the infinitesimal strain of a network of stations with displacements in x and y		InfStrain
9	FinStrain	Computes finite strain from an input displacement gradient tensor		CartToSph
10	PureShear	Computes displacement paths and progressive finite strain history for pure shear		
10	SimpleShear	Computes displacement paths and progressive finite strain history for simple shear		
10	GeneralShear	Computes displacement paths and progressive finite strain history for general shear		
10	Fibers	Determines the incremental and finite strain history of a fiber in a pressure shadow		
11	FaultBendFold	Plots the evolution of a simple step, Mode I fault-bend fold		SuppeEquation
11	SuppeEquation	Equation 11.8 for fault-bend folding	FaultBendFold, FaultBendFold Growth	
11	SimilarFold	Plots the evolution of a similar fold		
11	FixedAxisFPF	Plots the evolution of a simple step, fixed axis fault-propagation fold		
11	ParallelFPF			SuppeEquationTwo

(cont.)

Chapter	Function	Description	Called by	Calls Function(s)
11	SuppeEquationTwo	Plots the evolution of a simple step, parallel fault-propagation fold	ParallelFPF, ParallelFPFGrowth	
11	Trishear	Equation 11.20 for parallel fault-propagation folding		
11	VelTrishear	Plots the evolution of a 2D trishear fault-propagation fold		VelTrishear
11	FaultBendFoldGrowth	Symmetric, linear v_x trishear velocity field	Trishear, BackTrishear	
11	FixedAxisFPFGrowth	Plots the evolution of a simple step, Mode I fault-bend fold and adds growth strata		SuppeEquation
11	ParallelFPFGrowth	Plots the evolution of a simple step, fixed axis fault-propagation fold and adds growth strata		
11	TrishearGrowth	Plots the evolution of a simple step, parallel fault-propagation fold and adds growth strata		SuppeEquationTwo
11	BalCrossErr	Plots the evolution of a trishear fault-propagation fold and adds growth strata		VelTrishear
12	BedRealizations	Computes shortening error in area balanced cross sections	RMLMethod	
12	CorrSpher	Generates realizations of a bed using a spherical variogram and the Cholesky method	BedRealizations	CorrSpher
12	BackTrishear	Calculates correlation matrix for a spherical variogram		
12	InvTrishear	Retrodeforms bed for the given trishear parameters and returns sum of square of residuals	InvTrishear	VelTrishear
12	RMLMethod	Inverse trishear modeling using a constrained, gradient-based optimization method	RMLMethod	BackTrishear
12		Runs a Monte Carlo type, trishear inversion analysis for a folded bed		BedRealizations, InvTrishear

Table 1.1 List of MATLAB functions in the book

directions 180° apart, and dip direction must be fixed by specifying a geographic quadrant. Some geologists use the quadrant format such as "N 37 W 43 SW" or "E 15 N 22 S" for recording the strike and dip (Fig. 1.5). Though a charming anachronism, students should always be encouraged to eschew this terminology and instead use "azimuth" notation by citing an angle between zero and 360°, which is much less prone to error as well as being easier to program. The same bearings as above, but in azimuth format, are "323°" and "075°"; note that we always use three digits for horizontal azimuths to distinguish them from vertical angles.

Two methods of recording the orientation of a plane avoid the ambiguity that arises from dip direction. First, one can record the strike azimuth such that the dip direction is always clockwise from it, a convention known as the *right-hand rule*. This tends to be the convention of choice in North America because it is easy to determine using a standard Brunton compass. A second method is to record the dip and dip direction, which is more common in Europe where the Freiberg compass makes this measurement directly. Of course, the pole also uniquely defines the plane, but it cannot be measured directly with either type of compass.

Lines are generally recorded in one of two ways. Those associated with planes are commonly recorded by their orientation with respect to the strike of the plane, that is, their *pitch* or *rake* (Fig. 1.1). Although this way is commonly the most convenient in the field, it can lead to considerable uncertainty if one is not careful because of the ambiguity in strike, mentioned above, and the fact that pitch can be either of two complementary angles. We follow the convention that the rake is always measured from the given strike azimuth. For example, the southeast-plunging line in the northerly striking, east-dipping plane shown in Figure 1.5 would have a rake of greater than 90° because the strike of the plane using the right-hand rule is 015°. The second method – recording the trend and plunge directly – is completely unambiguous as long as the lower hemisphere is always treated as positive. Vectors that point into the upper hemisphere (e.g., paleomagnetic poles) can simply be given a negative plunge.

As for the conventions used in this book, unless explicitly stated otherwise, planes will be given as strike and dip using the right-hand rule and lines will be given as trend and plunge, with a negative plunge indicating a vector pointing upward.

1.4.2 Using stereonets

In the most common type of stereonet, the outermost circle, or primitive, corresponds to the horizontal plane; it is the top edge of the bowl in Figure 1.4a. Compass azimuths, or horizontal bearings, are measured along the primitive. To plot a vertical angle such as a dip or a trend, one counts the degrees inward from the primitive (horizontal) towards the center of the net (vertical) along one of the two straight lines in the net. On a stereonet plot, lines are represented as points and planes trace out great circles (Fig. 1.4).

By measuring angles downwards from the horizontal, structural geologists implicitly employ a coordinate system in which down is positive. This is why structural geologists use a lower hemisphere projection. The lower hemisphere stereonet is particularly well suited to standard measurements made in the field with compass and clinometer. However, there is no reason why the lower hemisphere must be used, or why the primitive must represent the horizontal. Mineralogists, by convention, plot on the upper hemisphere, and the primitive can represent any planar surface. For example, it is often instructive to plot structure data on a cross section where the primitive represents the vertical plane of the section. The user or reader can then immediately see whether the features being plotted lie in the plane (i.e., plot along the primitive)

1.4 Spherical projections

or are oblique to the plane of the section. The operation that makes this possible in modern computer stereonet programs is the coordinate transformation, which we will see in Chapter 3.

Figure 1.6 illustrates the stereonet solution to the problem first introduced above in Figure 1.3. Let's say we know the strike and true dip of the plane (056°, 35° SE). One first rotates the stereonet until its great circles parallel the strike of the desired plane (Fig. 1.5a). Note that the geographic directions, N, E, S, W, do not rotate with the net because they are fixed to the construction. Then, the true dip δ is measured inwards $35°$ from the primitive along the straight line perpendicular to the great circles of the net. Finally, we draw the great circle that contains

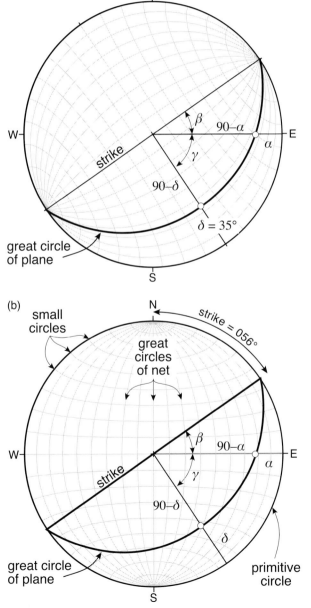

Figure 1.6 Same angular relations as in Figures 1.3 and 1.4 but now with the background of a typical equal area stereonet. To plot the plane of interest, we rotate the net (a) so that the great circles on the net are parallel to the strike of the plane of interest. In (b) we return the net so that its great circles are aligned with geographic north. Note the similarity of great circles with lines of longitude and small circles with lines of latitude. Symbols are as in Figures 1.3 and 1.4.

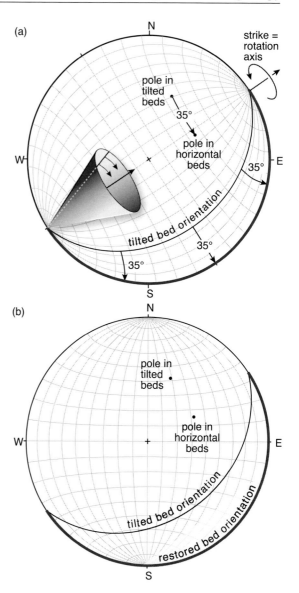

Figure 1.7 Rotating planes and lines on a stereonet. (a) Because the rotation axis is horizontal and parallel to the strike of bedding, we rotate the net so that its pole is parallel to the strike. The cone shows how a line oblique to the rotation axis sweeps out a conical trace as it rotates; small circles are conical sections on a sphere. (b) Same diagram, with the net restored so that its pole is parallel to the geographic poles.

both the strike and the true dip; because the true dip is 35° and the index great circles on the net in Figure 1.6 are spaced 10° apart, the great circle that represents our plane falls midway between the 30° and 40° index great circle. By restoring the net so that its great circles coincide with geographic north, we can now determine the apparent dip α along the EW straight line: 21°. You can verify that this is the correct answer by substituting the appropriate values into Equation 1.1.

Many structural geology lab manuals do an excellent job of describing step-by-step procedures for carrying out a great many operations using stereonets (e.g., Marshak and Mitra, 1988; Ragan, 2009) and we will not repeat those instructions here. There is, however, one operation that illustrates particularly well the power and limitations of stereonets for carrying out structural calculations: rotation of data. A line rotated about a rotation axis sweeps out a cone shape if the angle between the line and axis (the *apical angle*) is less than 90° (Fig. 1.7a inset)

1.4 Spherical projections

and sweeps out a plane if the angle is exactly 90°. The intersection of a sphere and a cone with its apex at the center of the sphere is a small circle and, as mentioned before, the intersection of a plane and the sphere is a great circle. This is, in fact, how computer stereonet programs draw the small circles and great circles that form the net. They take a point on the primitive at, say, 40° apical angle to the north–south axis and rotate it by equal increments through 180° until it arrives at the other side of the net, thus "drawing" the small circle at 40° to the pole. We will explore in more detail how to do this in Chapter 3, once we have developed an efficient way to do rotations.

The important concept for now is that points rotated on a stereonet follow small circle traces. Figure 1.7 shows bedding striking 056° and dipping 35° SE (same as before) but now we have added a line with a trend and plunge of 020, 45, which might, for example, represent a paleomagnetic pole measured in the rocks. For many reasons, we might want to see what the orientation of the paleomagnetic pole would have been when the rocks were horizontal, something practitioners call a fold test. To rotate bedding and everything else back to horizontal we define a horizontal rotation axis whose azimuth is parallel to the strike of bedding. Because the beds dip 35°, a 35° rotation about the strike will return the beds to horizontal. On a paper stereonet, we rotate the net with respect to the overlay until the great circles parallel the strike of the plane (Fig. 1.7a).

Here, however, we need to introduce an important formalism about the sign of the rotation and confront the ambiguity of the strike. By convention, positive rotations are clockwise when looking in the direction of the given azimuth of the rotation axis, and negative when counter-clockwise. The ambiguity arises because one can cite the strike as either 056° or 236° (that is, 180° away). Which do we use for the azimuth of our rotation axis? If you imagine looking in the direction 056° (Fig. 1.7), you can see that a clockwise rotation would produce a steeper dip, not a shallower dip (i.e., zero). So, we can specify −35° rotation about 056°, or +35° about 236°. On the stereonet, every point on the great circle that defines the plane moves 35° along the small circle until it reaches the primitive (and the bedding is horizontal). The paleomagnetic pole also moves 35° along the small circle that it occupies until it reaches the orientation that it would have had prior to the tilting of the rocks (assuming, of course that the magnetism happened before the folding!).

On a paper stereonet, one can only do rotations about horizontal axes because the small circles are concentric about the poles of the net. We use the small circles to determine graphically how any point will rotate. This was fine in the above problem because strike lines are by definition horizontal, but it creates headaches when rotation axes are not horizontal (or vertical). To get some feeling for the contortions required for rotations about inclined axes on paper stereonets, look at the description in any basic structure lab manual for how to determine bedding dip from three drill holes! In Chapter 3, we will develop the equations for how to do rotations about any axis.

1.4.3 How spherical projection works

As most practicing structural geologists now use a computer program to make spherical projections, we should ask the question, how does the computer know where to plot our precious data? Where on the x–y grid of a computer screen does the computer decide to plot the point that corresponds to a line we have measured in the field? How do spherical projections actually work? Recall that the purpose of a projection is to take data on a sphere and project them onto a two-dimensional piece of paper. There are two types of projections commonly in use in structural geology: the *equal angle* and *equal area* projections.

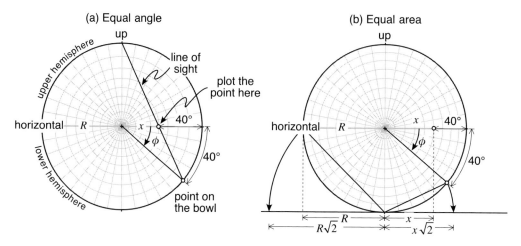

Figure 1.8 The angular relations involved in calculating the two spherical projections commonly used in structural geology: (a) equal angle (stereographic or Wulff) projection, and (b) equal area (Schmidt) projection. The primitives of both projections represent a vertical plane and thus these plots are perpendicular to a typical stereonet. R is the desired radius of the projection. The angle ϕ is the vertical angle we wish to plot (e.g., a plunge or a dip). Circles are at 10° increments; along the horizontal, their spacing is the same as it would be on a horizontal lower hemisphere projection.

The equal angle projection – also called a stereographic projection or Wulff net – is the simplest (Fig. 1.8a). We imagine that the viewer is at the top of the upper hemisphere looking straight down into the bowl of the lower hemisphere. We see where a line with a plunge of ϕ pierces the bowl and draw a straight line between the eye of the viewer and the point on the bowl. The point is plotted where that line of sight intersects the horizontal plane. The distance, x, from the center of the net of radius R is given by

$$x = R \tan\left(45 - \frac{\phi}{2}\right) \tag{1.5}$$

As you can see from Figure 1.8, this method of projection preserves angles perfectly and thus, on the horizontal, degrees are equally spaced. The preservation of angles has a downside: areas are distorted. Thus, for example, a ten by ten degree spherical cap plots as a smaller circle near the center of the net but is distorted to a larger circle near the edges (Fig. 1.9a). Distortion of areas was a significant problem when geologists tried to assess the density of points plotted on the projection, as was necessary back in the days of paper stereonets.

To address the area issue, the equal area, or Schmidt, net was introduced to structural geology (Fig. 1.8b). This construction produces point distributions that have the same density on the sphere as on the projection. A line with a plunge of ϕ plots a distance x from the center of a net of radius R, where x is given by

$$x = R\sqrt{2} \sin\left(45 - \frac{\phi}{2}\right) \tag{1.6}$$

The square root of 2 is a scaling factor to ensure that the original sphere and the projection have the same radius (Fig. 1.8b). The tradeoff, of course, is that angles are no longer preserved and conic sections, including great circles and small circles, are no longer true circles but fourth order quadrics (Fig. 1.9b). Because of the importance of analyzing concentrations of points,

1.4 Spherical projections

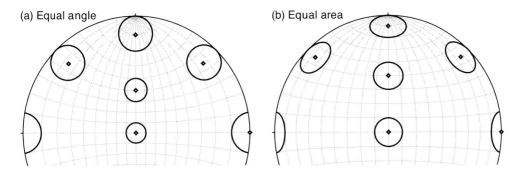

Figure 1.9 (a) The equal angle or Wulff net; (b) the equal area or Schmidt net. In both projections, spherical caps (small circles) of 10° radius have been plotted on various parts of the net to show the effect of the projection on the size and shape of the circle. Because both are lower hemisphere projections, a small circle that crosses the primitive plots on the opposite side of the net.

equal area nets have long been the stereonet of choice of the structural geologist. In reality, all modern stereonet programs contour on the sphere rather than on the projection so assessing densities is the same on both. Fortunately, the procedure to plot lines and planes is identical on both types of projection.

The MATLAB function **StCoordLine** below calculates Equations 1.5 and 1.6. It is followed by a commonly used helper function **ZeroTwoPi** whose sole purpose is to make sure that azimuths are always between 0 and 360° (zero and 2π in radians).

```
function [xp,yp] = StCoordLine(trd,plg,sttype)
%StCoordLine computes the coordinates of a line
%in an equal angle or equal area stereonet of unit radius
%
%   USE: [xp,yp] = StCoordLine(trd,plg,sttype)
%
%   trd = trend of line
%   plg = plunge of line
%   sttype = An integer indicating the type of stereonet. 0 for equal angle
%            and 1 for equal area
%   xp and yp are the coordinates of the line in the stereonet plot
%
%   NOTE: trend and plunge should be entered in radians
%
%   StCoordLine uses function ZeroTwoPi

% Take care of negative plunges
if plg < 0.0
    trd = ZeroTwoPi(trd+pi);
    plg = -plg;
end

% Some constants
piS4 = pi/4.0;
```

```
s2 = sqrt(2.0);
plgS2 = plg/2.0;

% Equal angle stereonet: From Equation 1.5 above
% Also see Pollard and Fletcher (2005), eq.2.72
if sttype == 0
    xp = tan(piS4 - plgS2)*sin(trd);
    yp = tan(piS4 - plgS2)*cos(trd);
% Equal area stereonet: From Equation 1.6 above
% Also see Pollard and Fletcher (2005), eq.2.90
elseif sttype == 1
    xp = s2*sin(piS4 - plgS2)*sin(trd);
    yp = s2*sin(piS4 - plgS2)*cos(trd);
end
end

function b = ZeroTwoPi(a)
%ZeroTwoPi constrains azimuth to lie between 0 and 2*pi radians
%
% b = ZeroTwoPi(a) returns azimuth b (from 0 to 2*pi)
% for input azimuth a (which may not be between 0 to 2*pi)
%
% NOTE: Azimuths a and b are input/output in radians

b=a;
twopi = 2.0*pi;
if b < 0.0
    b = b + twopi;
elseif b >= twopi
    b = b - twopi;
end
end
```

1.5 MAP PROJECTIONS

At first glance, the stereonet looks like our typical image of a globe and it has great circles and small circles that look, and in fact are, identical to lines of longitude and latitude. The tradeoffs we have just seen for stereonets – do we want to preserve areas or angles – are exactly those confronted when we want to make a flat map of a spherical body like the Earth. A full discussion of map projections and their subtleties is well beyond the scope of this book and there are excellent free sources of information available (Snyder, 1987). Nonetheless, given the importance of maps to geologists, and given their similarities to stereonets, they merit a brief mention here.

1.5.1 Map datum and projection

To make a map of the Earth, or some part of it, several considerations must be taken into account. Of prime importance is the fact that the Earth is not a sphere but an ellipsoid; its radius is 21 km smaller at the poles than at the Equator. Nor is the Earth a perfect ellipsoid but an irregular one that is best defined by the gravitational equipotential surface at sea level known as the *geoid*. The geoid is the reference level for elevations, not the ideal ellipsoid, but map coordinates are defined

1.5 Map projections   19

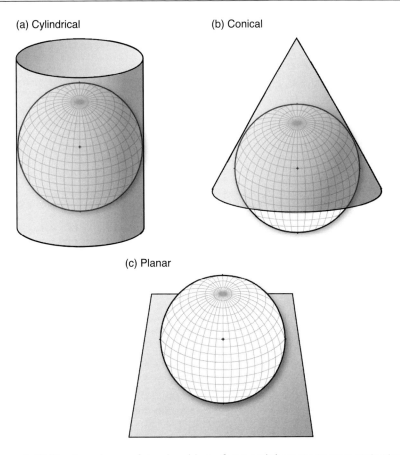

Figure 1.10 The three types of developable surfaces and their use in map projections. Within each category, there exist many different types of projections based on different mathematical formulae.

relative to an ideal ellipsoid. Because geoid anomalies deviate by no more than 100 m from an ideal ellipsoid, the difference is very small. As measurements of the shape of the Earth have improved over time, we have developed ellipsoids that more accurately represent the shape of the Earth. An ellipsoid model for part of the Earth, or the entire Earth, has to be matched with a *datum*, which is how you define horizontal position (latitude and longitude) and vertical elevation on the ellipsoid model. In the United States, we commonly use the 1980 Geodetic Reference System ellipsoid (GRS80), North American Datum of 1983 (NAD83) for the horizontal datum, and the National Geodetic Vertical Datum of 1929. These are being supplanted by the World Geodetic System 1984 standard (WGS84) with an ellipsoid slightly different than the GRS80 ellipsoid: The polar radius in the WGS84 system is 0.1 mm larger than that in GRS80!

The second major consideration is how to map the Earth to a flat surface. A *developable surface* is one that can be flattened without distortion; there are three that form the basis for most map projections (Fig. 1.10): a cylinder, a cone, or a plane (the first two, of course, must be sliced before they can be laid flat). Spherical projections, like the ones described in the last section, are projections onto a planar surface, sometimes also called *azimuthal projections*. For that reason, they can only show, at most, one hemisphere or the other. In addition to these types of developable surfaces, map projections may be calculated to preserve shape and have the

Projection	Type	Conformal	Equal area	Equi-distant	True direction	Scale
Globe	Sphere	yes	yes	yes	yes	World
Mercator	Cylindrical	yes			partly	Regional and smaller
Miller	Cylindrical					World
Orthographic	Azimuthal				partly	Hemisphere
Stereographic	Azimuthal	yes			partly	Hemisphere to local
Lambert azimuthal equal area	Azimuthal		yes			Hemisphere to regional
Albers equal area conic	Conic		yes			Sub-hemisphere to regional
Lambert conformal conic	Conic	yes			partly	Regional to local
Equidistant conic	Conic			partly		Regional

Table 1.2 Common map projections

same scale in every direction locally (conformal), preserve area (equal area), show the correct distance between a point at the center and any other point (equidistant), or show true directions locally. Planar maps cannot be both conformal and equal area, nor can they be equal area and equidistant. We see this in the case of our two structural projections: In the equal angle Wulff net (Fig. 1.9a), circles are true circles everywhere (conformal) but they get larger with distance from the center of the projection, even though they are the same area on the sphere (not equal area). In the Schmidt net (Fig. 1.9b), areas are the same everywhere (equal area), but true circles result only at the exact center of the net and are distorted everywhere else (not conformable).

Except for globes, which are inconvenient to carry around and are suitable only for continental or oceanic scale, all map projections represent a tradeoff on these characteristics. Therefore the choice of map projection depends on the needs of the mapmaker and user (see USGS Eastern Region, 2000). Table 1.2 summarizes some common projection types, their attributes, and their appropriate scale of usage.

1.5.2 The UTM projection

One type of projection merits special notice, particularly because this book focuses on rectangular Cartesian coordinate systems. The *Universal Transverse Mercator* (UTM) projection yields a map with rectangular coordinates in distance (meters). Except close to the poles, the Earth is divided into 60 zones, each 6° of longitude wide. Zone 1 lies between 180° and 174° W longitude, and the zones increase eastward. The $x - y$ coordinate system in each zone is defined by the central meridian – the longitude halfway between the edges of the zone – and the Equator. For example, zone 31 is located between 0° and 6° E; its central meridian is 3° E longitude (Fig. 1.11). For the Northern Hemisphere, the central meridian is assigned a value of 500 000 m in the x, or *eastings*, direction, and a point along the Equator is given a value of zero meters in the y, or *northings* direction. In our zone 31 example, above, a point lying 2500 km north of the Equator at 3° E would have an easting of 500 000 m and a northing of 2 500 000 m. For points in the Southern Hemisphere, the easting value is the same – i.e., 500 000 m along the central meridian of the zone – but the Equator is assigned a northing value of 10 000 000 m (Fig. 1.11b).

1.5 Map projections

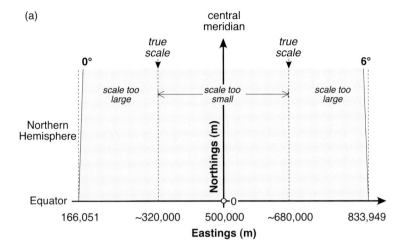

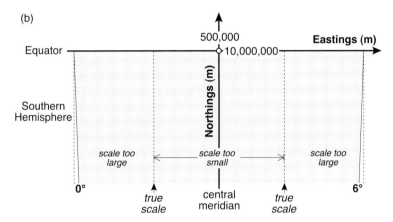

Figure 1.11 Relationship between geographic coordinates and UTM coordinates. The gray area shows UTM zone 31, located between 0 and 6° E longitude. (a) In the Northern Hemisphere, the origin defined by the central meridian (in this case, 3° E) and the Equator is assigned a value of (500 000 m, 0 m). (b) In the Southern Hemisphere, the central meridian still has a value of 500 000 m, but the Equator is assigned a y-value of 10 000 000 m, a false northing. In both hemispheres east and north are both positive directions. A zone narrows both northward and southward away from the Equator. The coordinates are reset for each zone so that, for example, 9° E longitude, the central meridian for zone 32, also has an x-value of 500 000 m.

If our point along the central meridian in zone 31 were 2500 km south of the Equator, it would have UTM coordinates 500 000 m east and 7 500 000 m north.

A couple of things might strike you as odd in this scheme: First, why isn't the origin of the coordinate system defined by the western edge of the zone? Why use the central meridian and assign it a seemingly arbitrary value of 500 000 m? The reason is that the distance between the lines of longitude that define a zone is not constant but is greatest at the Equator (668 km) and diminishes towards the poles; at 6° N a zone is about 3.5 km narrower than it is at the Equator. By using the central meridian and assigning it a large positive value, we ensure that the x, or easting, value within the zone is always positive. If you were located 250 km east of the central meridian at the Equator, you would still be in the zone, but if you were 250 km east of the same central meridian at, say, 60° N, you would actually be in the next zone to the east!

Second, why is the Equator assigned a value of 10 000 000 m for points in the Southern Hemisphere? This assignment enables us to assign positive y values (sometimes called *false northings*) to all points in the Southern Hemisphere because the distance from pole to the Equator is about 10 000 km. In both hemispheres, north and east are always positive. Because points in both Northern and Southern Hemispheres can have the same northing values, it is necessary to specify the hemisphere when converting between UTM and another coordinate system. Depending on the program used, one identifies the hemisphere in different ways. For example, a point in the Southern Hemisphere in Chile could be recorded as zone −19 or as zone 19S. Using the letters "S" or "N" to identify hemispheres has its pitfalls, however, because an older more complicated version of the UTM system uses letters to define latitudinal ranges. Thus, "19S" could mean Chile or it could mean a point between 32° and 40° N latitude, in the Atlantic Ocean offshore to the eastern United States!

Northings values are always less than 10 000 000 m because the UTM system is only applied to 84° N latitude and 80° S latitude. In the polar regions, the Universal Polar Stereographic (UPS) coordinate system is used, instead. The UTM zones are very regular around the globe except for the region of eastern Norway and the island of Spitsbergen, where they have been broadened to accommodate the local geography (Snyder, 1987).

The UTM coordinate system will be very useful when we want to calculate strains over a wide area. We will see the application, in particular, to determining strain rates in a geodetic GPS data set in Chapter 8. The equations for converting between longitude–latitude and UTM are relatively straightforward but tedious (Snyder, 1987). Fortunately, MATLAB has built-in functions to do the conversion for us. If you have the MATLAB Mapping Toolbox, you can use functions **mfwdtran** and **minvtran** to convert from lat-long to UTM and vice versa. Alternatively, you can check the MATLAB Central File Exchange, a website where MATLAB users share code. The Geodetic Toolbox by Michael R. Craymer, which can be downloaded from this site, has functions to do the conversions.

CHAPTER

TWO

Coordinate systems, scalars, and vectors

2.1 COORDINATE SYSTEMS

Virtually everything we do in structural geology explicitly or implicitly involves a coordinate system. When we plot data on a map each point has a latitude, longitude, and elevation. Strike and dip of bedding are given in azimuth or quadrant with respect to north, south, east, and west and with respect to the horizontal surface of the Earth. In the western United States, samples may be located with respect to township and range. We may not realize it, but more informal coordinate systems are used as well, particularly in the field. The location of an observation or a sample may be described as "1.2 km from the northwest corner fence post and 3.5 km from the peak with an elevation of 3150 m at an elevation of 1687 m."

A key aspect, but one that is commonly taken for granted, of all of these ways of reporting a location is that they are interchangeable. The sample that comes from near the fence post and the peak could just as easily be described by its latitude, longitude, and elevation or by its township, range, and elevation. Just because we change the way of reporting our coordinates (i.e., change our coordinate system), it does not mean that the physical location of the point in space has changed. This seems so simple as to be trivial, but we will see in Chapter 5 that this ability to change coordinate systems without changing the fundamental nature of what we are studying is essential to the concept of tensors.

2.1.1 Spherical versus Cartesian coordinate systems

As described in Chapter 1, because the Earth is nearly spherical, it is most convenient for structural geologists to record their observations in terms of spherical coordinates. Although a spherical coordinate system is the easiest to use for collecting data in the field, it is not the simplest for accomplishing a variety of calculations that we need to perform. One gets an inkling of this from the fact that, in continuum mechanics texts, spherical coordinates are usually

24 Coordinate systems, scalars, and vectors

presented and applied towards the back of the book or in an appendix – not exactly front page material. Far simpler, both conceptually and computationally, are rectangular Cartesian coordinates that are composed of three mutually perpendicular axes. Normally, one thinks of plotting a point by its distance from the three axes of the Cartesian coordinate system. As we will see below, a feature can equally well be plotted by the angles that a vector, connecting it to the origin, makes with the axes. If the portion of the Earth we are studying is sufficiently small so that our horizontal reference surface is essentially perpendicular to the radius of the Earth, then we can solve many different problems in structural geology, simply and easily, by expressing them in terms of Cartesian, rather than spherical, coordinates. Before we can do this, however, there is an additional aspect of coordinate systems that we must examine.

2.1.2 Right-handed and left-handed coordinate systems

The way that the axes of coordinate systems are labeled is not arbitrary. In the case of the Earth, it matters whether we consider a point that is below sea level to be positive or negative. "That's crazy," you say, everybody knows that elevations above sea level are positive! If that were the case, then why do structural geologists commonly measure positive angles downward from the horizontal? Why is it that mineralogists use an upper hemisphere stereographic projection whereas structural geologists use the lower hemisphere? The point is that it does not matter which is chosen so long as one is clear and consistent. Some simple conventions in the labeling of coordinate axes insure that consistency.

Coordinate systems can be of two types. Right-handed coordinates are those in which, if you hold your hand with the thumb pointed from the origin in the positive direction of the first axis, your fingers will curl from the positive direction of the second axis towards the positive direction of the third axis (Fig. 2.1a). A left-handed coordinate system would function the same except that the left hand is used. To make the coordinate system left handed, simply reverse the positions of the X_2 and X_3 axes as in Figure 2.1b. All this may seem academic and not very useful, but we will see in Chapter 3 that there are certain types of operations known as transformations which can change the sense of a coordinate system from right to left handed or vice versa. By convention, the preferred coordinate system is a right-handed one and that is the one we will use.

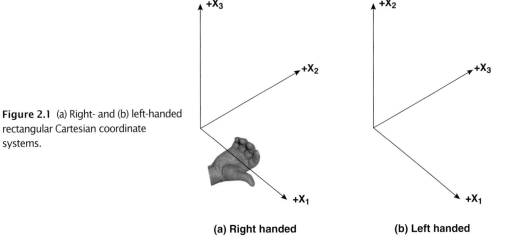

Figure 2.1 (a) Right- and (b) left-handed rectangular Cartesian coordinate systems.

2.3 Vectors

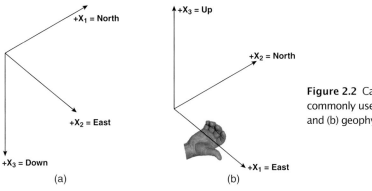

Figure 2.2 Cartesian coordinates commonly used in (a) structural geology and (b) geophysics and topography.

2.1.3 Cartesian coordinate systems in geology

Now we can return to the topic at the end of Section 2.1.1, that is, what Cartesian coordinate systems are appropriate to geology? Sticking with the right-handed convention, there are two obvious choices, the primary difference being whether one regards up or down as positive (Fig. 2.2).

In general, the north-east-down (**NED**) convention is more common in structural geology where positive angles are measured downwards from the horizontal. In geophysics, as well as in geographic maps, the east-north-up convention is more customary; after all, elevation above sea level is commonly treated as positive. Note that these are not the only possible right-handed coordinate systems. For example, west-south-up is also a perfectly good right-handed system, although this and all the other possible combinations are seldom used. In the rest of this book, unless otherwise stated, we will use the north-east-down convention. We will see how to convert between spherical and Cartesian coordinates in Section 2.3.7.

2.2 SCALARS

Scalars are the simplest physical component we will deal with. They are nothing more than the value – or magnitude – of some property at any particular point in space. Scalars are independent of coordinate system and furthermore they have the same value regardless of the coordinate system. As we will see in the following sections, this is not true for vector components. Common examples of scalar quantities are temperature, mass, density, volume, or energy. There is no direction associated with these properties, they simply exist in space and would have the same numerical value, regardless of whether one uses spherical or Cartesian coordinates or even Farmer Joe's northwest corner fence post.

2.3 VECTORS

Vectors form the basis for virtually all structural calculations so it is important to develop a very clear, intuitive feel for them. Vectors are physical quantities that have a magnitude and a direction; they can be defined only with respect to a given coordinate system, which is why we developed the idea of coordinate systems early in this chapter. Displacement, velocity, temperature gradient, and force are all common examples of vectors. For example, it does not make any sense to think about your velocity unless you know in what direction you are going.

2.3.1 Vectors vs. axes

All geological orientations have a direction in space with respect to a given coordinate system so all are vectors. However, for many calculations, it makes no difference on which end of the line you put the arrow. Thus, we make an informal distinction between vectors, which are lines with a direction (i.e., an arrow at one end of the line) and axes, which are lines with no directional significance. For example, think about the lineation that is made by the intersection between cleavage and bedding. That line, or axis, certainly has a specific orientation in space and is described with respect to a coordinate system, but there is no difference between one end of the line and the other. The hinge – or axis – of a cylindrical fold is another example of a line that has no directional significance. In both of these examples, we could be very systematic as to how we collected or calculated the data such that the arrow of the vector always pointed in a consistent direction, but it is seldom worth the trouble. Some common geological examples of vectors that cannot be treated as axes are the slip on a fault (i.e., displacement of piercing points), paleocurrent indicators (flute cast, etc.), and paleomagnetic poles.

When structural geologists use a lower hemisphere stereographic projection exclusively, *we are automatically treating all lines as axes*. To plot lines on the lower hemisphere, we arbitrarily assume that all lines point downwards. Generally, this is not an issue, but consider the problem of a series of complex rotations involving paleocurrent directions. At some point during this process, the current direction may point into the air (i.e., the upper hemisphere). If we force that line to point into the lower hemisphere, we have just reversed the direction in which the current flowed! Commonly, poles to bedding are treated as axes as, for example, when we make a π-diagram. This, however, is not strictly correct. There are really two bedding poles, the vector that points in the direction that strata become younger, and the vector that points towards older rocks.

Despite this difference between vectors and axes, there are few problems treating an axis as a vector for the purposes of the calculations that we will describe below, with a few exceptions (see Section 2.4.1). The potential problems are far greater treating vectors as axes than axes as vectors.

2.3.2 Basic vector notation

Clearly, with two different types of quantities around (scalars and vectors), we need a shorthand way to distinguish between them in equations. We will write scalars, and scalar components of vectors, in italics. Vectors in these notes are shown in lower case with bold face print (which is sometimes known as symbolic or Gibbs notation):

$$\vec{v} = \mathbf{v} = [\begin{matrix} v_1 & v_2 & v_3 \end{matrix}] \tag{2.1}$$

The above notation is common in linear algebra books but can be confusing because it seems to equate a vector with three scalars. Here is what it really means: Vectors in three-dimensional space can be described by three scalar components, indicated above as v_1, v_2, and v_3. In a Cartesian coordinate system, they give the magnitude of the vector in the direction of, or projected onto each of the three axes (Fig. 2.3). We will continue to use that notation; when the reader encounters it, they should interpret the equal sign as "has scalar components in the

2.3 Vectors

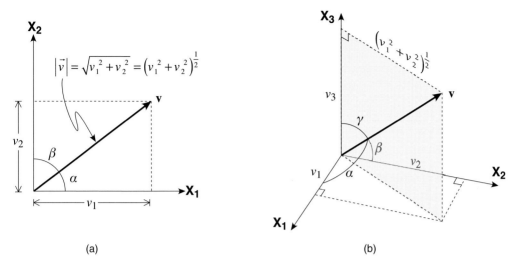

Figure 2.3 Components of a vector in Cartesian coordinates (a) in two dimensions and (b) in three dimensions.

current coordinate system of …" Because it is tedious to write out the three components all the time a shorthand notation, known as indicial notation, is commonly used:

$$v_i, \text{ where } i = 1, 2, 3 \tag{2.2}$$

The power of this sort of notation will be explained more fully in Chapter 4.

2.3.3 Magnitude of a vector

The magnitude of a vector is, graphically, just the length of the arrow. It is a scalar quantity and is, therefore, generally marked in regular weight, italicized type. If there is any ambiguity, then vertical bars will be used to definitively indicate the magnitude. In two dimensions (Fig. 2.3a), it is quite easy to see that the magnitude of vector **v** can be calculated from the Pythagorean Theorem (the square of the hypotenuse is equal to the sum of the squares of the other two sides). This is easily generalized to three dimensions (Fig. 2.3b), yielding the equation for the magnitude of a vector:

$$v = |\mathbf{v}| = \left(v_1^2 + v_2^2 + v_3^2\right)^{1/2} \tag{2.3}$$

2.3.4 Unit vector and direction cosines

A *unit vector* is just a vector with a magnitude of one and is indicated by a hat: $\hat{\mathbf{v}}$. Any vector can be converted into a unit vector parallel to itself by dividing the vector (and its components) by its own magnitude:

$$\hat{\mathbf{v}} = \frac{\mathbf{v}}{|\mathbf{v}|} = \left[\frac{v_1}{|\mathbf{v}|} \quad \frac{v_2}{|\mathbf{v}|} \quad \frac{v_3}{|\mathbf{v}|}\right] \tag{2.4}$$

If you carefully inspect Figure 2.3, you will see that the cosine of the angle that a vector makes with a particular axis is just equal to the component of the vector along that axis divided by the magnitude of the vector. Thus we get

$$\cos\alpha = \frac{v_1}{|\mathbf{v}|} \qquad \cos\beta = \frac{v_2}{|\mathbf{v}|} \qquad \cos\gamma = \frac{v_3}{|\mathbf{v}|} \tag{2.5}$$

Substituting Equation 2.5 into Equation 2.4 we see that a unit vector can be expressed in terms of the cosines of the angles that it makes with the axes. These scalars are known as *direction cosines*:

$$\hat{\mathbf{v}} = [\cos\alpha \quad \cos\beta \quad \cos\gamma] \tag{2.6}$$

2.3.5 Direction cosines and structural geology

The concept of a unit vector is particularly important in structural geology where we so often deal with orientations, but not sizes, of planes and lines. Any orientation can be expressed as a unit vector, whose components are the direction cosines. For example, in a north-east-down coordinate system, a line that has a 30° plunge due east (090°, 30°) would have the following components (Fig. 2.4):

$$\cos\alpha = \cos 90° = 0.0 \; (\alpha \text{ angle with respect to north})$$
$$\cos\beta = \cos 30° = \sqrt{3}/2 \; (\beta = \text{angle with respect to east})$$
$$\cos\gamma = \cos(90° - 30°) = 0.5 \; (\gamma = \text{angle with respect to down})$$

or simply

$$[\cos\alpha \quad \cos\beta \quad \cos\gamma] = [0.0 \quad \sqrt{3}/2 \quad 0.5]$$

For the third direction cosine, recall that the angle is measured with respect to the vertical, whereas plunge is given with respect to the horizontal. We will use direction cosines extensively to describe the orientation of lines in Cartesian coordinates, and then see how to convert from spherical to Cartesian coordinates in the sections that follow.

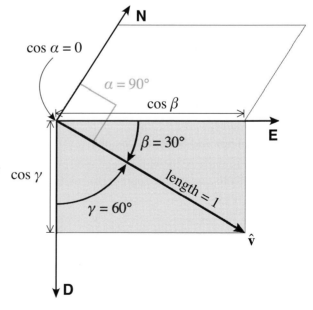

Figure 2.4 Orientation of a line (unit vector) lying in a vertical, east–west plane in gray and its representation by direction cosines.

2.3 Vectors

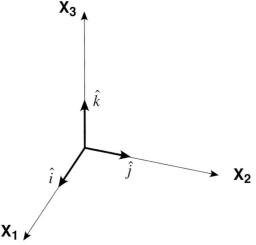

Figure 2.5 Three mutually perpendicular or "orthonormal" base vectors.

2.3.6 Base or reference vectors

Occasionally, it is convenient to represent the axes of a Cartesian coordinate system by three mutually perpendicular unit vectors known as base vectors (Fig. 2.5). Any vector can be expressed in terms of the base vectors for the coordinate systems by multiplying the components of the vector by the corresponding base vector:

$$\mathbf{v} = v_1 \hat{\mathbf{i}} + v_2 \hat{\mathbf{j}} + v_3 \hat{\mathbf{k}} \quad (2.7)$$

This equation is a more accurate way of describing the vector than Equation 2.1 because it clearly says that vector **v** is the sum of three unit vectors, each scaled by their respective scalar components.

2.3.7 Geologic features as vectors

Virtually all structural features can be reduced to two simple geometric objects: lines and planes. We commonly express more complex features, such as a deformed surface, as a series of measurements of lines or planes. For example, a fold is represented as a group of planar measurements (strikes and dips). The practice of dividing things into structural domains is an example of breaking something complex down into a series of simpler objects.

It takes no great challenge to see that lines can be treated as vectors. Likewise, because there is only one line that is perpendicular to a plane, poles to planes can also be treated as vectors. The question now is: How do we convert from orientations measured in spherical coordinates to Cartesian coordinates?

The relations between spherical and Cartesian coordinates are shown in Figure 2.6. Notice that the three angles α, β, and γ are measured along great circles between the point (which represents the vector) and the positive direction of the axis of the Cartesian coordinate system. Clearly, the angle γ is just equal to 90° minus the plunge of the line. Therefore (Fig. 2.7),

$$\cos \gamma = \cos(90 - plunge) = \sin(plunge) \quad (2.8a)$$

The relations between the trend and plunge and the other two angles are slightly more difficult to calculate. Recall that we are dealing just with orientations and therefore the vector of

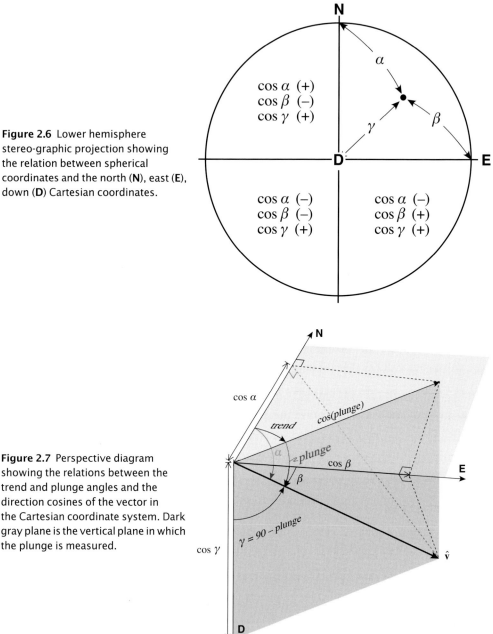

Figure 2.6 Lower hemisphere stereo-graphic projection showing the relation between spherical coordinates and the north (**N**), east (**E**), down (**D**) Cartesian coordinates.

Figure 2.7 Perspective diagram showing the relations between the trend and plunge angles and the direction cosines of the vector in the Cartesian coordinate system. Dark gray plane is the vertical plane in which the plunge is measured.

interest, $\hat{\mathbf{v}}$, is a unit vector. Therefore, from simple trigonometry the horizontal line that corresponds to the trend azimuth is equal to the cosine of the plunge. From here, it is just a matter of solving for the horizontal triangles in Figure 2.7:

$$\cos \alpha = \cos(trend) \cos(plunge) \qquad (2.8b)$$
$$\cos \beta = \cos(90 - trend) \cos(plunge) = \sin(trend) \cos(plunge) \qquad (2.8c)$$

2.3 Vectors

Axis	Direction cosines	Lines	Poles to planes (using right-hand rule)
North	$\cos\alpha$	$\cos(trend)\cos(plunge)$	$\sin(strike)\sin(dip)$
East	$\cos\beta$	$\sin(trend)\cos(plunge)$	$-\cos(strike)\sin(dip)$
Down	$\cos\gamma$	$\sin(plunge)$	$\cos(dip)$

Table 2.1 Conversion from spherical to Cartesian coordinates

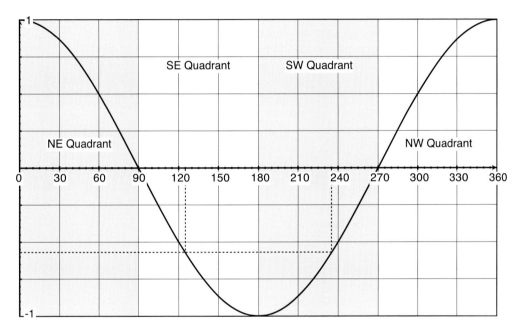

Figure 2.8 Graph of the cosine (vertical axis) for angles ranging from 0° to 360° (horizontal axis). For every positive (or negative) cosine, there are two possible azimuth values.

These relations, along with those for poles to planes, are summarized in Table 2.1.

Figures 2.6 and 2.8 show how the signs of the direction cosines vary with the quadrant. Although it is not easy to see an orientation expressed in direction cosines and immediately have an intuitive feel how it is oriented in space, one can quickly tell what quadrant the line dips in by the signs of the components of the vector. For example, the vector, [-0.4619, -0.7112, 0.5299], represents a line that plunges into the southwest quadrant (237°, 32°) because both $\cos\alpha$ and $\cos\beta$ are negative.

Understanding how the signs work is very important for another reason. Because it is difficult to get an intuitive feel for orientations in direction cosine form, after we do our calculations we will want to convert from Cartesian back to spherical coordinates. This can be tricky because, for each direction cosine, there will be two possible angles (due to the azimuthal range of 0–360°, Fig. 2.8). For example, if $\cos\alpha = -0.5736$, then $\alpha = 125°$ or $\alpha = 235°$. In order to tell which of the two is correct, one must look at the value of $\cos\beta$; if it is negative then $\alpha = 235°$, if positive then $\alpha = 125°$. When you use a calculator or a computer to calculate the inverse cosine, it will only give you one of the two possible angles (generally the smaller of the two). You must determine what the other one is by knowing the cyclicity of the cosine functions (Fig. 2.8).

The MATLAB® function `SphToCart`, below, carries out the conversions shown in Table 2.1.

```
function [cn,ce,cd] = SphToCart(trd,plg,k)
%SphToCart converts from spherical to cartesian coordinates
%
%   [cn,ce,cd] = SphToCart(trd,plg,k) returns the north (cn),
%   east (ce), and down (cd) direction cosines of a line.
%
%   k is an integer to tell whether the trend and plunge of a line
%   (k = 0) or strike and dip of a plane in right hand rule
%   (k = 1) are being sent in the trd and plg slots. In this
%   last case, the direction cosines of the pole to the plane
%   are returned
%
%   NOTE: Angles should be entered in radians

%If line (see Table 2.1)
if k == 0
    cd = sin(plg);
    ce = cos(plg) * sin(trd);
    cn = cos(plg) * cos(trd);
%Else pole to plane (see Table 2.1)
elseif k == 1
    cd = cos(plg);
    ce = -sin(plg) * cos(trd);
    cn = sin(plg) * sin(trd);
end
end
```

Of course, once we have calculated an answer in Cartesian coordinates, we commonly want the answer converted back to more familiar spherical coordinates. The following function `CartToSph` accomplishes this task. Because any cosine value can correspond to two possible angles between 0 and 360°, this routine uses code that checks the sign of the direction cosines to determine which angle is correct.

```
function [trd,plg] = CartToSph(cn,ce,cd)
%CartToSph Converts from cartesian to spherical coordinates
%
%   [trd,plg] = CartToSph(cn,ce,cd) returns the trend (trd)
%   and plunge (plg) of a line for input north (cn), east (ce),
%   and down (cd) direction cosines
%
%   NOTE: Trend and plunge are returned in radians
%
%   CartToSph uses function ZeroTwoPi

%Plunge (see Table 2.1)
plg = asin(cd);
```

2.3 Vectors

```
%Trend
%If north direction cosine is zero, trend is east or west
%Choose which one by the sign of the east direction cosine
if cn == 0.0
    if ce < 0.0
        trd = 3.0/2.0*pi; % trend is west
    else
        trd = pi/2.0; % trend is east
    end
%Else use Table 2.1
else
    trd = atan(ce/cn);
    if cn < 0.0
        %Add pi
        trd = trd+pi;
    end
    %Make sure trd is between 0 and 2*pi
    trd = ZeroTwoPi(trd);
end
end
```

2.3.8 Simple vector operations

To multiply a scalar times a vector, just multiply each component of the vector times the scalar:

$$x\mathbf{v} = \begin{bmatrix} xv_1 & xv_2 & xv_3 \end{bmatrix} \quad (2.9)$$

The most obvious application of scalar multiplication in structural geology is when you want to reverse the direction of the vector. For example, to change the vector from upper to lower hemisphere (or vice versa) just multiply the vector (i.e., its components) by −1. The resulting vector will be parallel to the original and will have the same length, but will point in the opposite direction.

To add two vectors together, you sum their components:

$$\mathbf{u} + \mathbf{v} = \mathbf{v} + \mathbf{u} = \begin{bmatrix} u_1 + v_1 & u_2 + v_2 & u_3 + v_3 \end{bmatrix} \quad (2.10)$$

Graphically, vector addition obeys the parallelogram law (Fig. 2.9a) whereby the resulting vector can be constructed by placing the two vectors to be added end-to-end.

Notice that the order in which you add the two vectors makes no difference. Vector subtraction is the same as adding the negative of one vector to the positive of the other (Fig. 2.9b). We will see an application of vector addition in Section 2.4.1.

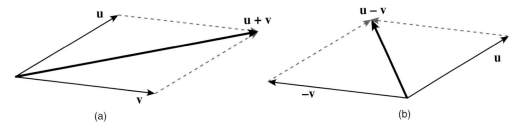

Figure 2.9 (a) Vector addition and (b) vector subtraction using the parallelogram rule.

2.3.9 Dot product and cross product

Vector algebra is remarkably simple, in part by virtue of the ease with which one can visualize various operations. There are two operations that are unique to vectors and that are of great importance in structural geology. If one understands these two, one has mastered the concept of vectors. They are the *dot product* and the *cross product*.

The dot product is also called the *scalar product* because this operation produces a scalar quantity. When we calculate the dot product of two vectors the result is the magnitude of the first vector times the magnitude of the second vector times the cosine of the angle between the two:

$$\mathbf{u} \cdot \mathbf{v} = \mathbf{v} \cdot \mathbf{u} = |\mathbf{u}|\,|\mathbf{v}| \cos\theta = u_1 v_1 + u_2 v_2 + u_3 v_3 \tag{2.11}$$

The physical meaning of the dot product is the length of $\mathbf{v}$ times the length of $\mathbf{u}$ as projected onto $\mathbf{v}$ (that is, the length of $\mathbf{u}$ in the direction of $\mathbf{v}$). Note that the dot product is zero when $\mathbf{u}$ and $\mathbf{v}$ are perpendicular (because in that case the length of $\mathbf{u}$ projected onto $\mathbf{v}$ is zero). The dot product of a vector with itself is just equal to the length of the vector, squared:

$$\mathbf{v} \cdot \mathbf{v} = |\mathbf{v}|^2 = v_1^2 + v_2^2 + v_3^2 \tag{2.12}$$

Equation 2.11 can be rearranged to solve for the angle between two vectors:

$$\cos\theta = \frac{u_1 v_1 + u_2 v_2 + u_3 v_3}{|\mathbf{u}|\,|\mathbf{v}|} \tag{2.13}$$

This last equation is particularly useful in structural geology. As stated previously, all orientations are treated as unit vectors. Thus, when we want to find the angle between any two lines, the product of the two magnitudes, $|\mathbf{u}|\,|\mathbf{v}|$, in Equations 2.11 and 2.13 is equal to one. Upon rearranging Equation 2.13, this provides a simple and extremely useful equation for calculating the angle between two lines:

$$\theta = \cos^{-1}(\cos\alpha_1 \cos\alpha_2 + \cos\beta_1 \cos\beta_2 + \cos\gamma_1 \cos\gamma_2) \tag{2.14}$$

The result of the cross product of two vectors is another vector. For that reason, you will often see the cross product called the *vector product*. The cross product is conceptually a little more difficult than the dot product, but is equally useful in structural geology. It is best illustrated with a diagram (Fig. 2.10), which relates to the Equations 2.15 to 2.17, below.

The cross product's primary use is when you want to calculate the orientation of a vector that is perpendicular to two other vectors. The resulting perpendicular vector is parallel to the unit vector and has a magnitude equal to the product of the magnitude of each vector times the sine of the angle between them. If $\mathbf{u}$ and $\mathbf{v}$ are both unit vectors, then the length of the resulting vector will be equal to the sine of the angle θ. The new vector obeys a right-hand rule with respect to the other two (Fig. 2.10):

$$\mathbf{v} \times \mathbf{u} = -\mathbf{u} \times \mathbf{v} = |\mathbf{v}|\,|\mathbf{u}| \sin\theta \,\hat{\ell} \tag{2.15}$$

and

$$\mathbf{v} \times \mathbf{u} = [(v_2 u_3 - v_3 u_2)\ \ (v_3 u_1 - v_1 u_3)\ \ (v_1 u_2 - v_2 u_1)] \tag{2.16}$$

which can also be written in terms of the base vectors of the coordinate system as

$$\mathbf{v} \times \mathbf{u} = (v_2 u_3 - v_3 u_2)\,\hat{i} - (v_3 u_1 - v_1 u_3)\,\hat{j} + (v_1 u_2 - v_2 u_1)\,\hat{k} \tag{2.17}$$

2.4 EXAMPLES OF STRUCTURE PROBLEMS USING VECTOR OPERATIONS

2.4.1 Example 1: Finding the mean of a group of vectors

A common problem in structural geology and geophysics is to determine the vector that statistically represents a group of individual vectors. For example, we may want to find the average of a

2.4 Examples of structure problems using vector operations

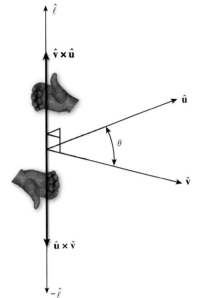

Figure 2.10 Diagram illustrating the meaning of the cross product, for the case of two unit vectors. The hand indicates the right-hand rule convention; for $\mathbf{v} \times \mathbf{u}$, the fingers curl from $\mathbf{v}$ towards $\mathbf{u}$ and the thumb points in the direction of the resulting vector, which is parallel to the unit vector $\hat{\ell}$. Note that $\mathbf{v} \times \mathbf{u} = -(\mathbf{u} \times \mathbf{v})$. The cross product can be calculated for any vectors, not just unit vectors.

group of paleomagnetic poles or the vector that best represents poles to bedding. This is a very easy operation using vector addition; it is much more difficult to do any other way. There are two things to be determined: (1) the orientation of a unit vector that is parallel to the average, or mean, of all of the individual vectors; and (2) an expression of how "concentrated" the vectors are.

The solution to this problem uses vector addition and is shown graphically in Figure 2.11. Numerically, the steps are given below; for a computer program to solve this problem, see function `CalcMV` at the end of this section. The solution is illustrated with a real problem: Determine the mean vector of the following four lines, given as trend and plunge: 026, 31; 054, 22; 037, 39; and 012, 47.

1. Convert all of your orientation data into direction cosines.

Trend and plunge	$\cos \alpha$	$\cos \beta$	$\cos \gamma$
026, 31	0.7704	0.3758	0.5150
054, 22	0.5450	0.7501	0.3746
037, 39	0.6207	0.4677	0.6293
012, 47	0.6671	0.1418	0.7314

2. Sum all of the individual components of the vectors, as in Equation 2.11. This will give you the resultant vector, **r**. If all the individual vectors have the same orientation, then the resultant vector will have a length that is equal to the number of vectors summed (in this example, $N = 4$); otherwise, it will always be less.

$$\sum \cos \alpha = 2.6032 \qquad \left(\sum \cos \alpha\right)^2 = 6.7767$$
$$\sum \cos \beta = 1.7354 \qquad \left(\sum \cos \beta\right)^2 = 3.0116$$
$$\sum \cos \gamma = 2.2503 \qquad \left(\sum \cos \gamma\right)^2 = 5.0639$$

Length of the resultant vector,

$$r = (6.7767 + 3.0116 + 5.0639)^{1/2} = 3.8539$$

3. Normalize the resultant vector by dividing each one of its components by the number of vectors summed together. The length of the normalized vector will always be less than or equal to 1. The closer it is to 1, the better the concentration.

   Note that $r = 0.9635$ indicates a reasonably strong preferred orientation.

$$\frac{\text{resultant length}}{N} = \frac{3.8539}{4} = 0.9635$$

4. Determine a unit vector, $\hat{\mathbf{m}}$, that is parallel to the resultant vector, $\mathbf{r}$. To do this, calculate the magnitude of the resultant vector (or the normalized resultant vector) and then divide the components by the magnitude (Eqs. 2.3 and 2.4). These components will now be in direction cosines.

$$\hat{\mathbf{m}} = \begin{bmatrix} \frac{2.6032}{3.8539} & \frac{1.7354}{3.8539} & \frac{2.2503}{3.8539} \end{bmatrix} = [0.6755 \ 0.4503 \ 0.5840]$$

5. Convert this final unit vector back to spherical coordinates.
   Trend and plunge of mean vector = 033.7°, 35.7°

This example points out one of the pitfalls of treating axes as vectors (Section 2.3.1). Suppose that we have two lines that plunge very gently into opposite quadrants (Fig. 2.12). If we deal with these lines as vectors, the sum of the two $(\mathbf{v} + \mathbf{u})$ is a very short, vertical vector that bisects the obtuse angle between the two (Fig. 2.12). This may be exactly what we want.

Commonly, however, the lines have no directional significance and are better thought of as axes. If this were the case then the result of averaging the two together would look very strange, indeed. After all, there is really very little difference in the orientation of the two lines. One possible way around this problem is to convert one of the two vectors to an upper hemisphere vector by multiplying it by –1 (–$\mathbf{u}$ in Fig. 2.12). Then the vector, $\mathbf{v} - \mathbf{u}$, is much more like what we probably had in mind (Fig. 2.12)! We will see a more elegant solution to the problem of how to determine the statistical average of a group of axes in Chapter 5.

The MATLAB script CalcMV, below, takes a group of n lines, whose trends and plunges are held in the arrays T(i), P(i), and calculates the mean vector. Additionally, the program calculates

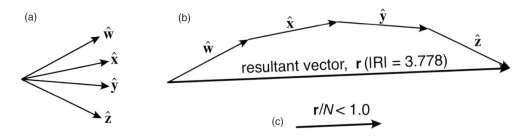

Figure 2.11 Example showing the use of vector addition to determine the mean vector. (a) The four original unit vectors, each of length = 1; (b) addition of the vectors using the parallelogram law to determine the resultant vector; (c) normalized resultant vector (i.e., resultant vector divided by the number of unit vectors) compared to a unit vector.

2.4 Examples of structure problems using vector operations

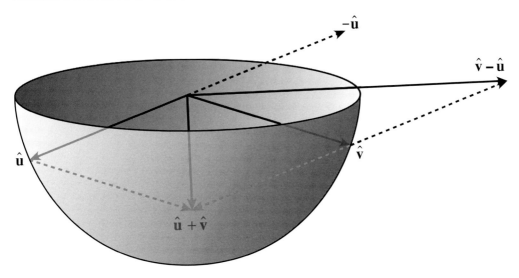

Figure 2.12 Perspective view of a lower hemisphere stereographic projection, showing the addition of vectors, **u** and **v**. Illustrates case in which addition of vectors can provide a misleading answer if the lines being analyzed are axes rather than vectors.

the Fisher statistics for the mean vector, which is the standard way to report uncertainties in paleomagnetic analyses. The variables, d99 and d95, are the cones of uncertainty at the 99 and 95% levels; that is, we are 99 and 95% certain that the mean vector lies within a cone of that apical angle. To solve Example 1 using `CalcMV` just type in MATLAB:

```
T=[26,54,37,12]*pi/180; %Lines trend
P=[31,22,39,47]*pi/180; %Lines plunge
[trd,plg,Rave,conc,d99,d95] = CalcMV(T,P); %Calculate Mean Vector

function [trd,plg,Rave,conc,d99,d95] = CalcMV(T,P)
%CalcMV calculates the mean vector for a given series of lines
%
%   [trd,plg,conc,d99,d95] = CalcMV(T,P) calculates the trend (trd)
%   and plunge (plg) of the mean vector, its normalized length, and
%   Fisher statistics (concentration factor (conc), 99 (d99) and
%   95 (d95) % uncertainty cones); for a series of lines whose trends
%   and plunges are stored in the vectors T and P
%
%   NOTE: Input/Output trends and plunges, as well as uncertainty
%   cones are in radians
%
%   CalcMV uses functions SphToCart and CartToSph

%Number of lines
nlines = max(size(T));

%Initialize the 3 direction cosines which contain the sums of the
%individual vectors (i.e. the coordinates of the resultant vector)
```

```
CNsum = 0.0;
CEsum = 0.0;
CDsum = 0.0;

%Now add up all the individual vectors
for i=1:nlines
    [cn,ce,cd] = SphToCart(T(i),P(i),0);
    CNsum = CNsum + cn;
    CEsum = CEsum + ce;
    CDsum = CDsum + cd;
end
%R is the length of the resultant vector and Rave is the length of
%the resultant vector normalized by the number of lines
R = sqrt(CNsum*CNsum + CEsum*CEsum + CDsum*CDsum);
Rave = R/nlines;
%If Rave is lower than 0.1, the mean vector is insignificant, return error
if Rave < 0.1
    error('Mean vector is insignificant');
%Else
else
    %Divide the resultant vector by its length to get the average
    %unit vector
    CNsum = CNsum/R;
    CEsum = CEsum/R;
    CDsum = CDsum/R;
    %Use the following 'if' statement if you want to convert the
    %mean vector to the lower hemisphere
    if CDsum < 0.0
        CNsum = -CNsum;
        CEsum = -CEsum;
        CDsum = -CDsum;
    end
    %Convert the mean vector from direction cosines to trend and plunge
    [trd,plg]=CartToSph(CNsum,CEsum,CDsum);
    %If there are enough measurements calculate the Fisher Statistics
    %For more information on these statistics see Fisher et al. (1987)
    if R < nlines
        if nlines < 16
            afact = 1.0-(1.0/nlines);
            conc = (nlines/(nlines-R))*afact^2;
        else
            conc = (nlines-1.0)/(nlines-R);
        end
    end
    if Rave >= 0.65 && Rave < 1.0
        afact = 1.0/0.01;
        bfact = 1.0/(nlines-1.0);
```

2.4 Examples of structure problems using vector operations

```
            d99 = acos(1.0-((nlines-R)/R)*(afact^bfact-1.0));
            afact = 1.0/0.05;
            d95 = acos(1.0-((nlines-R)/R)*(afact^bfact-1.0));
        end
end
end
```

2.4.2 Example 2: Calculating the rake of a line in a plane

Calculating the angle between any two lines is a common problem. The rake, or pitch, is an angle measured between a line of interest and the strike of the plane that contains the line. This example provides us with a perfect illustration of the use of the dot product (function **Angles** at the end of Section 2.4.3 includes code for this operation). Suppose we have a plane with a strike of 213° and a lineation within the plane has a trend and plunge of 278, 42; what is the rake of the lineation? The solution is easier than in the previous example:

1. Convert the data to direction cosines. Recall that the strike is just a line with zero plunge:

Trend and plunge	$\cos \alpha$	$\cos \beta$	$\cos \gamma$
213, 0	-0.8387	-0.5446	0.0
278, 42	0.1.34	-0.7359	0.6691

2. Then, just multiply the components together and calculate the inverse cosine as in Equation 2.14:

$$\theta = \cos^{-1}((-0.8387 \times 0.1034) + (-0.5446 \times -0.7359) + (0 \times 0.6691))$$
$$= \cos^{-1}(0.3140) = 71.7°$$

Note that the rake of 71.7° in this example is with respect to the given strike azimuth of 213°. If we had been given the other strike azimuth, 033°, then the pitch angle calculated would be the complement of the above, that is, 108.3°. It may also seem strange that, to solve this problem, we did not even need to know the dip or the dip direction of the plane. That would have been redundant information because the orientation of the plane is constrained by the two lines within it. In the next example, we will calculate the true dip of the plane.

2.4.3 Example 3: Determining a true dip from two apparent dips

Determining a line that is perpendicular to two other lines is one of the most common calculations in structural geology. For example, in analyzing a fault, the pole to the movement plane is perpendicular to the slip vector and the pole to the fault plane. In a little more familiar example, the pole to a plane is perpendicular to all of the lines within that plane. Thus, two apparent dip lines in a plane must be perpendicular to the pole of the plane. The previous example is just such a case; from the apparent dip and the strike line, both of which were given, we can calculate the pole to the plane and therefore the dip and dip direction. To accomplish this, we will use the cross product (function **Angles** at the end of this section implements this operation):

1. Convert the data to direction cosines. This was already done for us in the previous example.
2. Calculate the cross product from Equation 2.16. This will give us a vector that is parallel to the pole, **p**, but it will not be a unit vector because the lines are not perpendicular:

$$p_1 = (-0.7359 * 0.0) - (0.6691 * -0.5446) = 0.3644$$
$$p_2 = (0.6691 * -0.8387) - (0.1034 * 0.0) = -0.5612$$
$$p_3 = (0.1034 * -0.5446) - (-0.7359 * -0.8387) = -0.6736$$
$$\mathbf{p} = [0.3644, \ -0.5612, \ -0.6736]$$
$$|\mathbf{p}| = 0.9494$$

3. As you can see, the magnitude of **p** is not equal to 1, so it must be converted to a unit vector before we can determine the orientation using Equation 2.4. The components of the unit pole vector are

$$\hat{\mathbf{p}} = \left[\frac{0.3644}{0.9494}, \ \frac{-0.5612}{0.9494}, \ \frac{-0.6736}{0.9494}\right] = [0.3839, \ -0.5911, \ -0.7094]$$

4. Before going any farther, notice that the third component of $\hat{\mathbf{p}}$, the down direction cosine, is negative. Thus, the cross product we have calculated points upwards into the upper hemisphere (because down is positive in our north-east-down coordinate system). To calculate the lower hemisphere pole, multiply by -1:

$$-\hat{\mathbf{p}} = [-0.3839, \ 0.5911, \ 0.7094]$$

5. Now we can calculate the orientation of the pole to the plane in spherical coordinates:

$$\text{trend and plunge of pole} = 123°, 45.2°$$

The true dip of the plane is equal to 90 - 45.2 = 44.8°, and the dip direction is equal to 123 + 180 = 303°. Obviously, the dip direction is just 90° from the strike azimuth that we were given initially.

The function **Angles** below calculates either the angle between two lines if `ans0 = 'l'` is passed to it, or the strike and dip of a plane from two apparent dips in the plane if `ans0 = 'a'`. In the first case, the dot product is used and in the second, the cross product. If, instead, the user passes the strike and dip of two planes in the place of `trd1, plg1` and `trd2, plg2`, then the function will calculate either the intersection of two planes (`ans0 = 'i'`) or the angle between the two planes (`ans0 = 'p'`). To solve Example 2 using **Angles** just type in MATLAB:

`[ans1,ans2]=Angles(213*pi/180,0,278*pi/180,42*pi/180,'l');`

Example 3 can be solved by entering:

`[ans1,ans2]=Angles(213*pi/180,0,278*pi/180,42*pi/180,'a');`

```
function [ans1,ans2] = Angles(trd1,plg1,trd2,plg2,ans0)
%Angles calculates the angles between two lines, between two planes,
%the line which is the intersection of two planes, or the plane
%containing two apparent dips
%
%   [ans1,ans2] = Angles(trd1,plg1,trd2,plg2,ans0) operates on
%   two lines or planes with trend/plunge or strike/dip equal to
%   trd1/plg1 and trd2/plg2
```

2.4 Examples of structure problems using vector operations   41

```
%
%     ans0 is a character that tells the function what to calculate:
%
%         ans0 = 'a' -> the orientation of a plane given two apparent dips
%         ans0 = 'l' -> the angle between two lines
%
%         In the above two cases, the user sends the trend and plunge of two
%           lines
%
%         ans0 = 'i' -> the intersection of two planes
%         ans0 = 'p' -> the angle between two planes
%
%         In the above two cases the user sends the strike and dip of two
%         planes following the right-hand rule
%
%     NOTE: Input/Output angles are in radians
%
%     Angles uses functions SphToCart, CartToSph and Pole

%If planes have been entered
if ans0 == 'i' || ans0 == 'p'
    k = 1;
%Else if lines have been entered
elseif ans0 == 'a' || ans0 == 'l'
    k = 0;
end

%Calculate the direction cosines of the lines or poles to planes
[cn1,ce1,cd1]=SphToCart(trd1,plg1,k);
[cn2,ce2,cd2]=SphToCart(trd2,plg2,k);

%If angle between 2 lines or between the poles to 2 planes
if ans0 == 'l' || ans0 == 'p'
    % Use dot product = Sum of the products of the direction cosines
    ans1 = acos(cn1*cn2 + ce1*ce2 + cd1*cd2);
    ans2 = pi - ans1;
end

%If intersection of two planes or pole to a plane containing two
%apparent dips
if ans0 == 'a' || ans0 == 'i'
    %If the 2 planes or apparent dips are parallel, return an error
    if trd1 == trd2 && plg1 == plg2
        error('lines or planes are parallel');
    %Else use cross product
    else
        cn = ce1*cd2 - cd1*ce2;
        ce = cd1*cn2 - cn1*cd2;
```

```
            cd = cn1*ce2 - ce1*cn2;
            %Make sure the vector points down into the lower hemisphere
            if cd < 0.0
                cn = -cn;
                ce = -ce;
                cd = -cd;
            end
            %Convert vector to unit vector by dividing it by its length
            r = sqrt(cn*cn+ce*ce+cd*cd);
            % Calculate line of intersection or pole to plane
            [trd,plg]=CartToSph(cn/r,ce/r,cd/r);
            %If intersection of two planes
            if ans0 == 'i'
                ans1 = trd;
                ans2 = plg;
            %Else if plane containing two dips, calculate plane from its pole
            elseif ans0 == 'a'
                [ans1,ans2]= Pole(trd,plg,0);
            end
        end
    end
end
end
```

Function **Angles** calls function **Pole**, which calculates a plane, given its pole (k = 0) or a pole given the corresponding plane (k = 1).

```
function [trd1,plg1] = Pole(trd,plg,k)
%Pole returns the pole to a plane or the plane which correspond to a pole
%
%   k is an integer that tells the program what to calculate.
%
%   If k = 0, [trd1,plg1] = Pole(trd,plg,k) returns the strike
%   (trd1) and dip (plg1) of a plane, given the trend (trd)
%   and plunge (plg) of its pole.
%
%   If k = 1, [trd1,plg1] = Pole(trd,plg,k) returns the trend
%   (trd1) and plunge (plg1) of a pole, given the strike (trd)
%   and dip (plg) of its plane.
%
%   NOTE: Input/Output angles are in radians. Input/Output strike
%   and dip are in right-hand rule
%
%   Pole uses functions ZeroTwoPi, SphToCart and CartToSph

%Some constants
east = pi/2.0;

%Calculate plane given its pole
if k == 0
    if plg >= 0.0
```

```
            plg1 = east - plg;
            dipaz = trd - pi;
        else
            plg1 = east + plg;
            dipaz = trd;
        end
        %Calculate trd1 and make sure it is between 0 and 2*pi
        trd1 = ZeroTwoPi(dipaz - east);
    %Else calculate pole given its plane
    elseif k == 1
        [cn,ce,cd] = SphToCart(trd,plg,k);
        [trd1,plg1] = CartToSph(cn,ce,cd);
    end
end
```

2.5 EXERCISES

The following are a series of simple problems to be completed using vector algebra, exclusively, although you should report your results in spherical coordinates. The easiest way to do them is using the MATLAB functions provided above, although we recommend that you first solve the problems by hand. All can equally well be solved via a spreadsheet program.

1. A plane with a strike of 127° contains a line with a trend and plunge of 005°, 31°. What is the rake (pitch) of the line? What is the dip of the plane? Solve this problem first by hand and then by using the function **Angles**.
2. Two planes have the following orientations, given using the right-hand-rule format (RHR): 237, 25 and 056, 49. Calculate the orientation of the line of intersection between the two planes. Report your results in spherical coordinates. Solve this problem first by hand and then by using the function **Angles**.
3. A quarry has two vertical walls, one trending 002 and the other trending 135. The apparent dips of bedding on the faces are 40 N and 30 SE respectively. Calculate the strike and true dip of the bedding. Solve this problem first by hand and then by using the function **Angles**.
4. Two limbs of a chevron fold (A and B) have orientations (strike and dip) as follows: Limb A = 120, 40SW and limb B = 070, 60SE. Determine: (1) the trend and plunge of the hinge line of the fold; (2) the pitch of the hinge line in limb A; and (3) the pitch of the hinge line in limb B. Solve this problem using the function **Angles**.
5. Calculate the mean vector for the following group of lines. Report the magnitude and orientation (in spherical coordinates) of the mean vector. Solve this problem using either a spreadsheet or the function **CalcMV**.

    113.0, 73.0   081.0, 77.0
    076.0, 78.0   080.0, 58.0
    175.0, 71.0   058.0, 62.0
    229.0, 62.0   040.0, 57.0
    075.0, 62.0   042.0, 71.0
    111.0, 77.0   229.0, 23.0
    078.0, 85.0   110.0, 72.0
    316.0, 53.0   278.0, 61.0
    025.0, 78.0   264.0, 78.0
    021.0, 57.0

CHAPTER THREE

Transformations of coordinate axes and vectors

3.1 WHAT ARE TRANSFORMATIONS AND WHY ARE THEY IMPORTANT?

The word "transformation" looks imposing and mathematical but it is, in fact, quite a simple thing that we do commonly without thinking about it. Whenever we change coordinate systems, we do a coordinate transformation. Suppose we submit some samples of fossils and their locations in latitude, longitude, and elevation to a paleontologist for identification. The paleontologist writes back with the instructions that the locations in eastings and northings (i.e., UTM coordinates), not latitude and longitude, are required. Thus, a coordinate transformation is needed. This doesn't make us very happy because the change requires a long calculation that would be tedious to do by hand! In this chapter, we are interested only in transformations that can be precisely described mathematically, but one should realize that coordinate transformations are a very common thing. Basically, coordinate transformations are just another way of looking at the same thing.[1] In the above example, the specific numbers used to describe the location in the two coordinate systems are different but the physical location where the samples were collected has not changed. The change in numbers simply represents a change in the coordinate system not a change in the position or fundamental magnitude of the thing being described.

The concept of a transformation is very important and one with incredible power for a wide variety of structural applications. It is commonly necessary to look at a problem from two different points of view. For example, when studying continental drift (Fig. 3.1a), at least two different coordinate systems are commonly required, one in present-day geographic space and one attached to the continent at some time in the past when it was in a different place and orientation on the globe. Or, take the case of analysis of a fault (Fig. 3.1b). To understand what is

---

[1] Later in the book, we will use the word "transformation" in a different context to refer to changes brought about by deformation that takes place between an initial and a final state.

3.2 Transformation of axes

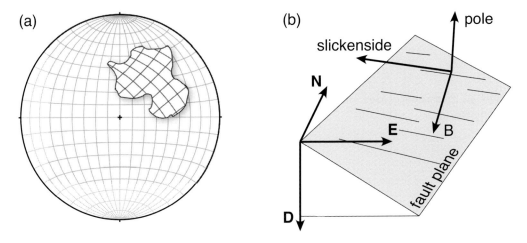

Figure 3.1 Examples of coordinate transformations in geology. (a) Continental drift; the continent has a coordinate system marked on it that corresponds to some time in the past. (b) A fault plane with two different coordinate systems.

going on from the perspective of the fault we need one coordinate system attached to the fault (e.g., with one axis perpendicular to the fault plane and another parallel to the slickensides on the fault). However, we also want to relate this to our more familiar geographic coordinate system; a transformation allows us to do that.

There is, however, an even more elemental reason for the importance of transformations. As intimated above, real, physical properties do not change when they are transformed from one coordinate system to another. As we will see in Chapter 5, this statement will be turned around to form the definition of an entire class of entities known as tensors. For right now, though, it is sufficient to be aware that the same thing can be described from many different viewpoints. Because the nature of something does not change when it is transformed, if we know its coordinates in one system we should be able to calculate its coordinates in any other system. This logic assumes that we know how the two coordinate systems are related to each other and that is our starting point.

3.2 TRANSFORMATION OF AXES

Before we can talk about transforming objects, we must consider the transformation that describes a change from one coordinate system to another. We will address only the change from one rectangular Cartesian coordinate system to another, which means the transformation is from one set of mutually perpendicular axes to another. As we will see in Section 3.2.3, this orthogonality makes our life very much easier.

3.2.1 Two-dimensional change in axes

The simplest type of transformation that you can think of is a two-dimensional shift or translation of axes without any rotation. Basically, we just establish the origin at a different place; it is simple to write equations that relate one set of axes to another. In the case of Figure 3.2,

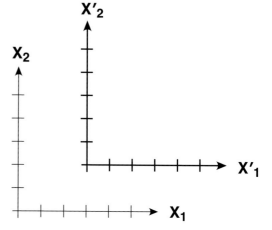

Figure 3.2 Translation of axes. The new axes are primed.

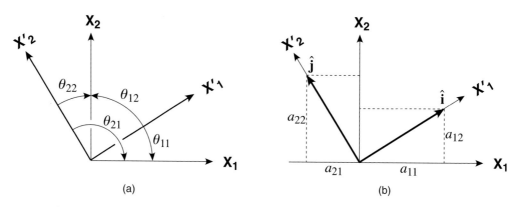

Figure 3.3 Rotation of axes in two dimensions. New axes are primed. (a) Shows the four angles, θ_{11}, θ_{12}, θ_{21}, and θ_{22}, that define the coordinate transformation. (b) Same transformation, but expressed in terms of base vectors and their direction cosines, a_{11}, a_{12}, a_{21}, and a_{22}.

$$X'_1 = X_1 - 3 \text{ and } X'_2 = X_2 - 2 \quad \text{(new in terms of old)}$$

and

$$X_1 = X'_1 + 3 \text{ and } X_2 = X'_2 + 2 \quad \text{(old in terms of new)}$$

We will come back to this example when we get to deformation (Chapter 7). Although this provides a useful starting point, it really doesn't provide any new information and therefore is not of great interest in our study of vectors. You can probably visualize that a vector will make the same angles with the axes in both the new and the old coordinates and, furthermore, the components of the vector will have the same magnitude in both coordinate systems. We have not really learned anything, yet.

More interesting is the case of rotation of a coordinate system. From Figure 3.3a you can see that, in two dimensions, there are four angles that define the transformation. Rather than give all of these a different letter, they are distinguished by double subscripts. As you can see by close inspection of the figure, the choice of subscripts is not arbitrary. The convention is that

3.2 Transformation of axes

the first subscript refers to the new (i.e., primed) axis, whereas the second subscript refers to the old (unprimed) axis. Thus, θ_{21} indicates the angle between the new, X'_2 axis and the old, X_1 axis.

Although there are, clearly, four angles, one can intuitively see that they are not all independent of each other. In fact, in two dimensions, we need only specify one of the four and the rest can be calculated from the first one. For example, $\theta_{11} = 90° - \theta_{12}$, $\theta_{21} = 90° + \theta_{22}$, etc. If we represent the axes by their base vectors, then you can see that the projection of the new axis onto the old axis is equal to the cosine of the angle between the two axes (Fig. 3.3b). For that reason, the relations between the two coordinate systems are commonly given in terms of the direction cosines between them: $a_{11} = \cos\theta_{11}$, $a_{12} = \cos\theta_{12}$, $a_{21} = \cos\theta_{21}$, and $a_{22} = \cos\theta_{22}$. By a simple application of the Pythagorean Theorem (see Fig. 2.4) and recalling that the length of a unit vector is 1 (i.e., $|\hat{i}| = 1$ in Fig. 3.3b), you can see that

$$a_{11}^2 + a_{12}^2 = 1 \text{ and } a_{21}^2 + a_{22}^2 = 1 \tag{3.1}$$

Furthermore, recall that the dot product of two perpendicular vectors is equal to zero (because the cosine of 90° is zero). Therefore, the dot product of the base vectors in the new system (Fig. 3.3) gives us a third constraint:

$$a_{11}a_{21} + a_{12}a_{22} = 0 \tag{3.2}$$

We have three equations, 3.1 and 3.2, and four unknowns, so only one of the direction cosines is independent. If you understand this two-dimensional case, extension to three dimensions is obvious.

3.2.2 Three-dimensional change in axes: The transformation matrix

The relations in three dimensions logically follow from those in two dimensions. There are three old axes and three new ones; hence, there will be nine angles that completely define the coordinate transformation (Fig. 3.4a). As before, we use double subscripts to identify the angles

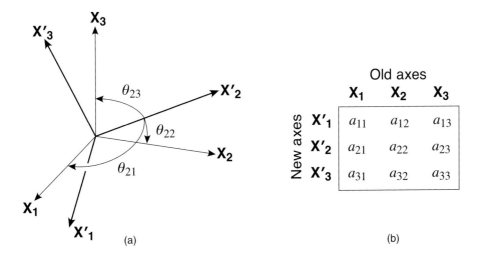

Figure 3.4 (a) A general, three-dimensional coordinate transformation. The new axes are primed; the old axes are in black. Only three of the nine possible angles are shown. (b) Graphic device for remembering how the subscripts of the direction cosines relate to the new and the old axes.

and their direction cosines, with the first subscript referring to the new axis and the second to the old axis (Fig. 3.4b). As before, not all nine of these angles are independent. Just visually, you can see that, given θ_{22} and θ_{23}, the third angle, θ_{21}, is fixed. Intuitively, you may be able to see that, to completely constrain the transformation, only one other angle between any of the other two new axes and any of the old axes is needed.

The array of direction cosines in Figure 3.4b is known as the *transformation matrix*. It is commonly written:

$$a_{ij} = \begin{pmatrix} a_{11} & a_{12} & a_{13} \\ a_{21} & a_{22} & a_{23} \\ a_{31} & a_{32} & a_{33} \end{pmatrix} \quad (3.3)$$

The way we have written Equation 3.3 uses some notation that we have not seen much of up to this point. We will see much more of matrices and indicial notation in the next chapter.

3.2.3 The orthogonality relations

Earlier, we began developing some general equations, 3.1 and 3.2, that described how the angles (or really their direction cosines) relate to one another. The development in three dimensions is an extension of the previous derivation. In three dimensions, the length of any vector is the square root of the sum of the squares of its three components (Eq. 2.3). If that vector is a unit vector, then the sum of the squares of the direction cosines will be equal to one. We showed in Section 3.2.1 that the components of the transformation matrix are nothing more than the direction cosines of the base vectors of the new coordinate system in the old coordinate system. Therefore, we can write the following three equations, which give the lengths (squared) of the three base vectors of the new coordinate system:

$$\begin{aligned} a_{11}^2 + a_{12}^2 + a_{13}^2 &= 1 \\ a_{21}^2 + a_{22}^2 + a_{23}^2 &= 1 \\ a_{31}^2 + a_{32}^2 + a_{33}^2 &= 1 \end{aligned} \quad (3.4)$$

Likewise, as stated above, the dot product of two perpendicular vectors is zero. Because each of the three base vectors of the new coordinate system is perpendicular to the others, we can write three additional equations:

$$\begin{aligned} a_{21}a_{31} + a_{22}a_{32} + a_{23}a_{33} &= 0 \\ a_{31}a_{11} + a_{32}a_{12} + a_{33}a_{13} &= 0 \\ a_{11}a_{21} + a_{12}a_{22} + a_{13}a_{23} &= 0 \end{aligned} \quad (3.5)$$

Equations 3.4 and 3.5 collectively form what are known as the *orthogonality relations*. Now, in three dimensions, we have six equations and nine unknowns (i.e., the nine direction cosines). This proves quantitatively what we already knew intuitively: There are only three independent direction cosines in the transformation matrix.

3.3 TRANSFORMATION OF VECTORS

Now that we have put the transformation of axes to rest, we'll look at something more practical: the transformation of vectors. As before, we'll start in two dimensions, where it is easier to get a feeling for the geometry. The two-dimensional transformation equations are derived by projecting the old components of the vector, v_1 and v_2, onto the new axes, $\mathbf{X}'_1$ and $\mathbf{X}'_2$. In Figure 3.5b,

3.3 Transformation of vectors

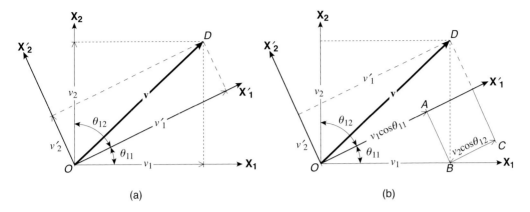

Figure 3.5 Transformation of vector **v** in two dimensions. (a) The components of the vector in the old coordinate system are v_1 and v_2; in the new coordinate system, the coordinates are v'_1 and v'_2. (b) Shows the geometry for deriving the v'_1 component of transformation equation (3.5) from triangles OAB and BCD.

you can see that v'_1 will be equal to the sum of line segments OA and BC, which can be calculated from the trigonometry of triangles OAB and BCD (Fig. 3.5b). We get

$$v'_1 = v_1 \cos \theta_{11} + v_2 \cos \theta_{12}$$

or, in terms of the direction cosines of the transformation matrix (and including without proof the equation for v'_2),

$$\begin{aligned} v'_1 &= v_1 a_{11} + v_2 a_{12} \\ v'_2 &= v_1 a_{21} + v_2 a_{22} \end{aligned} \tag{3.6}$$

Note that the above equations give the new components of the vector in terms of the old. By projecting the new components, v'_1 and v'_2, onto the old axes, $\mathbf{X_1}$ and $\mathbf{X_2}$, you can make the same geometric arguments and derive the reverse transformation, which is the old in terms of the new:

$$\begin{aligned} v_1 &= v'_1 a_{11} + v'_2 a_{21} \\ v_2 &= v'_1 a_{12} + v'_2 a_{22} \end{aligned} \tag{3.7}$$

There are some subtle, but important, changes between Equations 3.6 and 3.7. First, in the latter the primed components are on the right-hand side. Less obvious, but no less important, the positions of a_{12} and a_{21} have been switched or *transposed*. One of the nice things about vector algebra is that it is extremely symmetrical and logical!

Figure 3.6 shows a three-dimensional vector transformation. As before, notice that only the coordinates change, not the fundamental length or orientation of the vector, itself. Thus in Figure 3.6, $v_1 \neq v'_1$, $v_2 \neq v'_2$, and $v_3 \neq v'_3$ but the vector is just as long and points the same way in both coordinate systems.

The geometry in three dimensions is really the same as in two, only harder to visualize. Think about decomposing the vector into its three components parallel to the old axes, and then transforming those components along with the old axes into the new coordinate system. Thought of this way, the transformation of any vector is analogous to transforming the base vectors of the axes themselves, except that the components are not unit vectors. Three equations describe the three-dimensional vector transformation:

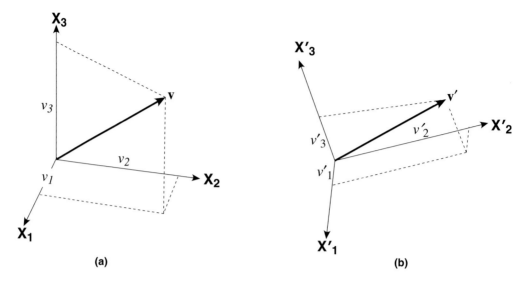

Figure 3.6 Vector **v** in two different coordinate systems. Note that the length and orientation of **v** on the page has not changed; only the axes have changed.

$$v'_1 = a_{11}v_1 + a_{12}v_2 + a_{13}v_3$$
$$v'_2 = a_{21}v_1 + a_{22}v_2 + a_{23}v_3 \qquad (3.8)$$
$$v'_3 = a_{31}v_1 + a_{32}v_2 + a_{33}v_3$$

and the reverse transformation (old in terms of new):

$$v_1 = a_{11}v'_1 + a_{21}v'_2 + a_{31}v'_3$$
$$v_2 = a_{12}v'_1 + a_{22}v'_2 + a_{32}v'_3 \qquad (3.9)$$
$$v_3 = a_{13}v'_1 + a_{23}v'_2 + a_{33}v'_3$$

Note that we have reversed the order of a and v in these equations from the earlier Equations 3.6 and 3.7, but that is perfectly okay because all terms are scalars. If you examine carefully Equations 3.8 and 3.9, it looks as though we have "flipped" the transformation matrix so that the rows are now columns and vice versa. Mathematically, "flipping" a matrix is known as taking the *transpose*, as we will see in a later chapter.

You can transform the coordinates of a point in space using the same equations that you would for vectors (3.8 and 3.9). That's because any point can be thought of as being connected to the origin of the coordinate system by a vector known as a *position vector*. The components of the vector are the same as the coordinates of the point. We will use this concept below in Section 3.4.1.

3.4 EXAMPLES OF TRANSFORMATIONS IN STRUCTURAL GEOLOGY

Generally, we give little thought to the fact that some of our most commonplace structural problems involve transformations of the type described in the previous section. That is because we are taught to do them with laborious manual methods, like orthographic projections, or on a stereonet. In this section, we will see how to solve two such problems using the methods developed in this chapter.

3.4 Examples of transformations in structural geology

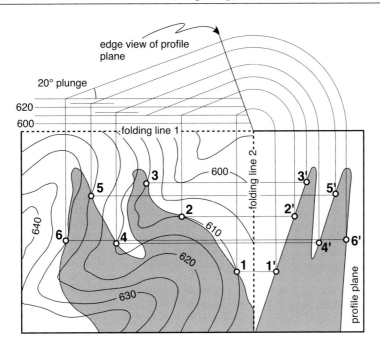

Figure 3.7 Graphical construction for drawing a down-plunge projection in a region with topography. The fold in this example plunges 20° east. The projection of six control points from the map onto the projection is shown. Modified from Ragan (2009, p. 461).

3.4.1 Down-plunge projection

To get the best sense of the "true" geometry of a cylindrical fold, geologists usually construct a *profile view* of the fold, a cross section of the fold perpendicular to the fold axis. When folds are cylindrical, all the points along all of their surfaces can be projected parallel to the fold axis onto this profile plane. This task is complicated by two facts: First, the surface of the Earth is irregular with hills and valleys and, second, folds commonly plunge oblique to the ground surface. The graphical method taught in virtually all structural geology lab manuals employs orthographic projection (Fig. 3.7). One chooses a horizontal folding line that is perpendicular to the fold axis and another that is horizontal and in the same plane as the fold axis. Then by swinging arcs with a compass and carefully drawing parallel straight lines you can construct the profile.

The construction is made more tedious by the fact that a separate folding line is needed for each elevation of the control points used. There is ample opportunity for error in the construction of the parallel lines as well as interpolation between widely separated control points.

There is, however, a different way of making a down-plunge projection that applies the methods we have seen earlier in this chapter. Specifically, we determine the geographic coordinates for a series of points along each bedding surface; that is, we digitize the bedding surfaces. Then we transform those points into the fold coordinate system (Fig. 3.8). All of this can be done on a computer much more rapidly than is possible by hand. Because we are dealing with geographic coordinates, our old, right-handed coordinate system will be X_1 = east, X_2 = north, and X_3 = up (or elevation); our new, right-handed system will be as shown in Figure 3.8, with X'_3 coinciding with the fold axis.

Now, we need to determine the transformation matrix in terms of the orientation (trend and plunge) of the fold axis. The angular relations are given in Figure 3.9. Some of the angles are

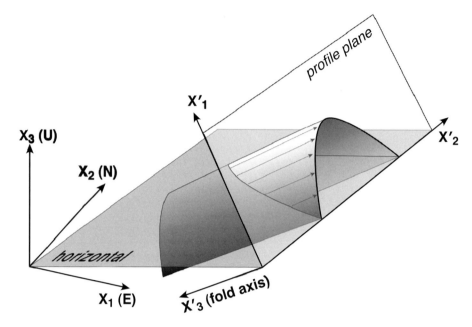

Figure 3.8 The down-plunge projection, showing the relation between its graphical construction and the right-hand coordinate system we will use below. The true profile plane is the plane that contains the X'_1 and X'_2 axes. The X'_2 axis corresponds to a folding line in the orthographic projection technique shown in Figure 3.7.

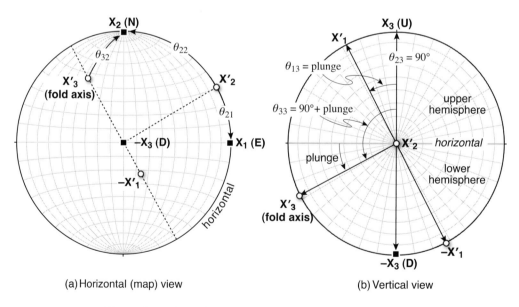

(a) Horizontal (map) view

(b) Vertical view

Figure 3.9 (a) Equal area lower hemisphere projection showing the angular relations between the two sets of axes in the down-plunge projection problem. ENU is the old coordinate system and the axes defined by the fold axis (X'_3) are the new coordinate system. Several of the angles that define the coordinate transformation (θ_{21}, θ_{22}, etc.) are shown. (b) The same coordinate transformation viewed in a vertical plane that contains the trend and plunge of the fold axis.

3.4 Examples of transformations in structural geology

obvious, as in the case of all of the new axes with respect to the old X_3 axis. For example, from Figure 3.9b it is clear that the angle between the new X'_3 and the old X_3 axes is equal to the fold axis plunge plus 90°. The angle between the X'_1 and X_3 axes is equal to the fold axis plunge, itself, and the angle between X'_2 and X_3 is just 90°. Thus, in terms of the direction cosines we can write

$$a_{13} = \cos(plunge)$$
$$a_{23} = \cos(90) = 0$$
$$a_{33} = \cos(90 + plunge) = -\sin(plunge)$$

Notice that, because all of these are with respect to one old axis, they are not all independent. If we can determine one more angle, we could use the orthogonality relations to calculate the rest. In fact, it will be easier to determine all of the angles directly in this example. The angles that X'_2 makes with the other two old axes are in a horizontal plane (Fig. 3.8) and therefore are just a function of the trend of the fold axis. Angle $(X'_2 X_1) = 360° -$ trend, and $(X'_2 X_2) =$ trend $- 270°$. This will give us two more direction cosines:

$$a_{21} = \cos(360 - trend) = \cos(trend)$$
$$a_{22} = \cos(trend - 270) = -\sin(trend)$$

The final direction cosines can be determined if we recall that they are nothing more than the direction cosines of the fold axis and its perpendicular (X'_1) in an east-north-up coordinate system. Thus, we can use the relations in Table 2.1 and modify them for the change in coordinate system. The direction cosines with respect to north and east will not change because $\cos(-plunge) = \cos(plunge)$. The cosine with respect to up will be equal to the $-\sin(plunge)$. Thus,

$$a_{31} = \sin(trend)\cos(plunge)$$
$$a_{32} = \cos(trend)\cos(plunge)$$

and the remaining direction cosines for the X'_1 axis can be calculated by projecting its negative into the lower hemisphere and then multiplying by -1:

$$a_{11} = -\sin(trend + 180)\cos(90 - plunge) = \sin(trend)\sin(plunge)$$
$$a_{12} = -\cos(trend + 180)\cos(90 - plunge) = \cos(trend)\sin(plunge)$$

Thus, we can combine all of the above equations and write out the transformation in shorthand form, as in Equation 3.3:

$$a_{ij} = \begin{pmatrix} \sin(trend)\sin(plunge) & \cos(trend)\sin(plunge) & \cos(plunge) \\ \cos(trend) & -\sin(trend) & 0 \\ \sin(trend)\cos(plunge) & \cos(trend)\cos(plunge) & -\sin(plunge) \end{pmatrix} \quad (3.10)$$

Now, to accomplish the down-plunge projection, substitute the direction cosines from Equation 3.10 into Equations 3.8 and coordinates in the new coordinate system can be calculated. In the actual projection, the v'_3 component is ignored because everything will be projected onto the $X'_1 X'_2$ plane. After that, it's just a matter of connecting the dots! The following MATLAB® function **DownPlunge** does the transformations for the down-plunge projection of a bed but it does not plot or connect the dots!

```
function dpbedseg = DownPlunge(bedseg,trd,plg)
%DownPlunge constructs the down plunge projection of a bed
%
% [dpbedseg] = DownPlunge(bedseg,trd,plg) constructs the down plunge
% projection of a bed from the X1 (East), X2 (North),
```

```
% and X3 (Up) coordinates of points on the bed (bedseg) and the
% trend (trd) and plunge (plg) of the fold axis
%
% The array bedseg is a two-dimensional array of size npoints x 3
% which holds npoints on the digitized bed, each point defined by
% 3 coordinates: X1 = East, X2 = North, X3 = Up
%
% NOTE: Trend and plunge of fold axis should be entered in radians

%Number of points in bed
nvtex = size(bedseg,1);

%Allocate some arrays
a=zeros(3,3);
dpbedseg = zeros(size(bedseg));

%Calculate the transformation matrix a(i,j). The convention is that
%the first index refers to the new axis and the second to the old axis.
%The new coordinate system is with X3' parallel to the fold axis, X1'
%perpendicular to the fold axis and in the same vertical plane, and
%X2' perpendicular to the fold axis and parallel to the horizontal. See
%equation 3.10
a(1,1) = sin(trd)*sin(plg);
a(1,2) = cos(trd)*sin(plg);
a(1,3) = cos(plg);
a(2,1) = cos(trd);
a(2,2) = -sin(trd);
a(2,3) = 0.0;
a(3,1) = sin(trd)*cos(plg);
a(3,2) = cos(trd)*cos(plg);
a(3,3) = -sin(plg);

%The east, north, up coordinates of each point to be rotated already define
%the coordinates of vectors. Thus we don't need to convert them to
%direction cosines (and don't want to either because they are not unit vectors)
%The following nested do-loops perform the coordinate transformation on the
%bed. The details of this algorithm are described in Chapter 4
for nv = 1:nvtex
    for i = 1:3
        dpbedseg(nv,i) = 0.0;
        for j = 1:3
            dpbedseg(nv,i) = a(i,j)*bedseg(nv,j) + dpbedseg(nv,i);
        end
    end
end
end
```

For example, say you want to construct the down-plunge projection of the contact between the white and gray units in Figure 3.7. Digitize the contact, and in a text editor make a file with

3.4 Examples of transformations in structural geology

the east, north, up coordinates of points on the contact, one point per line (coordinate entries can be separated by commas or spaces). Save this file as bedseg.txt. Now type in MATLAB:

```
load bedseg.txt; %Load bed
dpbedseg = DownPlunge(bedseg,90*pi/180,20*pi/180);% Down plunge projection
plot(dpbedseg(:,2),dpbedseg(:,1), 'k-'); %Plot bed
axis equal; %Make plot axes equal
```

You will get a chance to try this on a real structure in the exercises at the end of the chapter!

3.4.2 Rotation of orientation data

There are few operations more basic to structural geology than rotations. Unfolding lineations, paleomagnetic fold tests, and converting data measured on a thin section to its original geographic orientation all require rotations. The stereonet is a convenient graphic device for accomplishing rotations about a horizontal axis, but rotations about an inclined axis are more difficult. That is because points (lines) being rotated trace out small circles centered on the rotation axis. A stereonet only shows small circles centered on the horizontal. It can be done, but it is tedious.

A rotation is nothing more than a transformation of coordinate system and vectors. When we unfold linear elements, we are transforming from a geographic coordinate system to one pinned to bedding (or layering). Therefore, we should be able to use the mathematics developed in this chapter to determine the equations necessary to accomplish a general rotation about any axis in space. As before, we need to determine the transformation matrix that will allow us to transform the vectors representing our orientation measurements. The rotation axis is commonly specified by its trend and plunge, and the magnitude of rotation is given as an angle that is positive if the rotation is clockwise about the given axis (the old right-hand rule, again). The tricky part here is that the rotation axis does not generally coincide with the axes of either the new or the old coordinate system (unlike the previous example where the fold axis did define one of the new axes).

Ultimately we want to calculate the direction cosines for the transformation from the old axes to their new equivalents, rotated about the given rotation axis. Here we give the derivation for just one of the direction cosines, a_{22}; you can derive the rest yourself! In Figure 3.10, notice that, during the rotation, the X_2 axis tracks along a small circle centered on the rotation axis. The size of the circle, or in three dimensions the half-apical angle of the cone, is equal to the angle between the rotation axis and X_2, β. The angle between the new axis, X_2', and the rotation axis will also be β. Although, the points track along a small circle, the angle that we want to calculate is that between the new and old axes, θ_{22}, which is measured along a great circle (Fig. 3.10).

The simplest way to solve this problem is to use the law of cosines for spherical triangles. Notice that ω is the angle included between the two equal sides of the β-β-θ_{22} triangle (Fig. 3.10). Thus the appropriate formula to use is

$$\cos c = \cos a \cos b + \sin a \sin b \cos C$$

where $c = \theta_{22}$, $a = b = \beta$, and $C = \omega$. Substituting and rearranging, we get

$$\begin{aligned} a_{22} = \cos \theta_{22} &= \cos \beta \cos \beta + \sin \beta \sin \beta \cos \omega \\ &= \cos^2 \beta + (1 - \cos^2 \beta) \cos \omega \\ &= \cos \omega + \cos^2 \beta (1 - \cos \omega) \end{aligned} \quad (3.11a)$$

By the same reasoning the direction cosines for X_1–X_1' and X_3–X_3' are

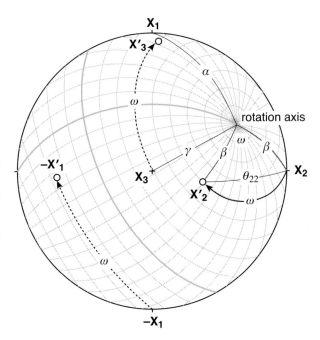

Figure 3.10 Lower hemisphere projection showing the geometry and angles involved in a general rotation about a plunging axis. The new axes are indicated by open circles and primed labels. The angle β is the angle between the rotation axis and the X_2 (east) axis, ω is the magnitude of the rotation, and θ_{22} is the angle between the old X_2 and the new X'_2 axes.

$$a_{11} = \cos\omega + \cos^2\alpha(1-\cos\omega)$$
$$a_{33} = \cos\omega + \cos^2\gamma(1-\cos\omega)$$
(3.11b)

We now have three equations and three independent unknowns. Therefore the remaining direction cosines can be calculated from the orthogonality relations or you can go through the somewhat more involved geometric derivation. They are given below without proof:

$$a_{12} = -\cos\gamma\sin\omega + \cos\alpha\cos\beta(1-\cos\omega)$$
$$a_{13} = \cos\beta\sin\omega + \cos\alpha\cos\gamma(1-\cos\omega)$$
$$a_{21} = \cos\gamma\sin\omega + \cos\beta\cos\alpha(1-\cos\omega)$$
$$a_{23} = -\cos\alpha\sin\omega + \cos\beta\cos\gamma(1-\cos\omega)$$
$$a_{31} = -\cos\beta\sin\omega + \cos\gamma\cos\alpha(1-\cos\omega)$$
$$a_{32} = \cos\alpha\sin\omega + \cos\gamma\cos\beta(1-\cos\omega)$$
(3.11c)

These equations give the direction cosines of the transformation matrix in terms of the direction cosines of the rotation axis and the magnitude of the rotation. All that is needed to accomplish a general rotation is to convert the trend and plunge of the rotation axis into direction cosines, and then use the transformation matrix in Equations 3.11 in the vector transformation Equations 3.8. Here is a MATLAB function, `Rotate`, to do a rotation about an arbitrary axis:

```
function [rtrd,rplg] = Rotate(raz,rdip,rot,trd,plg, ans0)
%Rotate rotates a line by performing a coordinate transformation on
%vectors. The algorithm was originally written by Randall A. Marrett
%
%   USE: [rtrd,rplg] = Rotate(raz,rdip,rot,trd,plg,ans0)
%
```

3.4 Examples of transformations in structural geology

```
%     raz = trend of rotation axis
%     rdip = plunge of rotation axis
%     rot = magnitude of rotation
%     trd = trend of the vector to be rotated
%     plg = plunge of the vector to be rotated
%     ans0 = A character indicating whether the line to be rotated is an axis
%     (ans0 = 'a') or a vector (ans0 = 'v')
%
%     NOTE: All angles are in radians
%
%     Rotate uses functions SphToCart and CartToSph

%Allocate some arrays
a = zeros(3,3); %Transformation matrix
pole = zeros(1,3); %Direction cosines of rotation axis
plotr = zeros(1,3); %Direction cosines of rotated vector
temp = zeros(1,3); %Direction cosines of unrotated vector

%Convert rotation axis to direction cosines. Note that the convention here
%is X1 = North, X2 = East, X3 = Down
[pole(1) pole(2) pole(3)] = SphToCart(raz,rdip,0);

% Calculate the transformation matrix
x = 1.0 - cos(rot);
sinRot = sin(rot); %Just reduces the number of calculations
cosRot = cos(rot);
a(1,1) = cosRot + pole(1)*pole(1)*x;
a(1,2) = -pole(3)*sinRot + pole(1)*pole(2)*x;
a(1,3) = pole(2)*sinRot + pole(1)*pole(3)*x;
a(2,1) = pole(3)*sinRot + pole(2)*pole(1)*x;
a(2,2) = cosRot + pole(2)*pole(2)*x;
a(2,3) = -pole(1)*sinRot + pole(2)*pole(3)*x;
a(3,1) = -pole(2)*sinRot + pole(3)*pole(1)*x;
a(3,2) = pole(1)*sinRot + pole(3)*pole(2)*x;
a(3,3) = cosRot + pole(3)*pole(3)*x;

%Convert trend and plunge of vector to be rotated into direction cosines
[temp(1) temp(2) temp(3)] = SphToCart(trd,plg,0);

%The following nested loops perform the coordinate transformation
for i = 1:3
    plotr(i) = 0.0;
    for j = 1:3
        plotr(i) = a(i,j)*temp(j) + plotr(i);
    end
end

%Convert to lower hemisphere projection if data are axes (ans0 = 'a')
```

```
if plotr(3) < 0.0 && ans0 == 'a'
    plotr(1) = -plotr(1);
    plotr(2) = -plotr(2);
    plotr(3) = -plotr(3);
end

%Convert from direction cosines back to trend and plunge
[rtrd,rplg]=CartToSph(plotr(1),plotr(2),plotr(3));
end
```

3.4.3 Graphical aside: Plotting great and small circles as a pole rotation

The transformation matrix we derived in the previous problem provides us with a simple and elegant way to draw great and small circles on any sort of spherical projection. The basic problem is, how to come up with a series of equally spaced points in the projection (lines in three dimensions) that one can connect with line segments to form the great or small circle. To solve this problem, we consider the pole to the great circle, or the axis of the conic section that defines the small circle, to be the rotation axis. Any vector perpendicular to the pole to the plane will, when rotated around the pole, trace out a plane that will intersect the projection sphere as a great circle. Likewise any vector that makes an angle of less than 90° will trace out a cone, which intersects the projection sphere as a small circle.

Thus, to make a program to draw great or small circles, you must first calculate the direction cosines of the pole to the plane or the center (axis) of the small circle. Then, pick a vector that lies somewhere on the great or small circle. If you are plotting a great circle, it is most convenient to choose the point where the circle intersects the primitive (i.e., the edge) of the projection. One of the main reasons for using a right-hand-rule format for specifying strike azimuths is that that vector will automatically trace out a lower hemisphere great circle when rotated 180° clockwise about the pole (a positive rotation). For small circles, you will probably want to choose the vector that has the minimum plunge (i.e., the vector with the same trend as the small circle axis and a plunge equal to the plunge of the axis minus the half apical angle of the small circle), unless the small circle intersects the edge of the stereographic projection, in which case the intersection is where you want to start.

From there, it is just a matter of rotating the vector a fixed increment and then drawing a line segment between the new and the old positions of the vector as projected on the net. This procedure is repeated until the total number of rotation increments equals 180° for a great circle or 360° for a small circle. On most computer screens, the resolution is such that 20 rotations in 9° increments (or something similar) will produce a reasonably smooth great circle. Smaller increments are time consuming and may actually produce a rougher great circle. The following MATLAB functions, **GreatCircle** and **SmallCircle**, use rotations to calculate the traces of great and small circles in equal area and equal angle projections:

```
function path = GreatCircle(strike,dip,sttype)
%GreatCircle computes the great circle path of a plane in an equal angle
%or equal area stereonet of unit radius
%
%   USE: path = GreatCircle(strike,dip,sttype)
%
%   strike = strike of plane
```

3.4 Examples of transformations in structural geology

```
%   dip = dip of plane
%   sttype = type of stereonet. 0 for equal angle and 1 for equal area
%   path = vector with x and y coordinates of points in great circle path
%
%   NOTE: strike and dip should be entered in radians.
%
%   GreatCircle uses functions StCoordLine, Pole and Rotate

%Compute the pole to the plane. This will be the axis of rotation to make
%the great circle
[trda,plga] = Pole(strike,dip,1);

%Now pick a line at the intersection of the great circle with the primitive
%of the stereonet
trd = strike;
plg = 0.0;

%To make the great circle, rotate the line 180 degrees in increments
%of 1 degree
rot=(0:1:180)*pi/180;
path = zeros(size(rot,2),2);
for i = 1:size(rot,2)
    %Avoid joining ends of path
    if rot(i) == pi
        rot(i) = rot(i)*0.9999;
    end
    %Rotate line
    [rtrd,rplg] = Rotate(trda,plga,rot(i),trd,plg,'a');
    %Calculate stereonet coordinates of rotated line and add to great
    %circle path
    [path(i,1),path(i,2)] = StCoordLine(rtrd,rplg,sttype);
end
end

function [path1,path2,np1,np2] = SmallCircle(trda,plga,coneAngle,sttype)
%SmallCircle computes the paths of a small circle defined by its axis and
%cone angle, for an equal angle or equal area stereonet of unit radius
%
%   USE: [path1,path2,np1,np2] = SmallCircle(trda,plga,coneAngle,sttype)
%
%   trda = trend of axis
%   plga = plunge of axis
%   coneAngle = cone angle
%   sttype = type of stereonet. 0 for equal angle and 1 for equal area
%   path1 and path2 are vectors with the x and y coordinates of the points
%   in the small circle paths
%   np1 and np2 are the number of points in path1 and path2,
%   respectively
%
```

```
%   NOTE: All angles should be in radians
%
%   SmallCircle uses functions ZeroTwoPi, StCoordLine and Rotate

%Find where to start the small circle
if (plga - coneAngle) >= 0.0
    trd = trda;
    plg = plga - coneAngle;
else
    if plga == pi/2.0
        plga = plga * 0.9999;
    end
    angle = acos(cos(coneAngle)/cos(plga));
    trd = ZeroTwoPi(trda+angle);
    plg = 0.0;
end

%To make the small circle, rotate the starting line 360 degrees in
%increments of 1 degree
rot=(0:1:360)*pi/180;
path1 = zeros(size(rot,2),2);
path2 = zeros(size(rot,2),2);
np1 = 0; np2 = 0;
for i = 1:size(rot,2)
    %Rotate line: Notice that here the line is considered as a vector
    [rtrd,rplg] = Rotate(trda,plga,rot(i),trd,plg,'v');
    % Add to the right path
    % If plunge of rotated line is positive add to first path
    if rplg >= 0.0
        np1 = np1 + 1;
        %Calculate stereonet coordinates and add to path
        [path1(np1,1),path1(np1,2)] = StCoordLine(rtrd,rplg,sttype);
    %If plunge of rotated line is negative add to second path
    else
        np2 = np2 + 1;
        %Calculate stereonet coordinates and add to path
        [path2(np2,1),path2(np2,2)] = StCoordLine(rtrd,rplg,sttype);
    end
end
end
```

Normally, stereonets are presented with the primitive equal to the horizontal (i.e., looking straight down). However, it is often convenient to construct a stereonet in another orientation. For example, one may want to plot data in the plane of a cross section (a view direction that is horizontal and perpendicular to the trend of the cross section), or in the down-plunge view of a cylindrical fold (a view direction parallel to the fold axis). The MATLAB function **GeogrToView** below enables one to calculate a stereonet looking in any direction, by transforming any point in the stereonet from **NED** coordinates to view direction coordinates.

3.4 Examples of transformations in structural geology

```
function [rtrd,rplg] = GeogrToView(trd,plg,trdv,plgv)
%GeogrToView transforms a line from NED to View Direction
%coordinates
%
%   USE: [rtrd,rplg] = Geogr To View(trd,plg,trdv,plgv)
%
%   trd = trend of line
%   plg = plunge of line
%   trdv = trend of view direction
%   plgv = plunge of view direction
%   rtrd and rplg are the new trend and plunge of the line in the view
%   direction.
%
%   NOTE: Input/Output angles are in radians
%
%   GeogrToView uses functions ZeroTwoPi, SphToCart and CartToSph

%Some constants
east = pi/2.0;

% Make transformation matrix between NED and View Direction
a = zeros(3,3);
[a(3,1),a(3,2),a(3,3)] = SphToCart(trdv,plgv,0);
temp1 = trdv + east;
temp2 = 0.0;
[a(2,1),a(2,2),a(2,3)] = SphToCart(temp1,temp2,0);
temp1 = trdv;
temp2 = plgv - east;
[a(1,1),a(1,2),a(1,3)] = SphToCart(temp1,temp2,0);

% Direction cosines of line
dirCos = zeros(1,3);
[dirCos(1),dirCos(2),dirCos(3)] = SphToCart(trd,plg,0);
% Transform line
nDirCos = zeros(1,3);
for i=1:3
    nDirCos(i) = a(i,1)*dirCos(1) + a(i,2)*dirCos(2)+ a(i,3)*dirCos(3);
end

% Compute line from new direction cosines
[rtrd,rplg] = CartToSph(nDirCos(1),nDirCos(2),nDirCos(3));

% Take care of negative plunges
if rplg < 0.0
    rtrd = ZeroTwoPi(rtrd+pi);
    rplg = -rplg;
end
end
```

Now we put all of the previous routines together in a function, **Stereonet**, that plots an equal area or equal angle stereonet in any view direction you want. This code is very short and efficient because it calls several of the previous functions in this chapter and Chapters 1 and 2.

```
function [] = Stereonet(trdv,plgv,intrad,sttype)
%Stereonet plots an equal angle or equal area stereonet of unit radius
%in any view direction
%
%   USE: Stereonet(trdv,plgv,intrad,stttype)
%
%   trdv = trend of view direction
%   plgv = plunge of view direction
%   intrad = interval in radians between great or small circles
%   sttype = An integer indicating the type of stereonet. 0 for equal angle,
%   and 1 for equal area
%
%   NOTE: All angles should be entered in radians
%
%   Example: To plot an equal area stereonet at 10 deg intervals in a
%   default view direction type:
%
%   Stereonet(0,90*pi/180,10*pi/180,1);
%
%   To plot the same stereonet but with a view direction of say: 235/42,
%   type:
%
%   Stereonet(235*pi/180,42*pi/180,10*pi/180,1);
%
%   Stereonet uses functions Pole, GeogrToView, SmallCircle and GreatCircle

% Some constants
east = pi/2.0;
west = 3.0*east;

% Plot stereonet reference circle
r = 1.0; % radius of stereonet
TH = (0:1:360)*pi/180; % polar angle, range 2 pi, 1 degree increment
[X,Y] = pol2cart(TH,r); % cartesian coordinates of reference circle
plot(X,Y,'k'); % plot reference circle
axis ([-1 1 -1 1]); % size of stereonet
axis equal; axis off; % equal axes, no axes
hold on; % hold plot

% Number of small circles
nCircles = pi/(intrad*2.0);
% Small circles
% Start at the North
trd = 0.0;
plg = 0.0;
```

3.4 Examples of transformations in structural geology 63

```
% If view direction is not the default (trd=0,plg=90), transform line to
% view direction
if trdv ~= 0.0 || plgv ~= east
    [trd,plg] = GeogrToView(trd,plg,trdv,plgv);
end
% Plot small circles
for i = 1:nCircles
    coneAngle = i*intrad;
    [path1,path2,np1,np2] = SmallCircle(trd,plg,cone Angle,sttype);
    plot(path1(1:np1,1),path1(1:np1,2),'b');
    if np2 > 0
        plot(path2(1:np2,1),path2(1:np2,2),'b');
    end
end

% Great circles
for i = 0:nCircles*2
    %Western half
    if i <= nCircles
        % Pole of great circle
        trd = west;
        plg = i*intrad;
    %Eastern half
    else
        % Pole of great circle
        trd = east;
        plg = (i-nCircles)*intrad;
    end
    % If pole is vertical, shift it a little bit
    if plg == east
        plg = plg * 0.9999;
    end
    % If view direction is not the default (trd=0,plg=90), transform line to
    % view direction
    if trdv ~= 0.0 || plgv ~= east
        [trd,plg] = GeogrToView(trd,plg,trdv,plgv);
    end
    % Compute plane from pole
    [strike,dip] = Pole(trd,plg,0);
    % Plot great circle
    path = GreatCircle(strike,dip,sttype);
    plot(path(:,1),path(:,2),'b');
end
hold off; %release plot
end
```

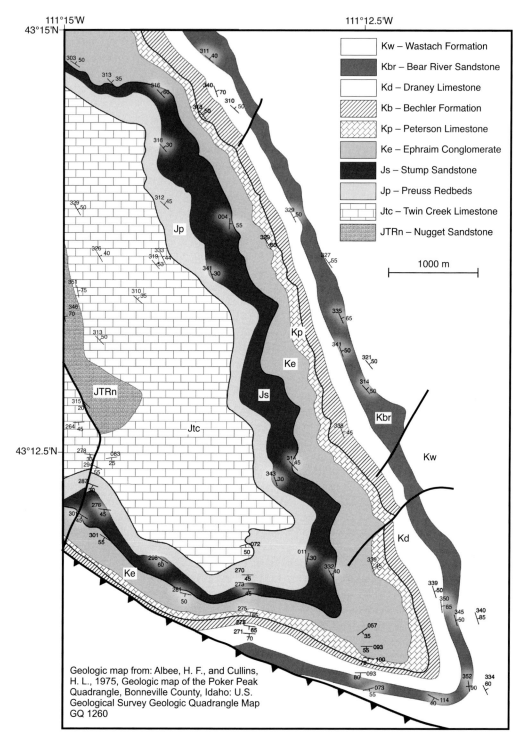

Figure 3.11 Simplified geologic map of the Big Elk anticline in southeastern Idaho, to accompany Exercise 8.

3.5 EXERCISES

1. Derive the equation for component a_{21} of the transformation matrix for the case of a general rotation.
2. Derive the transformation matrix of Equation 3.10, but this time as a vector transformation (Eq. 3.8) between the X_1 = north, X_2 = east, X_3 = up coordinate system and the fold axis based $X_1'-X_2'-X_3'$ coordinate system.
3. Derive the transformation matrix for a down-plunge projection in the right-handed coordinate system, X_1 = south, X_2 = west, X_3 = down.
4. Evaluate the problem of construction of a vertical section of a plunging cylindrical fold. Can this problem be carried out as a transformation of coordinates and points? If so, derive the transformation matrix; if not precisely state why not.
5. Construct the down-plunge projection of the contact between the gray and white units in Figure 3.7, using the MATLAB function **DownPlunge**.
6. Using the function **Stereonet**, plot equal area stereonets with 10° grid interval, and the following view directions: 123/42, 032/57, 245/21, 321/49.
7. Plot in MATLAB the following lines and planes in equal area stereonets with 10° grid interval, and view directions 000/90 and 214/56. Lines = 212/23, 014/56, 321/53. Planes = 211/24, 035/67, 238/76. Hint: Use functions **StCoordLine** (Chapter 1), **GreatCircle**, **GeogrToView**, and **Stereonet** (this chapter).
8. Figure 3.11 is a geologic map of the Big Elk anticline, located in the Mesozoic thrust belt in southeastern Idaho, United States (Albee and Cullins, 1975).
    a. The trend and plunge of the fold axis is 125/26. In Chapter 5, we will return to this example once you have learned how to calculate a best-fitting fold axis.
    b. Supplementary data file "Problem 3.8" contains the digitized contacts (east, north, up) of the top of the Jurassic Twin Creek Limestone (Jtc), the Jurassic Stump Sandstone (Js), and the Cretaceous Peterson Limestone (Kp). Using the equations and functions (e.g., **DownPlunge**) developed in this chapter, construct a down-plunge section of the Big Elk anticline.
    c. The Idaho–Wyoming thrust belt in which this structure occurs thrusts from west to east. What is the vergence (i.e., asymmetry) of the Big Elk anticline and does it agree with the general direction of thrusting? Do you note anything unusual about the sequence between Jtc and Js? This sequence contains the Preuss redbeds, which are known to contain evaporate minerals. Can you draw any conclusions with this additional information?

CHAPTER FOUR

Matrix operations and indicial notation

4.1 INTRODUCTION

Up to this point, we have successfully avoided introducing any unfamiliar mathematical concepts or strange symbology. All of the equations that have been presented are, individually, very simple, involving nothing more than addition and multiplication. There are a lot of them, however, and it gets tedious to keep rewriting very similar - and more importantly, predictable - equations over and over again. What we need is a shorthand way of writing things down that makes it easier on us while at the same time preserving, or even enhancing, the logic behind them. It should come as no surprise that such shorthand devices are readily available, and we will concentrate on two of them in this chapter: matrix notation and the indicial notation, including the Einstein summation convention. Although some of what follows may look exotic, just remember that the equations represented are no more complex than what we've seen before.

4.2 INDICIAL NOTATION

We have already been introduced, briefly, to indicial notation in Chapters 2 and 3. Instead of writing out components of, say, the transformation matrix in the previous chapter:

$$\mathbf{a} = \begin{pmatrix} a_{11} & a_{12} & a_{13} \\ a_{21} & a_{22} & a_{23} \\ a_{31} & a_{32} & a_{33} \end{pmatrix} \tag{4.1}$$

we can write it much more quickly as

$$\mathbf{a} = a_{ij} \quad (i, j = 1, 2, 3) \tag{4.2}$$

4.2 Indicial notation

where i and j refer to the new and the old axes, respectively. The expression in parentheses means that both indices can have values of 1, 2, or 3. Likewise, vectors can be written as

$$\mathbf{v} = v_i \quad (i = 1, 2, 3) \tag{4.3}$$

Generally, the expression in parentheses is omitted unless it is needed for clarity. In our three-dimensional Cartesian coordinate system, each index will always have a value of 1, 2, or 3; in two dimensions, 1 or 2.

There is a confusing variety in how the suffixes may be written depending on the author, the coordinate system, and the type of quantity being represented. All of the following may be encountered at one time or another:

$$V_i, T_i^j, f_{ij}, R^i{}_{jk}, \text{ etc.}$$

In this book, we will only have to deal with single or double subscripts; other formats will be avoided.

4.2.1 Einstein summation convention

Although the indicial notation saves us some time in writing down equations like (4.1), its real power lies in the ability to represent in a short space, long repetitive calculations. Many things that we do with vectors involve adding up their components in various ways. For example, take the equation for the magnitude of a vector in Chapter 2, which is

$$|\mathbf{v}| = \left(v_1^2 + v_2^2 + v_3^2\right)^{1/2}$$

Using a summation sign, Σ, we can write this equation somewhat more compactly as

$$|\mathbf{v}| = \left(\sum_{i=1}^{3} v_i^2\right)^{1/2} = \left(\sum_{i=1}^{3} v_i v_i\right)^{1/2}$$

But, even this is more than we need to do. We can state that, because the subscript i occurs twice on the right-hand side of the preceding equation, it is assumed that the summation occurs with respect to that index. This convention is known as the *Einstein summation convention*. Thus, we can write

$$|\mathbf{v}| = (v_i v_i)^{1/2} \tag{4.4}$$

Let's apply this to a more complex situation, the equations for the transformation of a vector, which were derived in the last chapter:

$$\begin{aligned} v'_1 &= a_{11} v_1 + a_{12} v_2 + a_{13} v_3 \\ v'_2 &= a_{21} v_1 + a_{22} v_2 + a_{23} v_3 \\ v'_3 &= a_{31} v_1 + a_{32} v_2 + a_{33} v_3 \end{aligned} \tag{4.5}$$

Notice that, on the right-hand side of each of the three equations, the second subscript of a and the single subscript of v increase systematically from 1 to 3. That suggests that we can write the equations as summations based on those subscripts:

$$v'_1 = \sum_{j=1}^{3} a_{1j}v_j$$
$$v'_2 = \sum_{j=1}^{3} a_{2j}v_j \qquad (4.6)$$
$$v'_3 = \sum_{j=1}^{3} a_{3j}v_j$$

Now that j occurs twice on the right-hand side of Equations 4.6, we can use our new-found summation convention so that:

$$v'_1 = a_{1j}v_j$$
$$v'_2 = a_{2j}v_j \qquad (4.7)$$
$$v'_3 = a_{3j}v_j$$

But, this is still too much work! The remaining indices 1, 2, and 3 now occur one on each side of Equations 4.7. So, we simply represent them as another letter index, in this case i, so that we can write the above three equations as a single one:

$$v'_i = a_{ij}v_j \qquad (4.8)$$

Clearly, we have saved ourselves a lot of tedious effort transcribing equations, not to mention potential errors, by reducing Equations 4.5 down to Equation 4.8. In general, in the Einstein summation convention, whatever subscript is repeated on one side of the equation is known as the *dummy suffix*; the summation within a single equation always occurs with respect to that suffix. The *free suffix* occurs only once on each side of the equation. There will be as many equations as there are values of the free suffix and each equation will have as many terms as there are dummy suffix values. So, assuming that $(i, j = 1, 2, 3)$, Equation 4.8 represents three separate equations (i is the free suffix) each of which has three terms (j is the dummy suffix). Note that in the case of Equation 4.8 we can reverse the order of the a and the v terms on the right-hand side without changing the meaning of the expression, but *we cannot change the order of the suffixes*:

$$v'_i = a_{ij}v_j = v_j a_{ij} \neq a_{ji}v_j \qquad (4.9)$$

It should be emphasized that, to be a dummy suffix, the subscript has to be repeated *within the same term* on the right-hand side of the equation. In the equation (which is not a real equation!)

$$v'_j = a_j + v_j$$

j is a free suffix, not a dummy suffix, and therefore three equations are indicated, one for each value of j.

4.2.2 Summation convention as a compact computer program

For those with some experience with computers, it is particularly useful to think of the summation convention as a kind of compact computer program. In the case of Equation 4.8, we have three arrays, two with dimensions of 1×3 (v_j and v'_i) and one with a dimension of 3×3 (a_{ij}). The summation about the dummy suffix, j, can be thought of as an inner loop and the free suffix, i, defines an outer. Thus, in MATLAB® we would program 4.8 as:

4.3 Matrix notation and operations

```
%vold is a 1 x 3 vector with old coordinates
%a is the 3 x 3 transformation matrix between old and new coordinates
vnew = zeros(1,3) %Initialize vector with new coordinates
for i=1:3 %i is the free suffix
      for j=1:3 %j is the dummy suffix
            vnew(i) = a(i,j)*vold(j) + vnew(i);
      end
end
```

Note that the indices of the arrays in this program appear in just the same order as the subscripts in Equation 4.8. Readers who have some programming experience should study this example carefully because it will make it easier to understand the more complex summation equations that we will encounter in the next chapter.

4.3 MATRIX NOTATION AND OPERATIONS

The summation convention introduced in the previous section will be used for most of our calculations because it is particularly easy to understand and because of the readiness with which it is translated into computer code. Also, the indices relate directly to the axes of our chosen coordinate system. However, there are some operations that are better expressed in matrix, rather than indicial notation. More importantly, one must make a distinction between a matrix as a mathematical concept and matrix notation. The latter, like indicial notation, is nothing more than simple shorthand though in many ways less immediately graspable. The use of a particular notation is commonly an either-or proposition, but in either case, we are fundamentally dealing with matrices.

Most structural geologists are conversant with the concept of matrices. In fact, we have already used the concept in our representation of the transformation matrix in the previous chapter (Section 3.2.2). In its simplest form, a matrix simply represents a rectangular table of numbers which may, or may not, represent a physical entity and may, or may not, be related to each other. In a computer program, any *array* of numbers is a matrix. The transformation matrix, **a**, contains nine numbers (Eq. 4.1), only three of which are independent of each other. The numbers in this case, however, do not represent any particular physical entity; they merely represent an arbitrary change of axes that is governed only by our whim and not by any set of physical conditions or constraints.

There are other matrices, however, which *do* represent tangible quantities that really exist. A vector can be described as a rectangular table of three numbers (in our three-dimensional Cartesian coordinate system):

$$v_i = [v_1 \quad v_2 \quad v_3] \tag{4.10}$$

The vector that represents the displacement of an element of matter, from one point to another, is something that is not dependent on our fancy but on something that really happened. Once the displacement has occurred, we can represent it with different numbers by changing the coordinate system, but we cannot change the fundamental nature of the displacement itself.

4.3.1 Notation and conventions

We use matrices to represent both the groups of numbers with, and without, physical significance. When a single letter is used to indicate a matrix, it appears in bold face, as on the left sides of Equations 4.1 and 4.2. In this book, if the matrix portrays a physical entity then its components are enclosed in square brackets, [ ], as in Equation 4.10, otherwise round brackets or parentheses, ( ), are used. This is the convention followed in Nye (1985) but it is by no means universal.

The only matrices that we will be concerned with in this book are simple, rectangular arrays of numbers. *Square matrices*, for example, have the same number of columns and rows; the first subscript of an individual component refers to its row number and the second to its column number. Thus, element a_{23} occurs in row two and column three, as in Equation 4.1. If we want to refer to the components in more general terms, it will be with indicial notation with i indicating row number and j the column number: a_{ij}. In a square matrix, the elements in which both subscripts are the same are collectively named the main or *principal diagonal*. If all of the elements in a matrix except the principal diagonal are equal to zero, then the matrix is called a *diagonal matrix*.

The *identity matrix* is a diagonal matrix in which the principal diagonal is entirely made up of ones:

$$\mathbf{I} = \delta_{ij} = \begin{pmatrix} 1 & 0 & 0 \\ 0 & 1 & 0 \\ 0 & 0 & 1 \end{pmatrix} \tag{4.11}$$

The indicial representation of this matrix, δ_{ij}, is given a special name, the *Kronecker delta*. As we will see below, the Kronecker delta has a number of useful applications. One of the handy properties of the Kronecker delta is to allow us to substitute one index of a vector for another:

$$v_i = \delta_{ij} v_j \tag{4.12}$$

Matrices can also have an unequal number of rows and columns. When we talk about an $m \times n$ matrix, the first letter (or number) tells us the number of rows and the second the number of columns. Equation 4.10 is a 1×3 matrix.

4.3.2 Elementary matrix operations

Multiplication by a single number

Any matrix can be multiplied by a single number simply by multiplying each one of its individual components by that number:

$$z\mathbf{P} = \begin{pmatrix} zP_{11} & zP_{12} & zP_{13} \\ zP_{21} & zP_{22} & zP_{23} \\ zP_{31} & zP_{32} & zP_{33} \end{pmatrix} \tag{4.13}$$

The multiplication of a vector by a scalar is an example of this sort of operation (Chapter 2).

Matrix addition

If two matrices have the same number of rows and columns, they can be added together by adding each component to its equivalent in the other matrix:

$$\mathbf{P} + \mathbf{Q} = \begin{pmatrix} (P_{11} + Q_{11}) & (P_{12} + Q_{12}) & (P_{13} + Q_{13}) \\ (P_{21} + Q_{21}) & (P_{22} + Q_{22}) & (P_{23} + Q_{23}) \\ (P_{31} + Q_{31}) & (P_{32} + Q_{32}) & (P_{33} + Q_{33}) \end{pmatrix} \tag{4.14}$$

This is the type of operation we do when we add two vectors together (Eq. 2.10).

4.3 Matrix notation and operations

Matrix multiplication

Two matrices can be multiplied together *only* if the number of columns in the first matrix matches the number of rows in the second. If two matrices have this property, they are said to be *conformable*. The resulting matrix has the same number of rows as the first matrix and the same number of columns as the second:

$$\mathbf{A}_{(m \times n)} \mathbf{B}_{(n \times k)} = \mathbf{C}_{(m \times k)} \tag{4.15}$$

With respect to the above equation, it is as if the n that the two matrices have in common is "canceled out." Notice that this operation is not reversible, that is, **AB** is conformable but **BA** is not because the number of columns of **B** (k) is not necessarily equal to the number of rows of **A** (m). The best way to describe how to carry out matrix multiplication is to write it in terms of the summation convention:

$$C_{ij} = A_{ik} B_{kj} \tag{4.16}$$

In this equation, k is the dummy suffix and i and j are the free suffixes. Thus, if the three suffixes each have values between 1 and 3, Equation 4.16 represents nine equations, each with three terms. The expansion for two of the terms is shown below:

$$C_{22} = A_{21}B_{12} + A_{22}B_{22} + A_{23}B_{32}$$
$$C_{31} = A_{31}B_{11} + A_{32}B_{21} + A_{33}B_{31}$$

In matrix notation, the dot product of two vectors can be calculated by representing the first vector as a 1×3 row matrix and the second vector as a 3×1 column matrix. This operation will yield a single number that is the sum of the products of the components of the two matrices:

$$\mathbf{u} \cdot \mathbf{v} = \mathbf{uv} = \begin{bmatrix} u_1 & u_2 & u_3 \end{bmatrix} \begin{bmatrix} v_1 \\ v_2 \\ v_3 \end{bmatrix} = [u_1 v_1 + u_2 v_2 + u_3 v_3] \tag{4.17}$$

The summation notation representation of Equation 4.17 is

$$\mathbf{uv} = v_i u_i \tag{4.18}$$

You can see that the result of Equation 4.17 is identical to Equation 2.11. Be very careful with this, however, because while $\mathbf{u} \cdot \mathbf{v} = \mathbf{v} \cdot \mathbf{u}$, it is clearly not the case with matrix notation. You can see that $\mathbf{uv} \neq \mathbf{vu}$, because the left side of this equation yields a single number (i.e., 1×1 matrix) whereas the right side of the equation yields a 3×3 matrix!

$$\mathbf{vu} = \begin{bmatrix} v_1 \\ v_2 \\ v_3 \end{bmatrix} \begin{bmatrix} u_1 & u_2 & u_3 \end{bmatrix} = \begin{bmatrix} v_1 u_1 & v_1 u_2 & v_1 u_3 \\ v_2 u_1 & v_2 u_2 & v_2 u_3 \\ v_3 u_1 & v_3 u_2 & v_3 u_3 \end{bmatrix} \tag{4.19}$$

In summation notation, we would write Equation 4.19 as

$$\mathbf{vu} = v_i u_j \tag{4.20}$$

Notice that the difference between Equations 4.18 and 4.20 is immediately obvious when written using indicial notation. The matrix multiplication of the two vectors, **vu**, as in Equations 4.19 and 4.20 is a special type of feature known as a tensor or *dyad product*, which we will see in the next chapter.

Transpose of a matrix

There are times, when dealing with square matrices, when it is necessary to interchange the columns and rows. This operation is called the *transpose* of a matrix and is denoted by a small superscript T.[1] For example, the transpose of the transformation matrix would be

$$\mathbf{a}^T = \begin{pmatrix} a_{11} & a_{21} & a_{31} \\ a_{12} & a_{22} & a_{32} \\ a_{13} & a_{23} & a_{33} \end{pmatrix} \quad (4.21)$$

If a square matrix is equal to its transpose, that is

$$\mathbf{C} = \mathbf{C}^T \text{ or } C_{ij} = C_{ji} \quad (4.22)$$

then the matrix is said to be *symmetric*. But, suppose we have the condition that $C_{ij} = -C_{ij}$. This can only be true if the principal diagonal of the matrix is all zeros (i.e., $C_{11} = -C_{11}$ only if $C_{11} = 0$, etc.). Matrices of this form are known as *antisymmetric* or *skew* matrices. The concept of symmetric and antisymmetric matrices will be very important in our discussion of strain later on in this book. An *orthogonal matrix* is one that, when multiplied by its transpose, is equal to the identity matrix. We will show below that the transformation matrix, **a**, has this property:

$$\mathbf{a}\mathbf{a}^T = \mathbf{I} \quad (4.23)$$

Although we gave a formula for the dot product in Equations 4.17 and 4.18, it is hardly the most logical way to write the expression because it is more natural to think of the two vectors as both row or both column vectors. The transpose gives us a way around this because the transpose of a row vector is a column vector. Thus, we can rewrite Equations 4.17 and 4.19 as

$$\mathbf{u} \cdot \mathbf{v} = \mathbf{u}\mathbf{v}^T \quad (4.24)$$
$$\mathbf{v} \otimes \mathbf{u} = \mathbf{v}^T\mathbf{u} \quad (4.25)$$

The $\otimes$ represents the dyad product of two vectors.

The matrix operations described in this chapter are one of MATLAB's specialties and thus can commonly be carried out with a single-line command. For people who understand linear algebra, this makes things very easy, but there is a great temptation to use the one-line commands as a black box. Thus, below we show the long way of carrying out these operations, as well as providing you with the one-line MATLAB equivalents. Those who wish to accomplish these operations in a different programming language will find the translations straightforward.

```
function c = MultMatrix(a,b)
%MultMatrix multiplies two conformable matrices
%
% USE: c = MultMatrix(a,b)
%
% Matrix a premultiplies matrix b to produce matrix c, as in the equation
% c = ab
```

---

[1] Nye (1985) denotes the transpose of a matrix by a small subscript t: $\mathbf{C}_t$.

4.3 Matrix notation and operations

```matlab
%
% NOTE: This function is only for illustration purposes. To multiply
% matrices MATLAB use the * operator (e.g. c = a*b)

aRow = size(a,1); %Number of rows in a
aCol = size(a,2); %Number of columns in a
bRow = size(b,1); %Number of rows in b
bCol = size(b,2); %Number of columns in b

%If the multiplication is conformable
if aCol == bRow
    %Initialize c
    c = zeros(aRow,bCol);
    for i = 1:aRow     % note the use of the nested loops
        for j = 1:bCol % to do the matrix multiplication
            for k = 1:aCol
                c(i,j) = a(i,k)*b(k,j) + c(i,j);
            end
        end
    end
%Else report an error
else
    error('Error: Matrices are not conformable');
end
end


function c = Transpose(a)
%Transpose calculates the transpose of a matrix
%
%   USE: c = Transpose(a)
%
%   The original matrix is a; the transpose of a is returned in c
%
%   NOTE: This function is only for illustration purposes. To get the
%   transpose of a matrix in MATLAB use the ' operator (e.g. c = a')

%Number of rows and columns in a
n = size(a,1);
m = size(a,2);
%Initialize c. Note the switch of number of rows and columns here
c = zeros(m,n);

for i = 1:n
    for j = 1:m
        c(j,i) = a(i,j); %Note the switch of indices, i & j here
    end
end
end
```

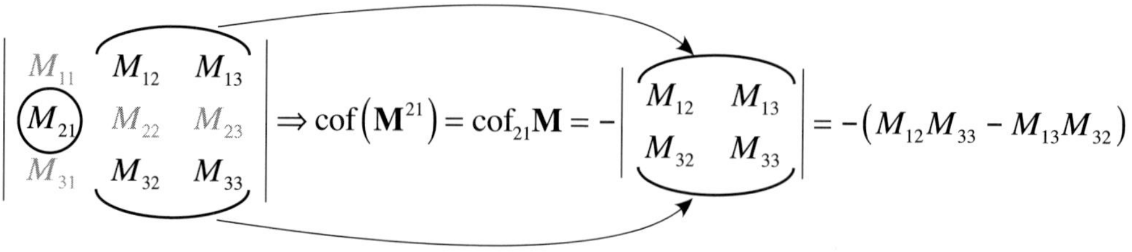

$$\begin{vmatrix} M_{11} & M_{12} & M_{13} \\ M_{21} & M_{22} & M_{23} \\ M_{31} & M_{32} & M_{33} \end{vmatrix} \Rightarrow \text{cof}(\mathbf{M}^{21}) = \text{cof}_{21}\mathbf{M} = -\begin{vmatrix} M_{12} & M_{13} \\ M_{32} & M_{33} \end{vmatrix} = -(M_{12}M_{33} - M_{13}M_{32})$$

Figure 4.1 How to construct the cofactor of element M_{21} of matrix **M**.

4.3.3 The determinant and inverse of a matrix

Determinant of a matrix

There is a single scalar function of square matrices, known as the *determinant*; it is represented by vertical lines on either side of the matrix or by the letters "*det*" preceding the matrix. For a simple 2×2 matrix, the determinant is easy to calculate:

$$\det \mathbf{C} = |\mathbf{C}| = \begin{vmatrix} C_{11} & C_{12} \\ C_{21} & C_{22} \end{vmatrix} = C_{11}C_{22} - C_{12}C_{21} \tag{4.26}$$

For larger matrices, calculating the determinant is considerably more difficult. In general, one finds the cofactors – that is, the determinants of subsets of the matrix – and multiplies them times their corresponding elements. For example, the cofactor, M_{21}, of a 3×3 matrix, **M**, is determined by taking the negative determinant of the sub-matrix that does not include either the row or column of the cofactor, itself. If $i + j$ is even, then you take the positive determinant of the sub-matrix. That is probably pretty obscure, but perhaps diagramming it out will help (Fig. 4.1):

The cofactor is negative because $i + j = 2 + 1 = 3$ is an odd number. Thus, we can define the determinant of the entire matrix by

$$|\mathbf{M}| = \begin{vmatrix} M_{11} & M_{12} & M_{13} \\ M_{21} & M_{22} & M_{23} \\ M_{31} & M_{32} & M_{33} \end{vmatrix} = M_{11}\text{cof}_{11}(\mathbf{M}) - M_{12}\text{cof}_{12}(\mathbf{M}) + M_{13}\text{cof}_{13}(\mathbf{M}) \tag{4.27}$$

Expanding the right side of this equation, we get

$$\begin{aligned} \det \mathbf{M} = |\mathbf{M}| &= M_{11}(M_{22}M_{33} - M_{23}M_{32}) \\ &+ M_{12}(M_{23}M_{31} - M_{21}M_{33}) \\ &+ M_{13}(M_{21}M_{32} - M_{22}M_{31}) \end{aligned} \tag{4.28}$$

or,

$$\begin{aligned} \det \mathbf{M} = |\mathbf{M}| &= M_{11}M_{22}M_{33} + M_{12}M_{23}M_{31} + M_{13}M_{21}M_{32} \\ &- M_{13}M_{22}M_{31} - M_{11}M_{23}M_{32} - M_{12}M_{21}M_{33} \end{aligned}$$

The cofactor method can be used to calculate the determinants of square matrices with orders higher than 3, but we will seldom need to do so in this book. Below, we show how to calculate the cofactors and determinant for a 3×3 matrix.

4.3 Matrix notation and operations

```
function cofac = CalcCofac(a)
%CalcCofac calculates all of the cofactor elements for a 3 x 3 matrix
%
%   USE: cofac = CalcCofac(a)
%
%   a is the matrix and cofac are the cofactor elements

%Number of rows and columns in a
n = size(a,1);
m = size(a,2);

%If matrix is 3 x 3
if n == 3 && m == 3
    %Initialize cofactor
    cofac = zeros(3,3);
    %Calculate cofactor. When i+j is odd, the cofactor is negative
    cofac(1,1) = a(2,2)*a(3,3) - a(2,3)*a(3,2);
    cofac(1,2) = -(a(2,1)*a(3,3) - a(2,3)*a(3,1));
    cofac(1,3) = a(2,1)*a(3,2) - a(2,2)*a(3,1);

    cofac(2,1) = -(a(1,2)*a(3,3) - a(1,3)*a(3,2));
    cofac(2,2) = a(1,1)*a(3,3) - a(1,3)*a(3,1);
    cofac(2,3) = -(a(1,1)*a(3,2) - a(1,2)*a(3,1));

    cofac(3,1) = a(1,2)*a(2,3) - a(1,3)*a(2,2);
    cofac(3,2) = -(a(1,1)*a(2,3) - a(1,3)*a(2,1));
    cofac(3,3) = a(1,1)*a(2,2) - a(1,2)*a(2,1);
else
    error('Matrix is not 3 x 3');
end
end

function [detA,cofac] = Determinant(a)
%Determinant calculates the determinant and cofactors for a 3 x 3 matrix
%
%   USE: [detA,cofac] = Determinant(a)
%
%   a is the matrix, detA is the determinant, and cofac are the
%   cofactor elements
%
%   Determinant uses function CalcCofac
%
%   NOTE: This function is only for illustration purposes. To get the
%   determinant of a square matrix of any size use the MATLAB function det
%   (e.g. detA = det(a))

%Number of rows and columns in a
n = size(a,1);
```

```
m = size(a,2);

%If matrix is 3 x 3
if n == 3 && m == 3
    %Calculate the array of cofactors for a. Note that this is not the most
    %efficient way of doing this because you will calculate six more
    %cofactors than you need. The time loss, however, is negligible
    cofac = CalcCofac(a);
    %Calculate the determinant of a as in equation 4.27, remembering that
    %the cofactor 1,2 from CalcCofac will already be negative
    detA = 0.0;
    for i = 1:3
        detA = a(1,i)*cofac(1,i) + detA;
    end
else
    error('Matrix is not 3 x 3');
end
end
```

Inverse of a matrix

We have already seen that multiplication of conformable matrices is possible, but suppose we have the equation

$$\mathbf{y} = \mathbf{M}\mathbf{x} \tag{4.29}$$

Can we solve this equation for **x** by dividing through by **M**? The answer, of course, is "no," dividing by a matrix has no meaning. We can, however, get around this limitation by defining the *inverse* of a matrix, denoted by the matrix symbol raised to the minus one power: $\mathbf{M}^{-1}$. A matrix, when multiplied by its inverse, is equal to the identity matrix, **I** (Eq 4.11):

$$\mathbf{M}\mathbf{M}^{-1} = \mathbf{M}^{-1}\mathbf{M} = \mathbf{I} = \delta_{ij} \tag{4.30}$$

It can be shown that, if a square matrix has a non-zero determinant, then the matrix has an inverse. Matrices of this type are called non-singular. We can then solve for **x** in Equation 4.29 as follows:

$$\mathbf{x} = \mathbf{M}^{-1}\mathbf{y} \tag{4.31}$$

The definition of the inverse of a matrix is simple but the actual calculation is not. The equation that one can use to find the inverse of matrix, **M**, is given below (see Nye, 1985, pp. 155-156, or Malvern, 1969, pp. 41-43, for the derivation):

$$\mathbf{M}^{-1} = \begin{pmatrix} \left(\frac{\text{cof}_{11}(\mathbf{M})}{|\mathbf{M}|}\right) & \left(\frac{\text{cof}_{21}(\mathbf{M})}{|\mathbf{M}|}\right) & \left(\frac{\text{cof}_{31}(\mathbf{M})}{|\mathbf{M}|}\right) \\ \left(\frac{\text{cof}_{12}(\mathbf{M})}{|\mathbf{M}|}\right) & \left(\frac{\text{cof}_{22}(\mathbf{M})}{|\mathbf{M}|}\right) & \left(\frac{\text{cof}_{32}(\mathbf{M})}{|\mathbf{M}|}\right) \\ \left(\frac{\text{cof}_{13}(\mathbf{M})}{|\mathbf{M}|}\right) & \left(\frac{\text{cof}_{23}(\mathbf{M})}{|\mathbf{M}|}\right) & \left(\frac{\text{cof}_{33}(\mathbf{M})}{|\mathbf{M}|}\right) \end{pmatrix} \tag{4.32}$$

You can see that, even for a 3×3 matrix, inverting it is not simple! Nonetheless, matrix inversion is the foundation of some very powerful algorithms in geophysics and structural

4.4 Transformations of coordinates and vectors revisited

geology that go under the general heading of "inverse methods." We will see an example of matrix inversion when we solve the problem of extracting strain rate from Global Positioning System (GPS) velocity vectors later on. For large matrices, numerical methods are commonly used. Below, we show how to calculate the inverse of a 3×3 matrix.

```
function aInv = Invert(a)
%Invert calculates the inverse of a 3 x 3 matrix
%
%    USE: aInv = Invert(a)
%
%    a is the matrix, and aInv is the inverse matrix
%
%    Invert uses function Determinant
%
%    NOTE: This function is only for illustration purposes. To get the
%    inverse of a square matrix of any size use the MATLAB function inv
%    (e.g. aInv = inv(a))

%Calculate the cofactors and determinant of a
[detA,cofac] = Determinant(a);

%Calculate the inverse matrix following equation 4.32
aInv = zeros(3,3); %Initialize aInv
for i = 1:3
    for j = 1:3
        aInv(i,j) = cofac(j,i)/detA; %Note the switch of i & j in cofac
    end
end
end
```

4.4 TRANSFORMATIONS OF COORDINATES AND VECTORS REVISITED

The transformation of a vector from one coordinate system to another is just one example of a whole general class of matrix algebra operations known as *linear transformations*. Anytime we have the same number of equations and unknowns they can be written as a set of simultaneous linear equations, and matrix concepts provide a simple way of solving these equations. Many textbooks, for example, describe solutions of linear equations using Cramer's Rule and there are other approaches for larger matrices. Most of these methods are beyond the scope of this book.

The reverse transformation (old in terms of new)

In Chapter 3, we introduced the concept of the transformation matrix, **a**, and the set of six equations that govern the relations between the direction cosines, which are known as the *orthogonality relations* (Eqs. 3.3 and 3.4). Now that we have indicial notation and some matrix concepts more fully in mind, we can reexamine these relations in a new light. Using the summation convention, we can rewrite Equations 3.4 as

$$a_{ik}a_{jk} = 1, \text{ if } i \text{ and } j \text{ are equal}$$

Likewise, Equations 3.5 can be rewritten as

$$a_{ik}a_{jk} = 0, \text{ if } i \text{ and } j \text{ are not equal}$$

Recall that the identity matrix (Eq. 4.11) has these same properties. If its two indices are equal, then they have a value of 1 (i.e., the principal diagonal is equal to 1), otherwise it has a value of 0 (i.e., the "off-diagonal" elements are equal to 0). Thus, we can write the orthogonality relations as a single equation in indicial notation:

$$a_{ik}a_{jk} = \delta_{ij} \tag{4.33}$$

or in matrix notation:

$$\mathbf{a}\mathbf{a}^T = \mathbf{I} \tag{4.34}$$

Thus, the transformation matrix is an orthogonal matrix, as described above.

The transformation of a vector (Eqs. 3.8 and 4.5) in matrix notation can be written simply as

$$\mathbf{v}' = \mathbf{a}\mathbf{v} \quad \text{or} \quad v'_i = a_{ij}v_j \tag{4.35}$$

Clearly, from the discussion of inverse matrices, if we want to solve for $\mathbf{v}$ (i.e., we want the reverse transformation of old in terms of new), we should be able to pre-multiply $\mathbf{v}'$ by the inverse of $\mathbf{a}$ (compare Eqs. 4.29 and 4.31):

$$\mathbf{v} = \mathbf{a}^{-1}\mathbf{v}' \tag{4.36}$$

However, if you compare Equation 4.23 with Equation 4.30, you will see that, for an orthogonal matrix, its transpose is equal to its inverse; that is, $\mathbf{a}^T = \mathbf{a}^{-1}$. Therefore, we can also write Equation 4.36 as

$$\mathbf{v} = \mathbf{a}^T\mathbf{v}' \quad \text{or} \quad v_i = a_{ji}v'_j \tag{4.37}$$

If you expand Equation 4.37, you will see that it is identical to Equations (3.9). This is a more elegant way of deriving the reverse transformation than we were able to do in Chapter 3.

Change of "handedness" of axes

Imagine looking at the image of a right-handed coordinate system in the mirror. What you would see is a left-handed coordinate system (Fig. 4.2). This is nothing more than a certain type of coordinate transformation. One particularly common place to encounter this type of transformation is in considering the symmetry of mineral crystals. It would be nice if there were some way to determine whether or not a particular transformation would produce a change in the handedness of axes of our coordinate system. As it turns out, there is (Nye, 1985, pp. 35-38). It is possible to show that the determinant of the transformation matrix, $\mathbf{a}$ (and of all orthogonal matrices), can only be equal to +1 or -1. A close inspection of Figure 4.2 shows that the only axis to change is X_2; in the mirror, it points in the opposite direction from the original axis whereas all of the others point essentially in the same direction (accounting for the perspective of the diagram). Thus the transformation matrix for the situation shown in Figure 4.2 is

4.5 Exercises

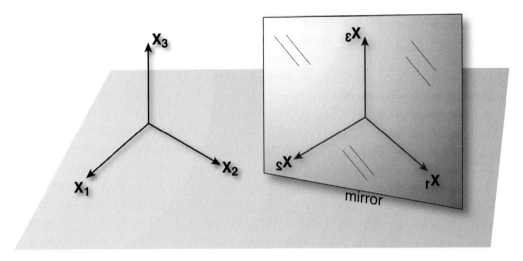

Figure 4.2 The reflection of a right-handed coordinate system in the mirror produces a left-handed coordinate system.

$$a_{ij} = \begin{pmatrix} 1 & 0 & 0 \\ 0 & -1 & 0 \\ 0 & 0 & 1 \end{pmatrix} \quad (4.38)$$

The determinant of this matrix, calculated from Equation 4.32, is

$$\begin{aligned} \det a = |a| &= a_{11}(a_{22}a_{33} - a_{23}a_{32}) + a_{12}(a_{23}a_{31} - a_{21}a_{33}) + a_{13}(a_{21}a_{32} - a_{22}a_{31}) \\ &= a_{11}(a_{22}a_{33} - a_{23}a_{32}) + 0 + 0 \\ &= a_{11}a_{22}a_{33} - 0 \\ &= (1)(-1)(1) = -1 \end{aligned}$$

Thus, you can show that, if the determinant of the transformation matrix, $|a|$, equals -1, then the transformation will produce a change in the hand of the axes. It is just as easy to show that a transformation that does not change the hand of the axes has a determinant of $+1$.

4.5 EXERCISES

1. Matrices **A** and **B** are given below. Calculate the following sums and products: (a) **A** + **B**, (b) **B** + **A**, (c) **AB**, and (d) **BA**.

$$\mathbf{A} = \begin{pmatrix} 2 & 6 \\ 4 & 9 \end{pmatrix} \quad \text{and} \quad \mathbf{B} = \begin{pmatrix} 3 & 14 \\ 7 & 10 \end{pmatrix}$$

2. What are the determinants of matrices **A** and **B** in Exercise 1?
3. In Equations 4.27 and 4.28, we determined the determinant of matrix **M** by expansion of the first row of the matrix. Show that expansion of the second row produces the same determinant for the matrix.

4. Solve for matrix **x** in the following equation:

$$\mathbf{y} = \mathbf{Mx}$$

$$\begin{bmatrix} 4 \\ 7 \end{bmatrix} = \begin{bmatrix} 9 & 1 \\ 3 & 6 \end{bmatrix} \begin{bmatrix} x_1 \\ x_2 \end{bmatrix}$$

5. Expand the following equation:

$$\frac{\partial u_i}{\partial x_j} = \bar{e}_{ij}$$

6. Expand the following equation:

$$\varepsilon_{ij} = \frac{1}{2}(e_{ij} + e_{ji})$$

7. Expand the terms e_{11} and e_{31} in the following equation:

$$e_{ij} = \frac{1}{2}\left[\frac{\partial u_i}{\partial x_j} + \frac{\partial u_j}{\partial x_i} + \frac{\partial u_k}{\partial x_i}\frac{\partial u_k}{\partial x_j}\right]$$

CHAPTER

FIVE

Tensors

5.1 WHAT ARE TENSORS?

Few things are more imposing to structural geologists than the concept of tensors. In most continuum mechanics textbooks you will find the formal definition of a tensor as a physical quantity that "transforms like a tensor" or that tensors transform in such a way that a "valid tensor equation in one coordinate system will be valid in any other coordinate system." These are rigorous definitions that are important to understand fully (we will come back to them in Section 5.3), but to someone meeting this concept for the first time they are not terribly illuminating! Yes, we know, or at least have been told at one point or another, that stress and strain are "tensors," but what does that statement really mean? Vectors we can handle, but tensors?

In fact, we have already used tensors extensively in this book. All vectors are a type of tensor quantity known as a first order or first rank tensor. Any physical quantity that is independent of a particular coordinate system – as we have already seen for vectors (Chapters 3 and 4) – is a tensor. We have already discussed two types of mathematical and physical entities and now we can add a third:

1. Scalar (zero order tensor): A quantity represented by a single number that is independent of the coordinate system (i.e., it has the same value, regardless of the coordinate system we choose). Some examples of scalars are:
   - temperature
   - mass
   - density.
2. Vector (first order tensor): A physical entity with a magnitude and direction represented by three numbers[1] whose values depend on the particular coordinate system. Although the

---

[1] In all of the following discussion, we assume a Cartesian coordinate system unless explicitly stated otherwise.

Tensor rank	Name	Quantities related
0	Scalar	Nothing
1	Vector	Two other scalars Scalar and a vector
2	2nd order tensor (commonly just "tensor")	Two vectors
3	3rd order tensor	Scalar and a tensor Vector and a 2nd order tensor
4	4th order tensor	Two 2nd order tensors

Table 5.1 Tensor rank and the types of related entities

magnitude of the numbers changes with coordinate system, the magnitude and direction of the vector is the same in all coordinate systems. A vector can relate a scalar and another vector. For example, in the equation **f** = m**a**, force and acceleration are vectors and mass is a scalar. Some other familiar examples of vectors are:
- velocity
- displacement
- temperature gradient.

3. Second order tensor: A physical quantity represented by nine numbers. The physical entity is independent of coordinate system. A second order tensor relates two vectors to each other or another second order tensor to a scalar. Some examples of second order tensors are:
- thermal conductivity
- stress
- strain.

We can continue this hierarchy of tensors virtually indefinitely. The order of a tensor simply equals the number of subscripts that it has. Vectors have one subscript so they are first order tensors, second order tensors have two subscripts, and so on. For convenience sake, when there is no ambiguity we will refer to second order tensors simply as "tensors" and first order tensors as "vectors" but you should be aware that both are members of a general class of physical quantities, independent of coordinate system, that we call tensors. In this chapter, we will deal with "generic" tensors and save the discussion of the most important tensors for structural geology – stress and strain – to the following chapters.

5.2 TENSOR NOTATION AND THE SUMMATION CONVENTION

Because second order tensors in three dimensions are represented by arrays of nine numbers, we treat them mathematically as 3×3 matrices. However, as was emphasized in Chapter 4, matrices may represent physical quantities like vectors and tensors or they may be totally artificial constructs such as the transformation matrix. In other words, all tensors are matrices but not all matrices are tensors.

5.2.1 Basic characteristics of a tensor

Like any 3×3 matrix, tensors are represented by bold face letters or by indicial notation with two subscripts, each of which can have values of 1, 2, or 3. When writing out the components of

5.2 Tensor notation and the summation convention

a tensor, we distinguish them from an arbitrary matrix by using square brackets rather than parentheses, just as we did with vectors:

$$\mathbf{T} = T_{ij} = \begin{bmatrix} T_{11} & T_{12} & T_{13} \\ T_{21} & T_{22} & T_{23} \\ T_{31} & T_{32} & T_{33} \end{bmatrix} \tag{5.1}$$

The nine components of the tensor – in this case a generic tensor, **T** – give the values of the tensor with reference to the three axes of the specific coordinate system. If we change the axes, then the nine components will change their values but, just like a vector, the fundamental nature of the tensor itself will not change. The exact nature of the relation between component and axis depends on the specific tensor. In the following chapters, we'll see two examples of this.

Like any matrix, tensors can be symmetric, asymmetric, or antisymmetric depending on the relations of the components to each other. If the tensor has nine independent components then it is *asymmetric*. If $T_{ij} = T_{ji}$, then there are only six independent components and the tensor is *symmetric*. Finally, *antisymmetric* (or *skew-symmetric*) tensors are those in which $T_{ij} = -T_{ji}$, in which case there are only three independent components. Any general asymmetric tensor can be decomposed into a symmetric tensor plus an antisymmetric tensor as follows:

$$T_{ij} = S_{ij} + A_{ij} \quad \text{where} \quad S_{ij} = \frac{T_{ij} + T_{ji}}{2} \quad \text{and} \quad A_{ij} = \frac{T_{ij} - T_{ji}}{2} \tag{5.2}$$

You can easily prove to yourself that S_{ij} is symmetric and A_{ij} is antisymmetric.

For all symmetric tensors, there is one set of coordinate axes where all the components, *except for those along the principal diagonal*, are zero. That is,

$$\mathbf{T} = T_{ij} = \begin{bmatrix} T_{11} & 0 & 0 \\ 0 & T_{22} & 0 \\ 0 & 0 & T_{33} \end{bmatrix} = \begin{bmatrix} T_2 & 0 & 0 \\ 0 & T_1 & 0 \\ 0 & 0 & T_3 \end{bmatrix} \tag{5.3}$$

The values along the principal diagonal, T_1, T_2, and T_3, are then known as the *principal axes* of the tensor. Note that, on the right side of Equation 5.3, we have purposefully *not* put T_1 in the T_{11} space, etc. This was done to emphasize a very important point: *A component with two subscripts refers to the axes of the coordinate system; a single subscript refers just to the magnitude of the component, not its position or orientation.* These three single subscripts define the major, intermediate, and minor magnitude axes of a three-dimensional surface known as the magnitude ellipsoid (Fig. 5.1). In the case of Equation 5.3, the largest component is specified

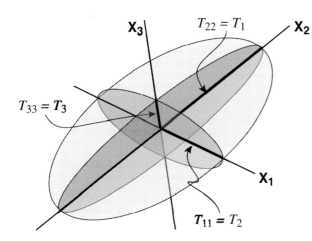

Figure 5.1 The magnitude ellipsoid and principal axes of a generic tensor, **T**. Note that the magnitude axes, T_1, T_2, and T_3, are parallel to $\mathbf{X}_2$, $\mathbf{X}_1$, and $\mathbf{X}_3$, respectively. Thus, $T_1 = T_{22}$, etc. Their orientations shown here correspond to those in Equation 5.3.

as being parallel to the X_2 axis of the coordinate system. Most structural geology students are already familiar with the "stress ellipsoid" and the "strain ellipsoid." The general equation for the magnitude ellipsoid (using our generic tensor, **T**, and assuming that principal axes of the tensor are parallel to the axes of the coordinate system with the same index, unlike in Eq. 5.3) is

$$\frac{X_1^2}{T_1^2} + \frac{X_2^2}{T_2^2} + \frac{X_3^2}{T_3^2} = 1 \tag{5.4}$$

We will see more about determining the principal axes of the tensor in Section 5.4.

5.2.2 Tensors relating two vectors

As stated above (Table 5.1), a tensor commonly relates two vectors; more formally, we can state that a tensor is a "linear vector operator" because the components of the tensor are the coefficients of a set of linear equations that relate two vectors. Suppose we have two vectors, **u** and **v**, that are related by tensor, **T**. In matrix or indicial notation, we write

$$\mathbf{u} = \mathbf{T}\mathbf{v} \quad \text{or} \quad u_i = T_{ij} v_j \tag{5.5}$$

With the summation convention, we can easily expand Equation 5.5, realizing that j is the dummy suffix and i is the free suffix:

$$\begin{aligned} u_1 &= T_{11} v_1 + T_{12} v_2 + T_{13} v_3 \\ u_2 &= T_{21} v_1 + T_{22} v_2 + T_{23} v_3 \\ u_3 &= T_{31} v_1 + T_{32} v_2 + T_{33} v_3 \end{aligned} \tag{5.6}$$

If you compare Equation 5.5 with Equation 4.8, you'll see that they have a very similar form and their expansion using the summation convention is also the same. That's because both sets of equations represent matrix multiplication involving a 3×3 and a 3×1 matrix. The similarity ends there, however. In Equation 4.8, the transformation matrix, **a**, is not a tensor; it is simply a linear operator describing the relationship between the same vector in two different coordinate systems. In Equation 5.5, **T** is a tensor and **u** and **v** are *different* vectors. You would program Equation 5.5 as follows:

```
% v (1 x 3 vector) and T (3 x 3 tensor) are previously declared
u = zeros(1,3); % initialize u (1 x 3 vector)
for i = 1:3 % i is the free suffix
    for j = 1:3 %j is the dummy suffix
        u(i) = T(i,j)*v(j) + u(i);
    end
end
```

The three iterations of the outer loop will produce three separate equations and the three iterations of the inner loop mean that each equation will have three terms (i.e., as in Equation 5.6). Again, note how similar the equation written using the summation convention is to a computer program.

In the previous chapter, we saw a somewhat different way of producing a tensor as a type of product of two vectors. This operation is a natural extension of the dot (or scalar) product and the cross (or vector) product. If we premultiply a column vector times a row vector, the operation is known as the dyad (or tensor) product and the result is a 3×3 matrix which is known as a dyad. The *dyad product* of two vectors, **u** and **v**, is

5.3 Tensor transformations

$$\mathbf{T} = \mathbf{u} \otimes \mathbf{v} = \begin{bmatrix} u_1 v_1 & u_1 v_2 & u_1 v_3 \\ u_2 v_1 & u_2 v_2 & u_2 v_3 \\ u_3 v_1 & u_3 v_2 & u_3 v_3 \end{bmatrix} \qquad (5.7a)$$

Using indicial notation, we would write

$$T_{ij} = u_i v_j \qquad (5.7b)$$

Coding Equation 5.7 for a computer is simpler because there is no summation involved; both i and j are free suffixes. Thus, we can write:

```
%u (1 x 3 vector) and v(1 x 3 vector) are previously declared
T = zeros(3,3); %Initialize T (3 x 3 tensor)
for i = 1:3 %i is a free suffix
     for j = 1:3 %j is also a free suffix
           T(i,j) = u(i)*v(j); %there is no summation here
     end
end
```

5.3 TENSOR TRANSFORMATIONS

Like vectors, second order tensors are physical quantities independent of a coordinate system. Therefore, if we know what the components of the tensor are in one coordinate system, we should be able to determine what they are in any other coordinate system, just as we did for vectors (Chapter 3). All we need to know is the transformation matrix, **a**. The equations for transforming a tensor are somewhat more complicated, however, because a tensor is a more complicated entity than a vector.

5.3.1 Derivation of the tensor transformation equations

To proceed, one must first realize that the tensor that relates two vectors in the new coordinate system is just the transformed version of the same tensor in the old coordinate system (as before, the primed quantities are in the new coordinate system):

$$u_i = T_{ij} v_j \qquad (5.8a)$$

and

$$u'_i = T'_{ij} v'_j \qquad (5.8b)$$

Therefore, we can derive an equation that relates **T** to **T'** by combining the transformation equations for **u** to **u'** and for **v** to **v'**:

$$u'_i = a_{ik} u_k \qquad (5.9a)$$

and

$$v_l = a_{jl} v'_j \qquad (5.9b)$$

Do not be confused by the fact that we are using some unfamiliar letters for subscripts. You can choose whatever letters you want as long as you don't confuse the free and dummy suffixes. All of these equations obey the summation convention rules. Note that we can just as easily write Equation 5.8a as

$$u_k = T_{kl} v_l \qquad (5.10)$$

Now, if we substitute Equation 5.9b into Equation 5.10, and then take that result and substitute it into Equation 5.9a, we can write

$$u'_i = a_{ik} a_{jl} T_{kl} v'_j \tag{5.11}$$

But, we also know from Equation 5.8b that $u'_i = T'_{ij} v'_j$, so the tensor transformation, given as the *new components in terms of the old*, is

$$\boxed{\begin{aligned} T'_{ij} &= a_{ik} a_{jl} T_{kl} \quad \text{(summation notation)} \\ \mathbf{T'} &= \mathbf{a}^T \mathbf{T} \mathbf{a} \quad \text{(matrix notation)} \end{aligned}} \tag{5.12}$$

By a similar series of steps, you can derive the reverse transformation, that is, the *old components in terms of the new*:

$$\boxed{\begin{aligned} T_{ij} &= a_{ki} a_{lj} T'_{kl} \quad \text{(summation notation)} \\ \mathbf{T} &= \mathbf{a} \mathbf{T'} \mathbf{a}^T \quad \text{(matrix notation)} \end{aligned}} \tag{5.13}$$

These transformations are the key to understanding tensors. The definition of a tensor is a physical quantity, independent of a specific coordinate system, which generally describes the relation between two linked vectors (or a scalar and another tensor). The test of a tensor is if it transforms from one coordinate system to another according to the above equations, 5.12 and 5.13, then it is a tensor. The whole point about tensors is that the nine coefficients simply correspond to a particular reference frame and change systematically by the above rules upon change of coordinate system. We can transform them to any other reference frame without changing the fundamental nature of the physical property that the tensor represents. Therein lies their power because it is often advantageous, or necessary, to change our view of things (i.e., our coordinate system) to understand them more clearly.

5.3.2 Tensor transformation as a computer program

Expanding Equations 5.12 and 5.13 is a tedious task, because there are two dummy suffixes, k and l, and two free suffixes, i and j. Equation 5.12 alone represents nine individual equations each with nine terms! For guidance on how to expand the equations by hand, see page 12 of Nye (1985). As before, for those with some computer programming experience it is easier to think about expanding them as a series of nested do-loops. The program fragment below carries out the summation in Equation 5.12:

```
%T_old (3 x 3) tensor and a (3 x 3 trans. matrix) are previously declared
T_new = zeros(3,3); %initialize T_new (3 x 3 tensor)
for i = 1:3 %Outer loops are controlled by the free suffixes i & j
      for j = 1:3
            for k = 1:3 %Inner loops are around the dummy suffixes k & L
                  for L = 1:3
                        T_new(i,j) = a(i,k)*a(j,L)*T_old(k,L)+T_new(i,j);
                  end
            end
      end
end
```

5.3 Tensor transformations

Notice that the dummy suffixes are always in the inner loops (the order in which you loop about k and L does not matter) and the free suffixes are in the outer loops. As you can see from the above, there will be nine sums for each T_new and there will be nine T_new's.

5.3.3 A special two-dimensional transformation

In this age of computers, carrying out tensor transformations numerically according to Equations 5.12 and 5.13 is quite straightforward. This was not always the case, as easily accessible computers have only come into being in the last 30 years. Furthermore, complex equations are commonly easier to visualize graphically; it's not easy to look at Equations 5.12 and 5.13 and immediately have an intuitive grasp of their significance! For simple two-dimensional transformations in which one of the three axes is the same before and after the transformation (i.e., a rotation of the coordinate system about one of its axes, Fig. 5.2), there is just such a graphical construction.

Consider the case where the axes of the old coordinate system are parallel to the principal axes of the symmetric tensor, **T**. Then, we wish to change the coordinate system to a different orientation by rotating about the intermediate axis (Fig. 5.2). The transformation matrix of this problem is

$$\mathbf{a} = \begin{pmatrix} \cos\theta & \cos 90 & \cos(90-\theta) \\ \cos 90 & \cos 0 & \cos 90 \\ \cos(90+\theta) & \cos 90 & \cos\theta \end{pmatrix} = \begin{pmatrix} \cos\theta & 0 & \sin\theta \\ 0 & 1 & 0 \\ -\sin\theta & 0 & \cos\theta \end{pmatrix} \quad (5.14)$$

The initial form of the tensor, **T**, in the old coordinate system is

$$\mathbf{T} = T_{ij} = \begin{bmatrix} T_1 & 0 & 0 \\ 0 & T_2 & 0 \\ 0 & 0 & T_3 \end{bmatrix} \quad (5.15)$$

Now, we can use the tensor transformation equation to calculate what the tensor is in the new coordinate system. Substituting Equations 5.14 and 5.15 into Equation 5.12 and carrying out the summation, we get

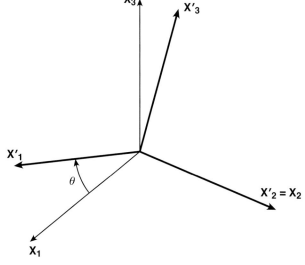

Figure 5.2 Coordinate transformation by a rotation about one of the axes of the coordinate system.

Figure 5.3 Mohr circle construction for a two-dimensional tensor transformation.

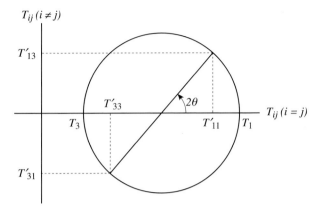

$$T'_{ij} = \begin{bmatrix} \left(T_1 \cos^2\theta + T_3 \sin^2\theta\right) & 0 & \left(-T_1 \sin\theta \cos\theta + T_3 \sin\theta \cos\theta\right) \\ 0 & 1 & 0 \\ \left(-T_1 \sin\theta \cos\theta + T_3 \sin\theta \cos\theta\right) & 0 & \left(T_1 \sin^2\theta + T_3 \cos^2\theta\right) \end{bmatrix} \quad (5.16)$$

The components of **T** in 5.16 can be put in a more useful form by using several trigonometric identities for double angles:

$$\sin 2\theta = 2\sin\theta \cos\theta \qquad \sin^2\theta = \frac{1 - \cos 2\theta}{2} \qquad \cos^2\theta = \frac{1 + \cos 2\theta}{2} \quad (5.17)$$

Substituting these equations into Equation 5.16 and rearranging, we get the following values for the components of the tensor in the new coordinate system:

$$T'_{11} = \left(\frac{T_1 + T_3}{2}\right) + \left(\frac{T_1 - T_3}{2}\right) \cos 2\theta$$

$$T'_{33} = \left(\frac{T_1 + T_3}{2}\right) - \left(\frac{T_1 - T_3}{2}\right) \cos 2\theta \quad (5.18)$$

$$T'_{13} = T'_{31} = -\left(\frac{T_1 - T_3}{2}\right) \sin 2\theta$$

Most structural geologists will recognize Equations 5.18 and the plot representing them (Fig. 5.3) as the Mohr circle. This construction, devised by the German engineer Otto Mohr in the late 1800s, is most commonly associated with the analysis of the stress tensor (i.e., Mohr circle for stress) but can be equally well applied to any symmetric second order tensor. Thus, we also have Mohr circle for infinitesimal strain, Mohr circle for finite strain in the deformed state, etc. These will be presented in following chapters.

5.4 PRINCIPAL AXES AND ROTATION AXIS OF A TENSOR

5.4.1 Magnitude ellipsoid and representation quadric

We stated above that a tensor is a linear vector operator, but it would be helpful if there were some way to visualize graphically how a tensor relates two vectors. The Mohr circle construction of the previous section is one such approach, but because it is plotted in "tensor space", rather than physical space, visualization is more difficult. We know that tensors can be

5.4 Principal axes and rotation axis of a tensor

represented by their magnitude ellipsoid and vectors by lines with arrows at one end; that's the type of thing that we need to get a physical understanding.

There is a surface that helps one visualize the angular relations between the two vectors related by a tensor, and it also helps to visualize how one calculates the orientation and magnitude of the principal axes. That surface is known as the *representation quadric*. This surface and its derivation are explained in more detail in Nye (1985, pp. 16-19 and 26-30), and we will only briefly touch on it here. Unlike the magnitude ellipsoid (Eq. 5.4), the representation quadric may have the geometric form of either an ellipsoid or hyperboloid defined by the following equation:

$$T_{ij}x_ix_j = 1 \qquad (5.19)$$

You can show that Equation 5.19 is a tensor by seeing whether it transforms according to Equation 5.12. Equation 5.19 can be written in terms of its principal axes, as follows:

$$T_1x_1^2 + T_2x_2^2 + T_3x_3^2 = 1 \qquad (5.20)$$

Note that, in the case of Equation 5.20 the principal axes are in the numerator, not in the denominator as they are in 5.4. Thus, when plotted on the same diagram the long axes of the magnitude ellipsoid and the quadric will be at right angles to each other (Fig. 5.4).

The relation between the representation quadric and the magnitude ellipsoid, as well as their major properties, are illustrated in Figure 5.4. This diagram is a principal section through the quadric and ellipsoid (i.e., a plane that contains the two principal axes of the tensor) for the relation $u_i = T_{ij}v_j$ (Eq. 5.5); it shows the angular relation between **u** and **v** in two dimensions. As **v** is rotated about the origin with a constant, unit length, the vector **u** traces out the surface of the magnitude ellipsoid of the tensor, **T**. The angle between the two vectors varies as a complex function of their position with respect to the principal axes of **T**. Vector **u** will always be perpendicular to the tangent to the representation quadric where the latter is intersected by **v**. This attribute is known as the radius-normal property of the representation quadric.

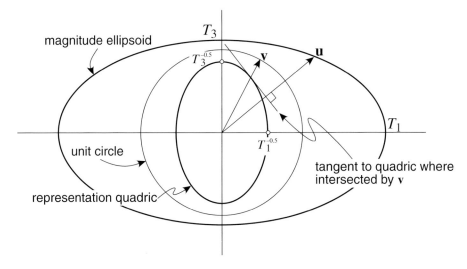

Figure 5.4 Geometric relations between the representation quadric and the magnitude ellipsoid of tensor **T**, and the two vectors that **T** relates, **u** and **v**.

5.4.2 Finding the magnitude and orientation of the principal axes

It is clear from the preceding discussion and Figure 5.4 that **u** and **v** are parallel only along the principal axes and nowhere else. At these positions, **u** has the same magnitude (and orientation of course) as the principal axis itself. If **u** and **v** are parallel, then they (i.e., their lengths) will be proportional to each other; one of the vectors, multiplied by a scalar, should be equal to the other. At a principal axis, the vector **v** will be parallel and equal to a unit vector, $\hat{\mathbf{x}}$, and **u** will simply be equal to a scalar, λ, times $\hat{\mathbf{x}}$ (Fig. 5.4):

$$\mathbf{v} = \hat{\mathbf{x}} \quad \text{and} \quad \mathbf{u} = \lambda \hat{\mathbf{x}} \tag{5.21}$$

Substituting the relations in 5.21 into Equation 5.5, we can write

$$\lambda \hat{\mathbf{x}} = \mathbf{T}\hat{\mathbf{x}} \tag{5.22a}$$

or using indicial notation,

$$\lambda x_i = T_{ij} x_j \tag{5.22b}$$

In Equations 5.21 and 5.22, λ is the unknown scalar constant - known as the *eigenvalue* - and **x** is an *eigenvector* of T_{ij}. Equation 5.22 can be solved by rearranging and using the substitution property of the Kronecker delta discussed in the previous chapter (Eq. 4.12):

$$x_i = \delta_{ij} x_j \;\Rightarrow\; T_{ij} x_j = \lambda \delta_{ij} x_j \;\Rightarrow\; (T_{ij} - \lambda \delta_{ij}) x_j = 0 \tag{5.23}$$

To solve for λ, take the determinant of this final equation (which is known as the *secular or characteristic equation*),

$$|T_{ij} - \lambda \delta_{ij}| = \begin{vmatrix} (T_{11} - \lambda) & T_{12} & T_{13} \\ T_{21} & (T_{22} - \lambda) & T_{23} \\ T_{31} & T_{32} & (T_{33} - \lambda) \end{vmatrix} = 0 \tag{5.24}$$

Expanding, we get a cubic polynomial in λ:

$$\lambda^3 - I\lambda^2 - II\lambda - III = 0 \tag{5.25}$$

The three roots of λ are the three eigenvalues; they will be the magnitudes of the three principal axes of the tensor. Note that all three roots will be real only if the tensor is symmetric; otherwise one or more will be imaginary. Once you know the three values of λ, you can then substitute each one in turn back into Equation 5.22 or 5.23 to solve for the three eigenvectors (*I*, *II*, and *III*) which give you the orientations of the principal axes.

In general, Equation 5.25 is solved numerically by computer using an algorithm known as the Jacobi Transformation or some more esoteric routine. Such routines can be found in Press *et al.* (1986). For a description of how to do this manually, see Nye (1985, Chapter IX). Thus, we can find the principal axes of any tensor in any general coordinate system by finding its eigenvectors and eigenvalues.

5.4.3 Invariants of a tensor

The three values of λ are scalars that correspond to the magnitudes of the principal axes of the tensor, which, of course, is independent of the coordinate system. Therefore, *I*, *II*, and *III*, the three coefficients of Equation 5.25, must also have the same values, regardless of the coordinate system we choose. Thus, they are known as the *invariants of the tensor* and their values, for any coordinate system, are given in Equations 5.26. If one happens to know the principal axes of the tensor, then they are particularly easy to calculate:

5.5 Example of eigenvalues and eigenvectors in structural geology

$$\begin{aligned}
I &= T_{11} + T_{22} + T_{33} = T_1 + T_2 + T_3 \\
II &= \frac{(T_{ij}T_{ij} - I^2)}{2} = -(T_1 T_2 + T_2 T_3 + T_3 T_1) \\
III &= \det \mathbf{T} = |T_{ij}| = T_1 T_2 T_3
\end{aligned} \quad (5.26)$$

We'll see that invariants of tensors have a number of uses, described in the following chapters.

5.4.4 Rotation axis of an antisymmetric tensor

An antisymmetric tensor (e.g., A_{jk} in Equation 5.2) is sometimes also known as an axial vector. To get the Cartesian coordinates, r_i, of that vector:

$$r_i = \frac{-b_{ijk}A_{jk}}{2} \quad (5.27)$$

b_{ijk} is a "permutation symbol" which is equal to $+1$ if the suffixes are cyclic, -1 if the suffixes are acyclic, and 0 if any two suffixes are repeated. The three components of vector $\mathbf{r}$, which give the orientation of the rotation axis, are

$$r_1 = \frac{-(A_{23} - A_{32})}{2}, \quad r_2 = \frac{-(-A_{13} + A_{31})}{2}, \quad r_3 = \frac{-(A_{12} - A_{21})}{2} \quad (5.28)$$

The amount of rotation in radians is just the length of the vector, $\mathbf{r}$:

$$|\mathbf{r}| = \sqrt{r_1^2 + r_2^2 + r_3^2} \quad (5.29)$$

5.5 EXAMPLE OF EIGENVALUES AND EIGENVECTORS IN STRUCTURAL GEOLOGY

The concepts discussed in this chapter form the basis for understanding the mechanics of structural geology. However, we defer their application to the next chapters where stress and strain are treated explicitly. Nonetheless, there is a very important type of problem, the solution to which relies heavily on the concept of eigenvalues and eigenvectors. This problem is: "how do we find the best-fit axes to a group of axial data that have no directional significance?" A more specific example is "how do we find the best-fit fold axis to a group of bedding poles?" As stated much earlier (Chapter 2), we cannot use the mean vector for this problem because the axes have no directional significance; we will commonly be plotting everything in the lower hemisphere.

5.5.1 Types of axial distributions

Before proceeding to the numerical solution to this problem, a digression into the types of line distributions in spherical space is needed. In general, lines can have three types of orientations (Fig. 5.5), which correspond to the three fundamental types of Euclidean geometric objects: lines (one-dimensional or 1D), planes (2D), and volumes (3D). A group of lines that are all parallel or sub-parallel to each other has a linear (1D) preferred orientation; if they were perfectly parallel to each other they would combine to form a single line, but more commonly there is some limited scatter (e.g., Fig. 5.5b). In this latter case, the ends of the lines define a surface whose shape fabric is approximately that of an elongate ellipsoid (a prolate or cigar-shaped ellipsoid). This type of distribution is commonly called a bipolar distribution. A girdle distribution (Fig. 5.5c) results when all of the lines are close to being coplanar. When this

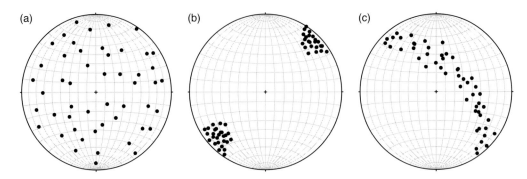

Figure 5.5 Three types of preferred orientations of linear elements displayed as points in an equal area projection. (a) Random, (b) bipolar, and (c) girdle.

happens, the ends of the lines trace out a surface of a flattened ellipsoid (an oblate or pancake-shaped ellipsoid). If the lines were all perfectly coplanar, then the oblate ellipsoid would be reduced to a two-dimensional circle. When all of the lines have a random distribution (Fig. 5.5a) their ends define a sphere.

5.5.2 Determination of "best-fit" axes

You can see that the problem we started out with is really a problem of finding the three mutually perpendicular axes of the ellipsoids referred to in the previous section. If we calculate three axes of nearly equal length, they define a sphere and point to a random distribution (Fig. 5.5a). Two short axes and one very long axis define a prolate ellipsoid and indicate a bipolar distribution (Fig. 5.5b). Likewise, two axes of equal length and one much shorter axis will define an oblate ellipsoid and a girdle distribution (Fig. 5.5c). In this last case, if we are trying to calculate the best-fit fold axis to a cylindrically folded surface, it is the shortest axis in which we are interested.

Suppose we are trying to calculate a fold axis, $\mathbf{f}$ (the derivation below follows that of Charlesworth et al., 1976). If the fold is perfectly cylindrical, then all of the bedding poles, $\mathbf{p}_{[n]}$, should be perpendicular to $\mathbf{f}$. As is commonly the case, to find the "best fit" to data with scatter, we want to find a model fit that reduces, by as much as possible, the sum of the squares of the deviations from this perfect case. If $\theta_{[i]}$ is the angle between the i'th bedding pole, $\mathbf{p}_{[i]}$, and the fold axis, then the cosine of that angle (which should be close to zero) can be used to represent the deviation. The cosine of $\theta_{[i]}$ is given by the dot product of $\mathbf{f}$ and $\mathbf{p}_{[i]}$. Treating both $\mathbf{f}$ and $\mathbf{p}_{[i]}$ as row vectors (unlike in Equation 4.17), we write the dot product as

$$\cos \theta_{[i]} = \mathbf{p}_{[i]} \mathbf{f}^T \tag{5.30}$$

Thus, we can express the sum of the squares of the deviations as

$$S = \sum_{i=1}^{n} \cos^2 \theta_{[i]} = \sum_{i=1}^{n} \left(\mathbf{p}_{[i]} \mathbf{f}^T \right)^2 \tag{5.31}$$

Because the dot product possesses commutability, we can write

$$\mathbf{p}_{[i]} \mathbf{f}^T = \mathbf{f} \mathbf{p}_{[i]}^T \tag{5.32}$$

and Equation 5.31 can be rewritten as

$$S = \sum_{i=1}^{n} \mathbf{f} \mathbf{p}_{[i]}^T \mathbf{p}_{[i]} \mathbf{f}^T = \mathbf{f} \mathbf{T} \mathbf{f}^T \tag{5.33}$$

5.5 Example of eigenvalues and eigenvectors in structural geology

where **T** is a matrix composed of the sums and products of the direction cosines ($p_i = [\cos\alpha \quad \cos\beta \quad \cos\gamma]$) of the individual lines:

$$\mathbf{T} = \sum_{i=1}^{n} \mathbf{p}_{[i]}^T \mathbf{p}_{[i]} = \sum_{i=1}^{n} (p_i p_j)_{[i]}$$

$$= \begin{bmatrix} \sum \cos^2\alpha_{[i]} & \sum \cos\alpha_{[i]}\cos\beta_{[i]} & \sum \cos\alpha_{[i]}\cos\gamma_{[i]} \\ \sum \cos\beta_{[i]}\cos\alpha_{[i]} & \sum \cos^2\beta_{[i]} & \sum \cos\beta_{[i]}\cos\gamma_{[i]} \\ \sum \cos\gamma_{[i]}\cos\alpha_{[i]} & \sum \cos\gamma_{[i]}\cos\beta_{[i]} & \sum \cos^2\gamma_{[i]} \end{bmatrix} \quad (5.34)$$

Matrix **T** is commonly known as the *orientation matrix* and it is used extensively in statistical treatment of orientation data that have Watson or Bingham distributions (see Fisher *et al.*, 1987). You can think of the orientation matrix as describing the ellipsoidal surface, depicted in the previous section, referred to an arbitrary coordinate system. To find the principal axes of ellipsoid, we need to calculate the eigenvalues and eigenvectors of matrix **T**. In the case of the problem we started out with, S in Equations 5.31 to 5.33 will correspond to the smallest eigenvalue. If the fold were perfectly cylindrical, S would be equal to zero (because all of the $\theta_{[i]}$'s would be 90°).

The calculation of the specific eigenvalues and eigenvectors is best left to any of a number of publicly available "canned" software packages. To get a better feeling about how these routines work in general, we highly recommend that you read Chapter 11 of Press *et al.* (1986). MATLAB® has a built-in function, **eig**, to solve the eigenvalue problem and thus it is particularly well suited to this problem. If, instead, you write your own code in a different language, you'll want to use two subroutines from *Numerical Recipes* (Press *et al.*, 1986) that calculate and sort the eigenvalues and eigenvectors (`Jacobi` and `Eigsrt`, respectively).

The following MATLAB function, **Bingham**, can be used to calculate the three mutually orthogonal axes of the orientation matrix and the uncertainty "cones" for the Bingham statistics. The statistical part is given here for information only because its complete description is beyond the scope of this book. For more information on the statistics, we suggest that you see Fisher *et al.* (1987), particularly Sections 6.3 to 6.6.

```
function [eigVec,confCone,bestFit] = Bingham (T,P)
%Bingham calculates and plots a cylindrical best fit to a pole distribution
%to find fold axes from poles to bedding or the orientation of a plane from
%two apparent dips. The statistical routine is based on algorithms in
%Fisher et al. (1987)
%
%    USE: [eigVec,confCone,bestFit] = Bingham (T,P)
%
%    T and P = Vectors of lines trends and plunges respectively
%
%    eigVec = 3 x 3 matrix with eigenvalues (column 1), and trends (column 2)
%    and plunges (column 3) of the eigenvectors. Maximum eigenvalue and
%    corresponding eigenvector are in row 1, intermediate in row 2,
%    and minimum in row 3.
%
%    confCone = 2 x 2 matrix with the maximum (column 1) and minimum
%    (column 2) radius of the 95% elliptical confidence cone around the
%    eigenvector corresponding to the largest (row 1), and lowest (row 2)
%    eigenvalue
%
```

```
%   besFit = 1 x 2 vector containing the strike and dip (right hand rule)
%   of the best fit great circle to the distribution of lines
%
%   NOTE: Input/Output trends and plunges, as well as confidence
%   cones are in radians. Bingham plots the input lines, eigenvectors and
%   best fit great circle in an equal area stereonet.
%
%   Bingham uses functions ZeroTwoPi, SphToCart, CartToSph, Stereonet,
%   StCoordLine and GreatCircle

%Some constants
east = pi/2.0;
twopi = pi*2.0;

%Number of lines
nlines = max(size(T));

%Initialize the orientation matrix
a=zeros(3,3);

%Fill the orientation matrix with the sums of the squares (for the
%principal diagonal) and the products of the direction cosines of each
%line. cn, ce and cd are the north, east and down direction cosines
for i = 1:nlines
    [cn,ce,cd] = SphToCart(T(i),P(i),0);
    a(1,1) = a(1,1) + cn*cn;
    a(1,2) = a(1,2) + cn*ce;
    a(1,3) = a(1,3) + cn*cd;
    a(2,2) = a(2,2) + ce*ce;
    a(2,3) = a(2,3) + ce*cd;
    a(3,3) = a(3,3) + cd*cd;
end

%The orientation matrix is symmetric so the off-diagonal components can be
%equated
a(2,1) = a(1,2);
a(3,1) = a(1,3);
a(3,2) = a(2,3);

%Calculate the eigenvalues and eigenvectors of the orientation matrix using
%MATLAB function eig. D is a diagonal matrix of eigenvalues and V is a
%full matrix whose columns are the corresponding eigenvectors
[V,D] = eig(a);

%Normalize the eigenvalues by the number of lines and convert the
%corresponding eigenvectors to the lower hemisphere
for i = 1:3
    D(i,i) = D(i,i)/nlines;
    if V(3,i) < 0.0
```

5.5 Example of eigenvalues and eigenvectors in structural geology

```
            V(1,i) = -V(1,i);
            V(2,i) = -V(2,i);
            V(3,i) = -V(3,i);
        end
end

%Initialize eigVec
eigVec = zeros(3,3);
%Fill eigVec
eigVec(1,1) = D(3,3); %Maximum eigenvalue
eigVec(2,1) = D(2,2); %Intermediate eigenvalue
eigVec(3,1) = D(1,1); %Minimum eigenvalue
%Trend and plunge of largest eigenvalue: column 3 of V
[eigVec(1,2),eigVec(1,3)] = CartToSph(V(1,3),V(2,3),V(3,3));
%Trend and plunge of intermediate eigenvalue: column 2 of V
[eigVec(2,2),eigVec(2,3)] = CartToSph(V(1,2),V(2,2),V(3,2));
%Trend and plunge of minimum eigenvalue: column 1 of V
[eigVec(3,2),eigVec(3,3)] = CartToSph(V(1,1),V(2,1),V(3,1));

%Initialize confCone
confCone = zeros(2,2);
%If there are more than 25 lines, calculate confidence cones at the 95%
%confidence level. The algorithm is explained in Fisher et al. (1987)
if nlines >= 25
    e11 = 0.0;
    e22 = 0.0;
    e12 = 0.0;
    d11 = 0.0;
    d22 = 0.0;
    d12 = 0.0;
    en11 = 1.0/(nlines* (eigVec(3,1) - eigVec(1,1))^2);
    en22 = 1.0/(nlines* (eigVec(2,1) - eigVec(1,1))^2);
    en12 = 1.0/(nlines* (eigVec(3,1) - eigVec(1,1))*(eigVec(2,1)...
        - eigVec(1,1)));
    dn11 = en11;
    dn22 = 1.0/(nlines* (eigVec(3,1) - eigVec(2,1))^2);
    dn12 = 1.0/(nlines* (eigVec(3,1) - eigVec(2,1))*(eigVec(3,1)...
        - eigVec(1,1)));
    vec = zeros(3,3);
    for i = 1:3
        vec(i,1) = sin(eigVec(i,3) + east)* cos(twopi - eigVec(i,2));
        vec(i,2) = sin(eigVec(i,3) + east)* sin(twopi - eigVec(i,2));
        vec(i,3) = cos(eigVec(i,3) + east);
    end
    for i = 1:nlines
        c1 = sin(P(i)+east)* cos(twopi-T(i));
        c2 = sin(P(i)+east)* sin(twopi-T(i));
        c3 = cos(P(i)+east);
        u1x = vec(3,1)* c1 + vec(3,2)* c2 + vec(3,3)* c3;
```

```
        u2x = vec(2,1)* c1 + vec(2,2)* c2 + vec(2,3)* c3;
        u3x = vec(1,1)* c1 + vec(1,2)* c2 + vec(1,3)* c3;
        e11 = u1x*u1x * u3x*u3x + e11;
        e22 = u2x*u2x * u3x*u3x + e22;
        e12 = u1x *u2x * u3x*u3x + e12;
        d11 = e11;
        d22 = u1x*u1x * u2x*u2x + d22;
        d12 = u2x * u3x * u1x*u1x + d12;
    end
    e22 = en22* e22;
    e11 = en11* e11;
    e12 = en12* e12;
    d22 = dn22* d22;
    d11 = dn11* d11;
    d12 = dn12* d12;
    d = -2.0*log(.05)/nlines;
    % initialize f
    f = zeros(2,2);
    if abs(e11*e22-e12*e12) >= 0.000001
        f(1,1) = (1/(e11*e22-e12*e12)) * e22;
        f(2,2) = (1/(e11*e22-e12*e12)) * e11;
        f(1,2) = -(1/(e11*e22-e12*e12)) * e12;
        f(2,1) = f(1,2);
        %Calculate the eigenvalues and eigenvectors of the matrix f using
        %MATLAB function eig. The next lines follow steps 1-4 outlined on
        %pp. 34-35 of Fisher et al. (1987)
        DD = eig(f);
        if DD(1) > 0.0 && DD(2) > 0.0
            if d/DD(1) <= 1.0 && d/DD(2) <= 1.0
                confCone(1,2) = asin(sqrt(d/DD(2)));
                confCone(1,1) = asin(sqrt(d/DD(1)));
            end
        end
    end
    % Repeat the process for the eigenvector corresponding to the smallest
    % eigenvalue
    if abs(d11*d22-d12*d12) >= 0.000001
        f(1,1) = (1/(d11*d22-d12*d12)) * d22;
        f(2,2) = (1/(d11*d22-d12*d12)) * d11;
        f(1,2) = -(1/(d11*d22-d12*d12)) * d12;
        f(2,1) = f(1,2);
        DD = eig(f);
        if DD(1) > 0.0 && DD(2) > 0.0
            if d/DD(1) <= 1.0 && d/DD(2) <= 1.0
                confCone(2,2) = asin(sqrt(d/DD(2)));
                confCone(2,1) = asin(sqrt(d/DD(1)));
            end
        end
    end
```

```
    end
end

%Calculate the best fit great circle to the distribution of points
bestFit=zeros(1,2);
bestFit(1) = ZeroTwoPi(eigVec(3,2) + east);
bestFit(2) = east - eigVec(3,3);

%Plot stereonet
Stereonet(0,90*pi/180,10*pi/180,1);

%Plot lines
hold on;
for i = 1:nlines
    [xp,yp] = StCoordLine(T(i),P(i),1);
    plot(xp,yp,'k.');
end

%Plot eigenvectors
for i = 1:3
    [xp,yp] = StCoordLine(eigVec(i,2),eigVec(i,3),1);
    plot(xp,yp,'rs');
end

%Plot best fit great circle
[path] = GreatCircle(bestFit(1),bestFit(2),1);
plot(path(:,1),path(:,2),'r');

%release plot
hold off;
end
```

5.6 EXERCISES

1. Decompose the following tensor, **T**, into symmetric and antisymmetric components. Then calculate the axial vector magnitude, orientation, and sense of rotation:

$$T_{ij} = \begin{bmatrix} 8 & -1 & -1 \\ 1 & 6 & 0 \\ -5 & 0 & 2 \end{bmatrix}$$

2. Expand Equation 5.12 for the terms T'_{11} and T'_{13}. The two equations should each have nine terms in them. It may help to follow by hand the example of the computer code given in Section 5.3.2.
3. Derive the transformation matrix given in Equation 5.14.
4. Use your expansion in Exercise 2 to derive the equations for the Mohr circle given in Equation 5.16, using the transformation matrix in Equation 5.14 and the initial form of the tensor in 5.15.
5. Is the orientation matrix a tensor? Explain your answer.
6. Use the function **Bingham** to calculate the "best-fit" great circle and the fold axis for the bedding poles in the Big Elk anticline (Fig. 3.11 and Exercise 8 in Chapter 3).

CHAPTER SIX

Stress

6.1 STRESS "VECTORS" AND STRESS TENSORS

There is much confusion amongst structural geology students regarding the concept of stress. This confusion remains even after one has got it straight that stress and strain are not interchangeable. The purpose of this chapter and the next is to examine these two fundamental concepts in light of the tools we have developed in the preceding five chapters. With this background we are now in a position to be much more precise about exactly what we mean by "stress" and "strain."

Most structural geologists learn fairly early on that stress is defined as a force, **f**, divided by the area of the plane, A, on which it acts:

$$\sigma = \frac{\mathbf{f}}{A} \tag{6.1}$$

This definition is a perfectly good one and conveys the meaning that stress is a measure of force "intensity." But, if you examine this equation carefully, you will see that force, **f**, is a vector and area, A, is a scalar. By this definition, "stress," σ, should also be a vector, just like force. Later on, in the same introductory course on structural geology, students learn that stress at a point can be represented by nine numbers and is, in fact, a tensor:

$$\sigma_{ij} = \begin{bmatrix} \sigma_{11} & \sigma_{12} & \sigma_{13} \\ \sigma_{21} & \sigma_{22} & \sigma_{23} \\ \sigma_{31} & \sigma_{32} & \sigma_{33} \end{bmatrix} \tag{6.2}$$

No wonder students are confused! These two types of stress are certainly related to each other, as we will see in the next section, but they are not, by any means, identical. In this book, the word "stress" by itself will refer to the *stress tensor* as in Equation 6.2. The quantity given by Equation 6.1 will be referred to as the *stress vector* or more correctly (following general usage in continuum mechanics and engineering literature) a *traction*.

6.2 Cauchy's Law

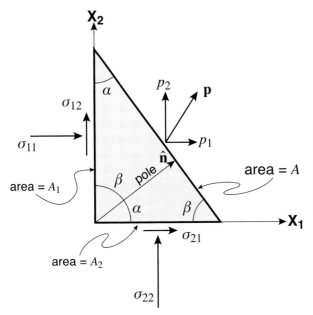

Figure 6.1 Tractions on the sides of a two-dimensional triangular element. Traction **p** acts on the inclined plane with area A; the other tractions act on the two planes perpendicular to the axes of the coordinate system, X_1 and X_2.

6.2 CAUCHY'S LAW

6.2.1 Stresses in two dimensions

We'll begin exploring the relation between stress and traction in two dimensions, where things are easier to visualize, and then expand the analysis to three dimensions. Suppose we have a triangular element as shown in Figure 6.1. Two sides of the element are perpendicular to our coordinate system and the third side, or "plane," is inclined to the two axes at some arbitrary angle; in this case the pole to the plane makes an angle of α with respect to the X_1 axis and β with respect to the X_2 axis. The stress vector, or traction, on the inclined plane with area A is **p**, which can be resolved into two vectors, p_1 parallel to the X_1 axis and p_2 parallel to the X_2 axis. The tractions on the sides of the triangle, which are perpendicular (and, of course, parallel) to the coordinate system, are labeled with the Greek letter sigma and two subscripts.

The first subscript tells you that the plane is perpendicular to that axis. For example, σ_{11} and σ_{12} both act on the plane that is perpendicular to the X_1 axis. The second subscript identifies the axis that is parallel to the vector of interest. Thus, σ_{12} and σ_{22} are both parallel to X_2. Tractions that act perpendicular to a plane are called *normal tractions* (or *normal stress vectors*); in the case of planes perpendicular to the coordinate axes, such tractions will always have two identical subscripts (e.g., σ_{11} and σ_{22}). *Shear tractions* (or *shear stress vectors*) parallel the plane and have unequal subscripts (e.g., σ_{12} and σ_{21}).

To derive the relations between the traction on the inclined plane and those acting on the planes perpendicular to the coordinate system, we need to do a balance of forces, not stresses. Therefore, we need to take into account the areas of the sides of the triangle. The relations between the areas are determined by the angles α and β. In particular, from similar triangles (Fig. 6.1) and some simple trigonometry you can see that A_1 and A_2 can be written as functions of A and the direction cosines of the pole to the inclined plane:

$$A_1 = A\cos\alpha \quad \text{and} \quad A_2 = A\cos\beta \tag{6.3}$$

Figure 6.2 Tractions on a tetrahedral element with three faces perpendicular to the three axes of the coordinate system and the fourth face (on the back side) inclined to all three. Note the naming convention for the subscripts: the first subscript shows to which axis the plane is perpendicular and the second subscript shows to which axis the vector is parallel.

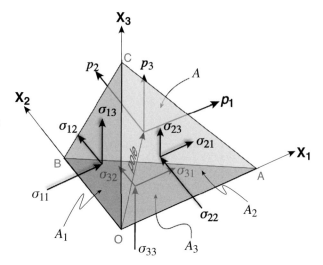

Now, we can write the force balance equation for the forces (i.e., tractions times area) parallel to X_1:

$$p_1 A = \sigma_{11} A_1 + \sigma_{21} A_2 = \sigma_{11} A \cos\alpha + \sigma_{21} A \cos\beta$$

Dividing through by the area of the inclined plane, A, we get

$$p_1 = \sigma_{11} \cos\alpha + \sigma_{21} \cos\beta = \sigma_{11} n_1 + \sigma_{21} n_2 \tag{6.4a}$$

where n_1 is the direction cosine that the pole to the plane makes with the X_1 axis and n_2 the direction cosine with the X_2 axis. Likewise, summing forces parallel to the X_2 axis we can write

$$p_2 = \sigma_{12} \cos\alpha + \sigma_{22} \cos\beta = \sigma_{12} n_1 + \sigma_{22} n_2 \tag{6.4b}$$

If the triangular element has no torques on it, then σ_{12} must equal σ_{21}. You can get an intuitive feel for why this must be so by imagining what would happen if the two tractions were not equal.[1] For example, in Figure 6.1, if σ_{12} were larger than σ_{21} the triangle would spin clockwise in the plane of the page. The only way that this will not happen is if $\sigma_{12} = \sigma_{21}$ and they both point either towards or away from the line of intersection of their mutual planes. This relationship is sometimes known as the *theorem of conjugate shear stresses*.

6.2.2 Stresses in three dimensions

The extension of these concepts to three dimensions is quite straightforward. Figure 6.2 shows the basic configuration, which follows all of the same basic conventions that we established in Figure 6.1. As before, we can balance the forces to see how the various tractions relate to one another; therefore we need to determine the areas on which those tractions act. There are several simple ways to determine this, with one of the most straightforward being to consider the volume of the tetrahedron:

$$V = \left(\frac{1}{3}\right)(\text{area of the base})(\text{height})$$

---

[1] For a more formal proof of this see Nye (1985), pp. 82–87.

6.2 Cauchy's Law

We can write the expression for the volume when each of the four sides is considered to be the base of the tetrahedron as follows (assuming in this case that the pole to the inclined plane is a unit vector):

$$V = \frac{1}{3}A = \frac{1}{3}A_1(\overleftrightarrow{OA}) = \frac{1}{3}A_2(\overleftrightarrow{OB}) = \frac{1}{3}A(\overleftrightarrow{OC})$$

and, from simple trigonometry:

$$A = \frac{A_1}{\cos \alpha} = \frac{A_2}{\cos \beta} = \frac{A_3}{\cos \gamma} \tag{6.5}$$

where α, β, and γ are the angles that the pole to the plane makes with the X_1, X_2, and X_3 axes, respectively.

Now, we can sum the forces parallel to the X_1 axis:

$$p_1 A = \sigma_{11}A_1 + \sigma_{21}A_2 + \sigma_{31}A_3 = \sigma_{11}A\cos\alpha + \sigma_{21}A\cos\beta + \sigma_{31}A\cos\gamma$$

Dividing through by the area of the inclined plane, A, we get

$$p_1 = \sigma_{11}\cos\alpha + \sigma_{21}\cos\beta + \sigma_{31}\cos\gamma = \sigma_{11}n_1 + \sigma_{21}n_2 + \sigma_{31}n_3 \tag{6.6a}$$

The expressions for the tractions parallel to the other axes are

$$p_2 = \sigma_{12}n_1 + \sigma_{22}n_2 + \sigma_{32}n_3 \tag{6.6b}$$
$$p_3 = \sigma_{13}n_1 + \sigma_{23}n_2 + \sigma_{33}n_3 \tag{6.6c}$$

where $n_1 = \cos\alpha$, $n_2 = \cos\beta$, and $n_3 = \cos\gamma$. The structure of these equations should look familiar to you. We can write them using our shorthand matrix notation as

$$\mathbf{p} = \mathbf{n}\boldsymbol{\sigma} = \boldsymbol{\sigma}^T\mathbf{n} \tag{6.7a}$$

or using the summation convention

$$p_i = \sigma_{ij}n_j \tag{6.7b}$$

With this exercise, we have just shown that the group of nine tractions, σ_{ij}, are in fact a tensor. The stress tensor relates two vectors, the traction on an arbitrary plane and the unit vector that describes the orientation of the pole to that plane. It is, as we described in Section 5.2.1, a linear vector operator, the coefficients in a set of three linear equations that describe the relations between these two vectors.

For the same reasons that we mentioned in the two-dimensional case, shear tractions on the adjoining faces of the block that parallel the coordinate axes must be equivalent. Thus,

$$\sigma_{12} = \sigma_{21}, \quad \sigma_{13} = \sigma_{31}, \quad \text{and} \quad \sigma_{32} = \sigma_{23}$$

Although there are nine different coefficients to the stress tensor, only six of them are independent. Stress, therefore, is a symmetric tensor in which the values above the principal diagonal of the matrix (Eq. 6.2) are the same as the values below. We know from Equation 4.20 that the transpose of a symmetric matrix is equal to itself so we can just as easily write Equations 6.7 as

$$\mathbf{p} = \boldsymbol{\sigma}\mathbf{n} \quad \text{or} \quad p_i = \sigma_{ij}n_j \tag{6.8}$$

Equation 6.8 is known as *Cauchy's Law*. Understanding it is the key to grasping why stress is a tensor; it is also the key to solving a large number of continuum mechanics problems in geology.

The MATLAB® function `Cauchy`, below, calculates the tractions on a plane of any orientation in any coordinate system. Thus, the axes do not have to be in a north-east-down coordinate system. Function **DirCosAxes**, which can be found immediately following **Cauchy**, calculates the direction cosines of the axes with respect to the **NED** coordinate system. Notice that to completely define the orientation of the orthogonal X_1, X_2, and X_3 axes, it is just necessary to give the trend and plunge of one axis (e.g., X_1), and the trend of a second axis (e.g., X_3).

```
function [T,pT] = Cauchy(stress,tX1,pX1,tX3,strike,dip)
%Given the stress tensor in a X1,X2,X3 coordinate system of any
%orientation, Cauchy computes the X1,X2,X3 tractions on an arbitrarily
%oriented plane
%
% USE: [T,pT] = Cauchy(stress,tX1,pX1,tX3,strike,dip)
%
% stress = Symmetric 3 x 3 stress tensor
% tX1 = trend of X1
% pX1 = plunge of X1
% tX3 = trend of X3
% strike = strike of plane
% dip = dip of plane
% T = 1 x 3 vector with tractions in X1, X2 and X3
% pT = 1 x 3 vector with direction cosines of pole to plane transformed
%        to X1,X2,X3 coordinates
%
% NOTE = Plane orientation follows the right hand rule
%         Input/Output angles are in radians
%
%Cauchy uses functions DirCosAxes and SphToCart

%Compute direction cosines of X1,X2,X3
dC = DirCosAxes(tX1,pX1,tX3);

%Calculate direction cosines of pole to plane
p = zeros(1,3);
[p(1),p(2),p(3)] = SphToCart(strike,dip,1);

%Transform pole to plane to stress coordinates X1,X2,X3
%The transformation matrix is just the direction cosines of X1,X2,X3
pT = zeros(1,3);
for i = 1:3
    for j = 1:3
        pT(i) = dC(i,j)*p(j) + pT(i);
    end
end

%Convert transformed pole to unit vector
r = sqrt(pT(1)*pT(1)+pT(2)*pT(2)+pT(3)*pT(3));
for i = 1:3
    pT(i) = pT(i)/r;
end
```

6.2 Cauchy's Law

```
%Calculate the tractions in stress coordinates X1,X2,X3
T = zeros(1,3); %Initialize T
%Compute tractions using Cauchy's law (Eq. 6.7b)
for i = 1:3
    for j = 1:3
        T(i) = stress(i,j)*pT(j) + T(i);
    end
end
end

function dC = DirCosAxes(tX1,pX1,tX3)
%DirCosAxes calculates the direction cosines of a right handed, orthogonal
%X1,X2,X3 cartesian coordinate system of any orientation with respect to
%North-East-Down
%
% USE: dC = DirCosAxes(tX1,pX1,tX3)
%
% tX1 = trend of X1
% pX1 = plunge of X1
% tX3 = trend of X3
% dC = 3 x 3 matrix containing the direction cosines of X1 (row 1),
%      X2 (row 2), and X3 (row 3)
%
% Note: Input angles should be in radians
%
% DirCosAxes uses function SphToCart

%Some constants
east = pi/2.0;
west = 1.5*pi;

%Initialize matrix of direction cosines
dC = zeros(3,3);

%Direction cosines of X1
[dC(1,1),dC(1,2),dC(1,3)] = SphToCart(tX1,pX1,0);

%Calculate plunge of axis 3
%If axis 1 is horizontal
if pX1 == 0.0
    if abs(tX1-tX3) == east || abs(tX1-tX3) == west
        pX3 = 0.0;
    else
        pX3 = east;
    end
%Else
else
    %From Equation 2.14 and with theta equal to 90 degrees
```

```
        pX3 = atan(-(dC(1,1)*cos(tX3)+dC(1,2)*sin(tX3))/dC(1,3));
end

%Direction cosines of X3
[dC(3,1),dC(3,2),dC(3,3)] = SphToCart(tX3,pX3,0);

%Compute direction cosines of X2 by the cross product of X3 and X1
dC(2,1) = dC(3,2)*dC(1,3) - dC(3,3)*dC(1,2);
dC(2,2) = dC(3,3)*dC(1,1) - dC(3,1)*dC(1,3);
dC(2,3) = dC(3,1)*dC(1,2) - dC(3,2)*dC(1,1);
% Convert X2 to a unit vector
r = sqrt(dC(2,1)*dC(2,1)+dC(2,2)*dC(2,2)+dC(2,3)*dC(2,3));
for i = 1:3
    dC(2,i) = dC(2,i)/r;
end
end
```

6.3 BASIC CHARACTERISTICS OF STRESS

Because stress is not a characteristic of a material itself (e.g., thermal conductivity), but is imposed on a material (like an electric field) it is called a *field tensor*. It is one of the simplest tensors we will deal with in structural geology, and for that reason is a much better place, mathematically, to start than with the various tensors related to deformation.

6.3.1 Principal axes of stress

Like any symmetric, second order tensor, the stress tensor can be expressed in terms of its principal axes, where only the principal diagonal of the corresponding matrix has non-zero values (e.g., Eq. 5.2). The principal stresses are merely the tractions that comprise the stress tensor when the coordinate system has a unique orientation. Conventionally, the principal stresses are written with just a single subscript:

$$\sigma_{ij} = \begin{bmatrix} \sigma_1 & 0 & 0 \\ 0 & \sigma_2 & 0 \\ 0 & 0 & \sigma_3 \end{bmatrix} \tag{6.9}$$

Again, let us remind you that, although the principal stresses are written in Equation 6.9 so that they are parallel to an axis of the same number, there is no reason why it has to be that way. Equally important, although they bear superficial similarity, σ_1, σ_2, and σ_3 are most definitively *not* the scalar components of a single vector. Single subscripts indicate magnitude only (technically, the three eigenvalues of the stress tensor) and not orientation in our given coordinate system. By convention, the largest principal stress is σ_1 and the smallest is σ_3. In geology, a common convention is that compression is positive, reflecting the fact that virtually all stresses inside the Earth are compressions except at very shallow levels in the crust. Engineering follows the opposite convention where tensions are positive.

If we know the six independent components of stress in any arbitrary coordinate system, we can find a coordinate system in which the axes are parallel to the principal stresses. In Section 5.4.2, we already saw how to solve this problem for our generic tensor, **T**, but it is worth going over it again for the case of stress. If a plane is perpendicular to a principal stress, there will be no shear stress on the plane because all of the off-diagonal components of the

6.3 Basic characteristics of stress

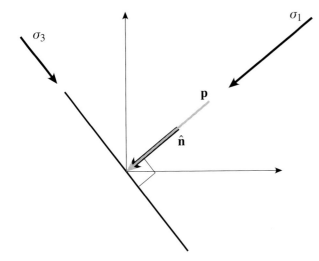

Figure 6.3 Illustration of stresses on a plane perpendicular to a principal stress. When the plane is perpendicular to a principal stress the traction on the plane, **p**, is parallel to the pole to the plane, **n̂**, the traction has the same magnitude and orientation as the principal stress (**p** = σ_1), and there are no shear tractions parallel to the plane.

stress matrix are zero. Remember, we argued above that the tractions with unequal subscripts, $i \neq j$, in the stress tensor are shear tractions and those with equal subscripts, $i = j$, are normal tractions. In Equation 6.9, all the components with $i \neq j$ are zero. Therefore, we want to find a plane, and the traction on that plane, where there is no shear stress. The only case where this is true is when the traction is parallel to the pole of the plane (Fig. 6.3).

Assuming that the principal stress has some unknown magnitude, λ, we can express the parallelism of the traction and the pole as

$$\mathbf{p}^{(\hat{n})} = \lambda \hat{\mathbf{n}} \tag{6.10}$$

Substituting Equation 6.10 into Cauchy's Equation 6.8 we get

$$\lambda \hat{\mathbf{n}} = \boldsymbol{\sigma} \hat{\mathbf{n}} \quad \text{or} \quad \lambda n_i = \sigma_{ij} n_j \tag{6.11}$$

Using the substitution property of the Kronecker delta (Eq. 4.12), we know that $n_i = \delta_{ij} n_j$ so that Equation 6.11 can be rearranged as

$$(\sigma_{ij} - \lambda \delta_{ij}) n_j = 0 \tag{6.12}$$

To solve for λ, take the determinant of the part in parentheses, above, and set it equal to zero, which will give us our familiar cubic in λ as in Equation 5.25. Again, this equation is generally solved numerically. The three eigenvalues, λ, are the three magnitudes of the principal stresses and the corresponding eigenvectors give the orientations of the principal axes of stress. Below are two MATLAB functions that deal with these problems. Function **TransformStress** transforms the stress tensor from one Cartesian system to another of other orientation. Function **PrincipalStress** calculates the principal stresses and their orientations for a given stress tensor in a Cartesian coordinate system of any orientation. **PrincipalStress**, below, relies on the MATLAB function **eig** to do the eigenvalue problem. If you are coding this from scratch in a normal programming language, you will need to call subroutines such as **Jacobi** and **Eigsrt** from *Numerical Recipes* (Press et al., 1986).

```
function nstress = TransformStress(stress,tX1,pX1,tX3,ntX1,npX1,ntX3)
%TransformStress transforms a stress tensor from old X1,X2,X3 to new X1'
%,X2',X3' coordinates
%
```

```
% USE: nstress = TransformStress(stress,tX1,pX1,tX3,ntX1,npX1,ntX3)
%
% stress = 3 x 3 stress tensor
% tX1 = trend of X1
% pX1 = plunge of X1
% tX3 = trend of X3
% ntX1 = trend of X1'
% npX1 = plunge of X1'
% ntX3 = trend of X3'
% nstress = 3 x 3 stress tensor in new coordinate system
%
% NOTE: All input angles should be in radians
%
% TransformStress uses function DirCosAxes

%Direction cosines of axes of old coordinate system
odC = DirCosAxes(tX1,pX1,tX3);

%Direction cosines of axes of new coordinate system
ndC = DirCosAxes(ntX1,npX1,ntX3);

%Transformation matrix between old and new coordinate system
a = zeros(3,3);
for i = 1:3
    for j = 1:3
        %Use dot product
        a(i,j) = ndC(i,1)*odC(j,1) + ndC(i,2)*odC(j,2) + ndC(i,3)*odC(j,3);
    end
end

%Transform stress tensor from old to new coordinate system (Eq. 5.12)
nstress = zeros(3,3);
for i = 1:3
    for j = 1:3
        for k = 1:3
            for L = 1:3
                nstress(i,j) = a(i,k)*a(j,L)*stress(k,L)+nstress(i,j);
            end
        end
    end
end

function [pstress,dCp] = PrincipalStress(stress,tX1,pX1,tX3)
%Given the stress tensor in a X1,X2,X3 coordinate system of any
%orientation, PrincipalStress calculates the principal stresses and their
%orientations (trend and plunge)
%
```

6.3 Basic characteristics of stress

```
% USE: [pstress,dCp] = PrincipalStress(stress,tX1,pX1,tX3)
%
% stress = Symmetric 3 x 3 stress tensor
% tX1 = trend of X1
% pX1 = plunge of X1
% tX3 = trend of X3
% pstress = 3 x 3 matrix containing the magnitude (column 1), trend
%           (column 2), and plunge (column 3) of the maximum (row 1),
%           intermediate (row 2), and minimum (row 3) principal stresses
% dCp = 3 x 3 matrix with direction cosines of the principal stress
%       directions: Max. (row 1), Int. (row 2), and Min. (row 3)
%
% NOTE: Input/Output angles are in radians
%
% PrincipalStress uses functions DirCosAxes and CartToSph

%Compute direction cosines of X1,X2,X3
dC = DirCosAxes(tX1,pX1,tX3);

%Initialize pstress
pstress = zeros(3,3);

%Calculate the eigenvalues and eigenvectors of the stress tensor. Use
%MATLAB function eig. D is a diagonal matrix of eigenvalues
%(i.e. principal stress magnitudes), and V is a full matrix whose columns
%are the corresponding eigenvectors (i.e. principal stress directions)
[V,D] = eig(stress);

%Fill principal stress magnitudes
pstress(1,1) = D(3,3); %Maximum principal stress
pstress(2,1) = D(2,2); %Intermediate principal stress
pstress(3,1) = D(1,1); %Minimum principal stress

%The direction cosines of the principal stress tensor are given with
%respect to X1,X2,X3 stress coordinate system, so they need to be
%transformed to the North-East-Down coordinate system (e.g. Eq. 3.9)
tV = zeros(3,3);
for i = 1:3
    for j = 1:3
        for k = 1:3
            tV(j,i) = dC(k,j)*V(k,i) + tV(j,i);
        end
    end
end

%Initialize dCp
dCp = zeros(3,3);

%Trend and plunge of maximum principal stress direction
```

```
dCp(1,:) = [tV(1,3),tV(2,3),tV(3,3)];
[pstress(1,2),pstress(1,3)] = CartToSph(tV(1,3),tV(2,3),tV(3,3));

%Trend and plunge of intermediate principal stress direction
dCp(2,:) = [tV(1,2),tV(2,2),tV(3,2)];
[pstress(2,2),pstress(2,3)] = CartToSph(tV(1,2),tV(2,2),tV(3,2));

%Trend and plunge of minimum principal stress direction
dCp(3,:) = [tV(1,1),tV(2,1),tV(3,1)];
[pstress(3,2),pstress(3,3)] = CartToSph(tV(1,1),tV(2,1),tV(3,1));
end
```

6.3.2 Mohr circle for stress

Mohr circle for stress, like any other Mohr circle, is a graphical calculator which allows us to determine the normal and shear stress on any plane that is parallel to one of the principal stresses and can make any angle with respect to the other two principal stresses. As described in Section 5.3.3, the Mohr circle is derived by making a rotation about one of the principal axes of a tensor. In Figure 6.4, the old axes are parallel to the principal axes of the tensor, σ_{ij}, and the rotation is around the σ_2 axis. By choosing our new coordinate system so that it is parallel to the pole to the plane, the components of the tensor in its new configuration, σ'_{ij}, will automatically give us the normal (σ'_{11}) and shear (σ'_{13}) stresses on the plane. Thus the old form of the stress tensor and the transformation matrix (**a**) are, respectively,

$$\sigma_{ij} = \begin{bmatrix} \sigma_1 & 0 & 0 \\ 0 & \sigma_2 & 0 \\ 0 & 0 & \sigma_3 \end{bmatrix} \quad \text{and} \quad a_{ij} = \begin{pmatrix} \cos\theta & 0 & \sin\theta \\ 0 & 1 & 0 \\ -\sin\theta & 0 & \cos\theta \end{pmatrix}$$

Using the identities $\cos(90 - \theta) = \sin\theta$ and $\cos(90 + \theta) = -\sin\theta$, the new form of the tensor, σ'_{ij}, is

$$\sigma'_{ij} = \begin{bmatrix} (\sigma_1\cos^2\theta + \sigma_3\sin^2\theta) & 0 & ((\sigma_3 - \sigma_1)\sin\theta\cos\theta) \\ 0 & \sigma_2 & 0 \\ -((\sigma_1 - \sigma_3)\sin\theta\cos\theta) & 0 & (\sigma_1\sin^2\theta + \sigma_3\cos^2\theta) \end{bmatrix} \quad (6.13)$$

Rearranging using the double angle formulas, we get the familiar equations for the Mohr circle:

$$\sigma'_{11} = \frac{(\sigma_1 + \sigma_3)}{2} + \frac{(\sigma_1 - \sigma_3)}{2}\cos 2\theta \quad (6.14a)$$

$$\sigma'_{13} = -\frac{(\sigma_1 - \sigma_3)}{2}\sin 2\theta \quad (6.14b)$$

The graphical representation of the Mohr circle is shown in Figure 6.5. Note that, in some introductory structural geology textbooks, you will see the angle 2θ in the Mohr circle diagram measured clockwise from σ_3. Those authors have taken as a convention that the angle θ is measured between σ_1 and the plane itself; in our derivation, above, θ is the angle between the *pole* to the plane and σ_1 (Fig. 6.4a). There is nothing particularly wrong with measuring θ from the plane rather than the pole because the two angles are complementary. Constructing the Mohr circle this way, however, tends to obscure its origin as a tensor transformation.

A useful property of all Mohr circle constructions is the concept of the pole to the Mohr circle (point P in Fig. 6.5a), which can help one to relate the Mohr circle diagram to the physical orientation of the vectors it represents (Ragan, 2009). Lines drawn from the pole to the Mohr

6.3 Basic characteristics of stress

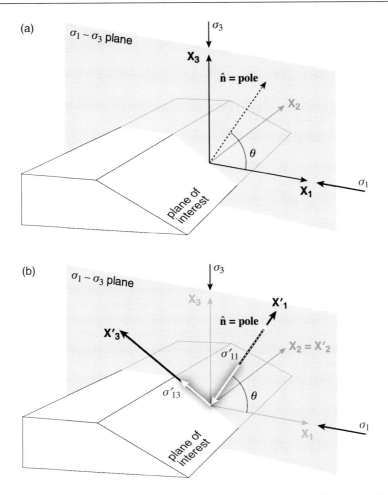

Figure 6.4 Coordinate systems and stress vectors for the Mohr circle for stress. (a) The "old" coordinate system, which is parallel to the principal axes of the stress tensor. Note that the X_2 axis, which is parallel to σ_2, is contained within the plane of interest. X_1, X_3, σ_1, σ_3, and the pole to the plane of interest are all coplanar. (b) The "new" coordinate system has now been transformed into the coordinate frame of the plane of interest. Note that X_2 has not changed (i.e., $X_2 = X'_2$) but X_1 and X_3, which are in the plane perpendicular to X_2, have been transformed to X'_1 and X'_3.

circle to the stress of interest on the circle are parallel to the physical orientation of that vector in space. For example, the long dashed lines in Figure 6.5a are parallel to the principal stress vectors in the gray block of material in Figure 6.5b.

Clearly, we can carry out the tensor transformation by rotating about any of the three principal stresses. Thus, there are three Mohr circles for any given state of stress. The above example is for the state of stress on planes that contain the σ_2 axis. The other two circles will be for planes that contain the σ_1 axis and planes that contain the σ_3 axis. The three transformations together give us a set of three nested circles (Fig. 6.6). Defined this way, all possible stresses in the body must plot in the region between the smallest and largest circles (the shaded region in Fig. 6.6). We are not limited to finding only those tractions that plot on one of the circles but, indeed, can find the stress on a plane of any orientation with respect to the coordinate and principal stress axes.

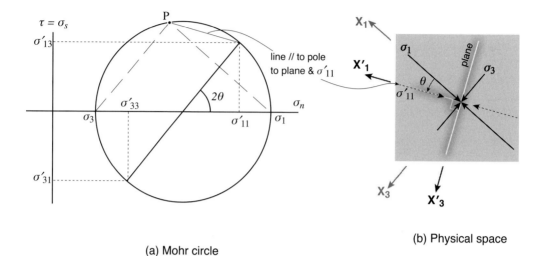

(a) Mohr circle

(b) Physical space

Figure 6.5 (a) Mohr circle for stress, where θ is the angle between the pole to the plane and the maximum principal stress, σ_1. Geological convention of compression positive is followed. Point P is the pole to the Mohr circle for stress. (b) The physical setting for the state of stress shown in part (a). X_1 and X_3 represent the old coordinate system parallel to the principal stresses, whereas X'_1 and X'_3 are the new coordinate system parallel to the pole to the plane and the plane itself, respectively. Note how the physical orientations of the stress vectors in (b) are parallel to the lines drawn between the stresses and point P on the Mohr circle in (a).

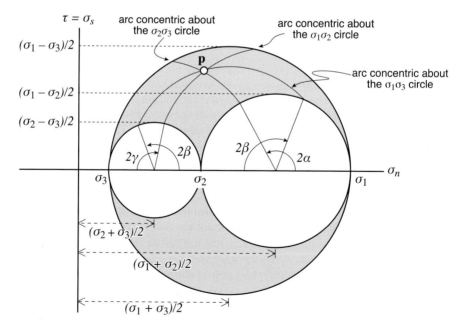

Figure 6.6 The three-dimensional Mohr circle for stress. All possible states of stress must plot within the shaded region; those that include a principal plane of stress plot on one of the margins of the three circles.

6.3 Basic characteristics of stress

Figure 6.6 shows the construction for finding the normal and shear tractions on a plane that makes angles of $\alpha = 59.5°$, $\beta = 55°$, and $\gamma = 50°$ with the $\mathbf{X_1}$ (σ_1), $\mathbf{X_2}$ (σ_2), and $\mathbf{X_3}$ (σ_3) axes, respectively. The tractions on that plane can be read off the diagram at the point of intersection of the three arcs (point **p**, Fig. 6.6). For those interested in the mathematical background for these relations, see Malvern (1969, pp. 94–101) or Jaeger and Cook (1979, pp. 27–30). Where you measure the double angles from is most conveniently remembered by recalling which axis they relate to. Thus, 2γ is measured from σ_3 on the Mohr circle, 2β from σ_2, and 2α from σ_1 (Fig. 6.6).

6.3.3 Special states of stress

There are several special types of stress that can be precisely defined with our understanding of the stress tensor. They are particularly easy to recognize when the coordinate axes are parallel to the principal axes of the tensor. They are listed below and several are illustrated with Mohr circles in Figure 6.7.

Uniaxial stress has only one non-zero principal stress. Nye (1985) gives as an example the state of stress in a vertical rod with a weight hung on one end. Uniaxial stress (Fig. 6.7c) has the form

$$\sigma_{ij} = \begin{bmatrix} \sigma_1 & 0 & 0 \\ 0 & 0 & 0 \\ 0 & 0 & 0 \end{bmatrix} \tag{6.15}$$

Biaxial stress (Fig. 6.7a) has two non-zero principal stresses:

$$\sigma_{ij} = \begin{bmatrix} \sigma_1 & 0 & 0 \\ 0 & \sigma_2 & 0 \\ 0 & 0 & 0 \end{bmatrix} \tag{6.16}$$

Triaxial stress (Fig. 6.7b) is the most general type of stress tensor. It has three non-zero principal stresses as in Equation 6.17:

$$\sigma_{ij} = \begin{bmatrix} \sigma_1 & 0 & 0 \\ 0 & \sigma_2 & 0 \\ 0 & 0 & \sigma_3 \end{bmatrix} \tag{6.17}$$

When two of the principal stresses are equal and the third is different, it is known as a *cylindrical state of stress* (Fig. 6.7c). In this case, only the direction of the different principal stress is unique; the other two principal stresses can have any orientation in the plane that is perpendicular to the third. Uniaxial stress, above, is a special case of cylindrical stress where the two equal stresses are zero.

Of particular import to structural geology is a *spherical state of stress*. This occurs when all three principal stresses have the same value (Fig. 6.7d). When this is the case, any direction in the body can be a principal axis (i.e., the stress magnitude ellipsoid is a sphere) and, therefore, *there are no planes that have shear traction acting on them*. When all three principal stresses are equal, the Mohr circle plots as a single point on the horizontal axis (Fig. 6.7d). Clearly, this point has a shear stress, $\sigma_s = 0$, and therefore there are no shear tractions in the body. This condition is also known as *hydrostatic stress* because, as long as a fluid is not moving, the pressure is equal in all directions and it can support no shear tractions.

Finally, a *pure shear stress* is one in which two of the principal stresses are equal and opposite in sign and the third is zero (Fig. 6.7e). The tensor looks like

$$\sigma_{ij} = \begin{bmatrix} \sigma & 0 & 0 \\ 0 & 0 & 0 \\ 0 & 0 & -\sigma \end{bmatrix} \tag{6.18}$$

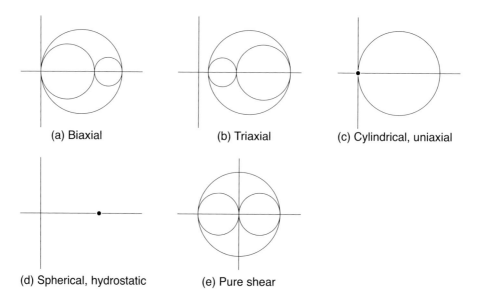

Figure 6.7 Several special states of stress as shown on general, three-dimensional Mohr circles. The solid black dot represents a single point. (a) $\sigma_3 = 0$, σ_1 and $\sigma_2 \neq 0$; (b) σ_1, σ_2, and $\sigma_3 \neq 0$; (c) $\sigma_3 = \sigma_2 = 0$, $\sigma_1 \neq 0$; (d) $\sigma_1 = \sigma_2 = \sigma_3$; (e) $\sigma_3 = -\sigma_1$, $\sigma_2 = 0$.

6.4 THE DEVIATORIC STRESS TENSOR

The concept of hydrostatic stress allows us to introduce an even more fundamental type of stress tensor which is very useful in structural geology. We can define the mean normal stress as the arithmetic average of the three normal tractions (i.e., the principal diagonal) of any stress tensor:

$$p = \frac{(\sigma_{11} + \sigma_{22} + \sigma_{33})}{3} \tag{6.19}$$

Note that the mean stress will be the same regardless of the coordinate system because the sum of tractions along the principal diagonal is just the first invariant of the stress tensor. From Section 4.3.2, we know that any matrix - and all tensors are matrices - can be expressed as the sum of two other matrices. Therefore we can write the stress tensor as

$$\sigma_{ij} = \begin{bmatrix} p & 0 & 0 \\ 0 & p & 0 \\ 0 & 0 & p \end{bmatrix} + \begin{bmatrix} \sigma_{11} - p & \sigma_{12} & \sigma_{13} \\ \sigma_{21} & \sigma_{22} - p & \sigma_{23} \\ \sigma_{31} & \sigma_{32} & \sigma_{33} - p \end{bmatrix} \tag{6.20a}$$

or, in indicial notation

$$\sigma_{ij} = p\delta_{ij} + s_{ij} \tag{6.20b}$$

The matrix on the left, $p\delta_{ij}$, is the spherical or hydrostatic stress tensor and the matrix on the right, s_{ij}, is the deviatoric stress tensor. As you might imagine, the hydrostatic stress tensor exerts uniform pressure all around the body of interest. This may cause the body to shrink or expand (i.e., change volume) but, on first glance, it is difficult to see how it would change the shape of a body. There is a way this can occur, however. If the body on which the stress is applied is anisotropic with respect to its material properties (e.g., it is stronger along certain planes than along others) then even uniform hydrostatic pressure will cause it to deform. Nonetheless, the deviatoric stress tensor is commonly responsible for the vast majority of shape changes; it is, after all, the tensor that has all of the shear stress associated with it.

The principal axes of the deviatoric stress tensor have the same orientation as those of the stress tensor itself and the magnitudes differ only by a factor of p:

$$s_1 = \sigma_1 - p, \quad s_2 = \sigma_2 - p, \quad \text{and} \quad s_3 = \sigma_3 - p \tag{6.21}$$

For arbitrary coordinate axes it is generally easier to determine the principal axes of the deviatoric tensor than for the stress tensor itself. This is because, for the former, there is an analytical solution to the eigenvalue problem. For the deviatoric stress tensor, the solution to the characteristic equation has the form:

$$\lambda^3 - II_{[s]}\lambda - III_{[s]} = 0 \tag{6.22}$$

The analytical solution (Malvern, 1969, p. 92) to this equation is

$$s_i = 2\cos\alpha_i \left(\frac{II_{[s]}}{3}\right)^{\frac{1}{2}} \tag{6.23a}$$

where

$$\cos 3\alpha_1 = \frac{III_{[s]}}{2}\left(\frac{3}{II_{[s]}}\right)^{\frac{3}{2}}, \quad \alpha_2 = \alpha_1 + \frac{2\pi}{3}, \quad \text{and} \quad \alpha_3 = \alpha_1 - \frac{2\pi}{3} \tag{6.23b}$$

and $II_{[s]}$ and $III_{[s]}$ are the second and third invariants of the stress tensor as described in Equation 5.26.

6.5 A PROBLEM INVOLVING STRESS

Many clever graphical and analytical methods have been developed to determine the magnitude and orientation of maximum shear stress on an arbitrarily oriented plane. This problem is particularly germane to any question involving faulting and fracturing of rocks in the upper crust. For example, during the 1980s, there was substantial interest in methods for finding a "best-fit" stress tensor for a group of fault plane–slickenside measurements. The orientation and magnitude of maximum shear stress on a plane is of key importance.

Rather than looking for the shortest solution, ours is designed to illustrate in a clear and organized way the principles developed in this and previous chapters. It relies on no graphical construction, simply a couple of tensor transformations. The rotations carried out in the graphical methods are conceptually the same as tensor transformations. There are three coordinate systems to deal with (Fig. 6.8): (1) The geographic coordinate system, **NED**, is what the data will be entered in and also the coordinate system in which we will want our final answers. (2) The second coordinate system is defined by the principal stress axes, $\sigma_1\sigma_2\sigma_3$. All three of these axes, including their magnitude and orientation are known in advance. (3) The third set of coordinates is determined by the fault plane itself. These are the pole to the plane, **n** (the first axis), the line in the fault plane along which there is zero shear traction, **b**, and the line in the plane that has the maximum shear traction, **s** (the third axis). Of these final three axes, we only know, at the beginning, the orientation of the pole.

Our solution to this problem will follow these basic steps: transform everything into principal stress coordinates; calculate the traction vector on the plane; use that vector to

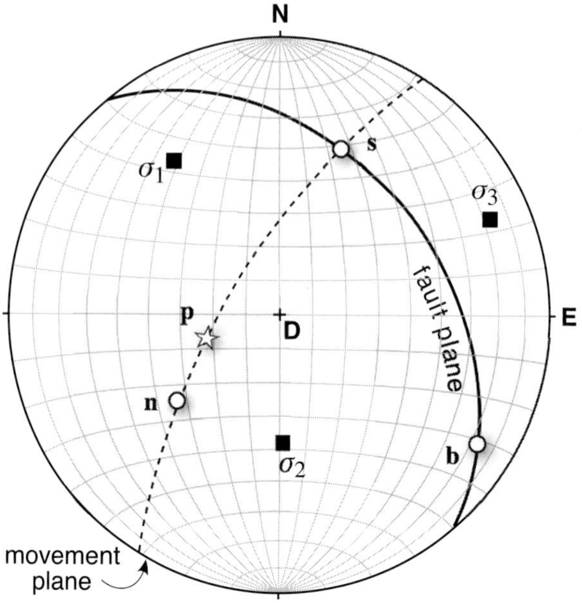

Figure 6.8 Lower hemisphere, equal area projection showing the three coordinate systems involved in determining the maximum shear traction on the plane. See text for description of axes.

determine the other two axes, **s** and **b**, of the third coordinate system; use a tensor transformation from $\sigma_1\sigma_2\sigma_3$ to **nbs** to calculate the magnitudes of the stresses on the plane; and, finally, do a vector transformation to get the orientations of **s** and **b** in geographic coordinates. Each of these steps is elaborated below.

6.5.1 Data entry and transformation to principal stress coordinates

The original data – the orientations and magnitudes of the principal stress axes and the orientation of the plane – are generally entered in geographic coordinates, as trend and plunge or strike and dip. The direction cosines of the principal stress axes will form the transformation matrix, **a**, for the **NED** (old) to $\sigma_1\sigma_2\sigma_3$ (new) transformation. To keep things straight, one only need enter the orientation of σ_1 and σ_3; calculating σ_2 as the cross product, $\sigma_3 \times \sigma_1$, will insure that the second coordinate system is right-handed. The first transformation matrix is

$$a_{ij} = \begin{pmatrix} \cos\alpha_{[\sigma_1]} & \cos\beta_{[\sigma_1]} & \cos\gamma_{[\sigma_1]} \\ \cos\alpha_{[\sigma_2]} & \cos\beta_{[\sigma_2]} & \cos\gamma_{[\sigma_2]} \\ \cos\alpha_{[\sigma_3]} & \cos\beta_{[\sigma_3]} & \cos\gamma_{[\sigma_3]} \end{pmatrix} = \begin{pmatrix} CN_{[\sigma_1]} & CE_{[\sigma_1]} & CD_{[\sigma_1]} \\ CN_{[\sigma_2]} & CE_{[\sigma_2]} & CD_{[\sigma_2]} \\ CN_{[\sigma_3]} & CE_{[\sigma_3]} & CD_{[\sigma_3]} \end{pmatrix} \quad (6.24)$$

The pole to the fault plane, **n**, is also entered in geographic coordinates. It, too, must be transformed into principal stress coordinates but, because the pole is not parallel to either the new or the old axes, its orientation in the new (principal stress) coordinate system is given by a vector transformation:

$$n'_i = a_{ij} n_j \quad (6.25)$$

6.5.2 Calculate the traction vector on the plane

Now that we know the pole to the plane in stress coordinates, we can calculate the traction on the plane **p**′ in principal stress coordinates, from Cauchy's Law:

6.5 A problem involving stress

$$p'_i = \sigma_{ij} n'_j \quad \text{where} \quad \sigma_{ij} = \begin{bmatrix} \sigma_1 & 0 & 0 \\ 0 & \sigma_2 & 0 \\ 0 & 0 & \sigma_3 \end{bmatrix} \quad (6.26)$$

Equations 6.24 to 6.26 are solved by the function `Cauchy` introduced in Section 6.2.

6.5.3 Determine the orientations of s' and b'

We need to know the orientations of **s'** and **b'** (the orientations of **s** and **b** in principal stress coordinates) so that the second transformation matrix, from principal stress to fault plane coordinates, can be determined. There are many different ways to do this. One of the simplest is to rely on the relationship that the maximum shear traction **s'** on a plane is also coplanar with the traction and the pole, **p'** and **n'**. This plane, which contains **p'**, **n'**, and **s'** (see Fig. 6.8), is perpendicular to the fault plane and in faulting analysis is called the *movement plane*. The pole to the movement plane is also **b'**, the second axis of our third coordinate system. Because we know the orientation of **n'** and **p'** already, **b'** can be determined by the cross product of those two, and then **s'** can be determined from the cross product of **n'** and **b'**:

$$\mathbf{b'} = \mathbf{n'} \times \mathbf{p'} \quad \text{and} \quad \mathbf{s'} = \mathbf{n'} \times \mathbf{b'} \quad (6.27)$$

The direction cosines of **n'**, **b'**, and **s'** *in principal stress coordinates* define our second transformation matrix, **c**. Note that the above cross products do not give us unit vectors, so the above must be divided by their magnitudes in order to get the direction cosines. The second transformation matrix is

$$c_{ij} = \begin{pmatrix} \hat{n}'_1 & \hat{n}'_2 & \hat{n}'_3 \\ \hat{b}'_1 & \hat{b}'_2 & \hat{b}'_3 \\ \hat{s}'_1 & \hat{s}'_2 & \hat{s}'_3 \end{pmatrix} \quad (6.28)$$

6.5.4 Vector transformation to get the geographic orientations

At this point, if we are just interested in the orientation of maximum shear on the plane, all that is needed is to transform **p'**, **b'**, and **s'** back to geographic coordinates (we already know what the pole, **n**, is). This transformation is from the *new* principal stress coordinate system back to the *old* geographic system, so the order of the subscripts of the transformation matrix, **a**, is reversed (i.e., transposed) from what it was in Equation 6.24:

$$s_i = a_{ji} s'_j \quad b_i = a_{ji} b'_j \quad \text{and} \quad p_i = a_{ji} p'_j \quad (6.29)$$

These vectors in geographic coordinates will probably have magnitudes different from one. Before we can convert them back into more familiar trends and plunges, they must be converted to unit vectors by dividing each of their components by their magnitudes.

6.5.5 Tensor transformation to get the magnitude of shear and normal tractions

To get the normal and shear tractions on the fault plane, we need to transform the stress tensor from the principal stress coordinate system (now, the old system) to the fault plane coordinates (new). The standard tensor transformation,

$$\sigma'_{ij} = c_{ik}c_{jl}\sigma_{kl} \tag{6.30}$$

will give us exactly what we need. The pole to the plane is the first axis so the stresses on the plane in the new coordinate system will have a first suffix of 1. The maximum shear direction in that plane, **s'**, was defined as the third axis so the shear stress on the plane will have a second subscript of 3. In summary,

- $\sigma'_{11} \equiv$ normal traction on the plane,
- $\sigma'_{13} \equiv$ shear traction on the plane,
- $\sigma'_{12} \equiv$ traction parallel to **b** $= 0$.

The equations for these three tractions, which result from Equation (6.30), are

$$\begin{aligned}\sigma'_{11} &= c_{11}c_{11}\sigma_1 + c_{12}c_{12}\sigma_2 + c_{13}c_{13}\sigma_3 \\ \sigma'_{12} &= c_{11}c_{21}\sigma_1 + c_{12}c_{22}\sigma_2 + c_{13}c_{23}\sigma_3 = 0 \\ \sigma'_{13} &= c_{11}c_{31}\sigma_1 + c_{12}c_{32}\sigma_2 + c_{13}c_{33}\sigma_3\end{aligned} \tag{6.31}$$

By setting the second equation in 6.31 to zero and using the orthogonality relations (Eqs. 3.3 and 4.28), we can derive an important quantity called the *principal stress ratio*, R (Gephart, 1990):

$$R = \frac{(\sigma_2 - \sigma_1)}{(\sigma_3 - \sigma_1)} = \frac{c_{13}c_{23}}{c_{12}c_{22}} \tag{6.32}$$

When $R = 1$, σ_2 is equal to σ_3; when $R = 0$, σ_2 is equal to σ_1. This ratio is of key importance to the problem of deriving stress from fault slip data (the inverse equivalent of the forward problem that we solved above). Fault reactivation is likewise critically dependent on the principal stress ratio. Figure 6.9 illustrates, for a single example, how the orientations and magnitudes of tractions vary with R; you can see that the rake of the potential directions of slip on a pre-existing fault plane can vary by 90°. In essence, Equation 6.32 shows that inversion of fault slip data for stress can yield only four independent quantities: the ratio of principal stresses R, and three independent angles (or direction cosines) that uniquely define the orientations of those principal stress axes. The fourth direction cosine in Equation 6.32 is dependent on the other three by the orthogonality relations. Thus, as shown by Gephart (1990), it is impossible to determine the magnitudes of the principal stresses from fault slip data. The importance of the principal stress ratio was first realized by Bott (1959). Note that Angelier (1984) defines a variation on the principal stress ratio:

$$\Phi = \frac{\sigma_2 - \sigma_3}{\sigma_1 - \sigma_3} \tag{6.33}$$

In this case, if $\Phi = 0$, then $\sigma_2 = \sigma_3$, and if $\Phi = 1$, then $\sigma_2 = \sigma_1$. Thus, $\Phi = 1 - R$. The MATLAB function `ShearOnPlane` below carries out all the calculations in this section.

```
function [TT,dCTT,R] = ShearOnPlane(stress,tX1,pX1,tX3,strike,dip)
%ShearOnPlane calculates the direction and magnitudes of the normal
%and shear tractions on an arbitrarily oriented plane
%
% USE: [TT,dCTT] = ShearOnPlane(stress,tX1,pX1,tX3,strike,dip)
%
% stress = 3 x 3 stress tensor
% tX1 = trend of X1
% pX1 = plunge of X1
```

6.5 A problem involving stress

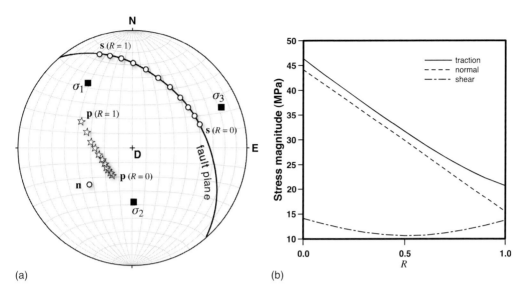

Figure 6.9 Illustration of the importance of the stress ratio, R (Eq. 6.32), for determining the direction and magnitude of shear on an arbitrarily oriented plane (which has the same orientation as that in Fig. 6.8). The orientations of the principal stresses and the plane are held constant and the values of $\sigma_1 = 50$ MPa and $\sigma_3 = 10$ MPa are also constant. All that varies is the value of R (i.e., σ_2 relative to σ_1 and σ_3). (a) Lower hemisphere, equal area projection showing the orientations of the principal stresses (solid squares), the plane of interest (great circle) and its pole (**n**), and the variation in orientation of the shear tractions on the plane, **s**, as well as the traction vector, **p**, with $0.0 \leq R \leq 1.0$. (b) Graph showing how, for the same example as (a), the magnitudes of the traction, normal, and shear vectors vary with R.

```
% tX3 = trend of X3
% strike = strike of plane
% dip = dip of plane
% TT = 3 x 3 matrix with the magnitude (column 1), trend (column 2) and
%      plunge (column 3) of: normal traction on the plane (row 1),
%      minimum shear traction (row 2), and maximum shear traction (row 3)
% dCTT = 3 x 3 matrix with the direction cosines of unit vectors parallel
%      to: normal traction on the plane (row 1), minimum shear traction
%      (row 2), and maximum shear traction (row 3)
% R = Stress ratio
%
% NOTE = Input stress tensor does not need to be along principal stress
%         directions
%         Plane orientation follows the right hand rule
%         Input/Output angles are in radians
%
%   ShearOnPlane uses functions PrincipalStress, Cauchy and CartToSph

%Initialize TT and dCTT
TT = zeros(3,3);
dCTT = zeros(3,3);
```

```
%Compute principal stresses and principal stress directions
[pstress,dCp] = PrincipalStress(stress,tX1,pX1,tX3);

%Update stress vector so that it is along principal stress directions
stress = zeros(3,3);
for i = 1:3
    stress(i,i) = pstress(i,1);
end

%Compute tractions on plane in principal stress direction (Eqs. 6.24-6.26)
[T,pT] = Cauchy(stress,pstress(1,2),pstress(1,3),pstress(3,2),strike,dip);

%Find the B axis by the cross product of T cross pT and convert to
%direction cosines (Eq. 6.27)
B = zeros(1,3);
B(1) = T(2)*pT(3) - T(3)*pT(2);
B(2) = T(3)*pT(1) - T(1)*pT(3);
B(3) = T(1)*pT(2) - T(2)*pT(1);

%Find the shear direction by the cross product of pT cross B. This will
%give S in right handed coordinates (Eq. 6.27)
S = zeros(1,3);
S(1) = pT(2)*B(3) - pT(3)*B(2);
S(2) = pT(3)*B(1) - pT(1)*B(3);
S(3) = pT(1)*B(2) - pT(2)*B(1);

%Convert T, B and S to unit vectors
rT = sqrt(T(1)*T(1)+T(2)*T(2)+T(3)*T(3));
rB = sqrt(B(1)*B(1)+B(2)*B(2)+B(3)*B(3));
rS = sqrt(S(1)*S(1)+S(2)*S(2)+S(3)*S(3));
for i = 1:3
    T(i) = T(i)/rT;
    B(i) = B(i)/rB;
    S(i) = S(i)/rS;
end

%Now we can write the transformation matrix from principal stress
%coordinates to plane coordinates (Eq. 6.28)
a = zeros(3,3);
a(1,:) = [pT(1),pT(2),pT(3)];
a(2,:) = [B(1),B(2),B(3)];
a(3,:) = [S(1),S(2),S(3)];

%Calculate stress ratio (Eq. 6.32)
R = (stress(2,2) - stress(1,1))/(stress(3,3)-stress(1,1));

%Calculate magnitude of normal and shear tractions (Eq. 6.31)
for i = 1:3
```

```
    TT(i,1) = stress(1,1)*a(1,1)*a(i,1) + stress (2,2)*a(1,2)*a(i,2) +...
    stress(3,3)*a(1,3)*a(i,3);
end

%To get the orientation of the tractions in north-east-down coordinates, we
%need to do a vector transformation between principal stress and
%north-east-down coordinates. The transformation matrix is just the
%direction cosines of the principal stresses in north-east-down coordinates
%(Eq. 6.29)
for i = 1:3
    for j = 1:3
        dCTT(1,i) = dCp(j,i)*pT(j) + dCTT(1,i);
        dCTT(2,i) = dCp(j,i)*B(j) + dCTT(2,i);
        dCTT(3,i) = dCp(j,i)*S(j) + dCTT(3,i);
    end
end

%Trend and plunge of traction on plane
[TT(1,2),TT(1,3)] = CartToSph(dCTT(1,1),dCTT(1,2),dCTT(1,3));
%Trend and plunge of minimum shear direction
[TT(2,2),TT(2,3)] = CartToSph(dCTT(2,1),dCTT(2,2),dCTT(2,3));
%Trend and plunge of maximum shear direction
[TT(3,2),TT(3,3)] = CartToSph(dCTT(3,1),dCTT(3,2),dCTT(3,3));
end
```

6.6 EXERCISES

1. Using the Mohr circle, perform a tensor transformation on the tensor shown in Equation 6.18 by a 45° rotation around the σ_2 axis. Discuss your results.
2. Show why there is no quadratic term, λ^2, in Equation 6.22.
3. Show that Equations 6.31 follow from 6.30 when the old coordinate system is parallel to the principal stress axes.
4. Derive Equation 6.32 from Equations 6.31 and the orthogonality relations.
5. A state of stress with the following principal stress magnitudes, $\sigma_1 = 40$ MPa, $\sigma_2 = 20$ MPa, $\sigma_3 = 10$ MPa, has a σ_1 axis oriented vertically, σ_2 aligned in a horizontal E-W direction, and σ_3 in a horizontal N-S direction. Calculate the magnitude and orientation of the normal and maximum shear stress acting on a plane striking 60° and dipping 55° SE. Hint: Use function **ShearOnPlane**.
6. In the Oseberg field, North Sea, the principal stresses are oriented $\sigma_1 = 080/00$, $\sigma_2 = 000/90$, and $\sigma_3 = 170/00$. If at 2 km depth, $\sigma_1 = 50$ MPa, $\sigma_2 = 40$ MPa, and $\sigma_3 = 30$ MPa, what is the normal and shear stress on a plane oriented (strike and dip, right-hand rule) 040/65? Hint: Use function **ShearOnPlane**.

CHAPTER SEVEN

Introduction to deformation

7.1 INTRODUCTION

A famous structural geologist once remarked, "As a structural geologist, I don't believe in stress." It is true that we never really see stress (because it is a field tensor) or can ever measure stress directly. All we can observe is the end product of imposing stress on a material, and that is what is broadly known as deformation. We may observe deformation while the accompanying stress is still present, as in the case of seismic waves generated by earthquakes or rock bursts in a quarry or, more commonly, we may observe rocks that were distorted by stress some hundreds of millions of years ago. Stress is instantaneous; deformation is what we see in the rocks.

In our study of deformation, we are embarking on new territory in one very important aspect: We will be comparing the states of material at two different points in time. When we did stress tensor transformations in the last chapter, we were simply taking two different looks at the same state of stress at the same instant in time. Studying deformation requires that we establish both temporal and spatial frames of reference.

Deformation, which most structural geologists choose to concentrate on, is in fact a far more complicated topic than stress. There is a plethora of different symmetric and asymmetric tensors, some quite messy, which describe deformation. In this chapter, we introduce some fundamental concepts about this topic and then in Chapter 8 we'll take a look at some simplifying mathematical assumptions that make deformation almost as easy as stress to deal with. In Chapter 9, we'll see just how messy it can get with an introduction to finite strain.

In the first chapter, we wrote that deformation is the product of strain (distortion), rotation, and translation; we are about to find out what that means. Presentations of strain in most structural geology texts start with some simple one-dimensional measures of strain:

$$\text{extension}: e = \frac{l_f - l_i}{l_i} \tag{7.1a}$$

7.2 Deformation and displacement gradients

$$\text{stretch: } S = \frac{l_f}{l_i} \qquad (7.1b)$$

$$\text{quadratic elongation: } \lambda = S^2 = \left(\frac{l_f}{l_i}\right)^2 \qquad (7.1c)$$

where l_f is the final length and l_i is the initial length. The presence of the l_i in the denominator of Equations 7.1 signals an implicit assumption that the initial state is the reference state. One could equally well choose the final state as the reference condition. On a more profound level, it is not clear from the above equations what the nature of the extension, stretch, and quadratic elongations are because there is no explicit coordinate system. We now have the tools to rectify that shortcoming, thereby enabling a deeper understanding of strain.

Before addressing the topic of strain, however, we need to develop a precise understanding for some simpler concepts: coordinate transformations, deformation gradients, and displacement gradients.

7.2 DEFORMATION AND DISPLACEMENT GRADIENTS

Though we commonly struggle to determine displacements in geology – the slip on a fault, the translation of a continent across the globe – individual displacements tell us virtually nothing about deformation itself. To determine the deformation of a region, we need to know how the displacement of one part of the region compares with displacement of several other parts of the region. The displacement vectors at several different points define a displacement field; deformation is the *gradient* of the displacement field.

To get a feeling for deformation and displacement gradients, take the following real-world, if oversimplified, example. Salt Lake City, Utah, and Carson City, Nevada, are located on opposite sides of the Basin and Range Province in the western United States (Fig. 7.1). The two cities are presently located about 700 km apart but, before extension occurred in the Basin and Range Province, they (or, strictly, the spots of ground they now occupy) were only about 350 km apart. The town of Austin, Nevada, lies more-or-less on the same line of section and is, at present, about 240 km from Carson City. We would like to find not only the distance between Carson City and Austin prior to extension, but also to have a convenient way of calculating where any other spot on the line was before the deformation. This problem clearly oversimplifies the Basin and Range extensional history, but is useful for developing an intuitive feel for deformation and displacement gradients.

You can think of there being two different coordinate systems in this problem, both of which have their origin at Carson City, Nevada (Fig. 7.2). We will refer to the axis prior to extension with a capital **X** and the present day axis with a small **x**. Both **X** and **x** have the same units, but the positions of the cities along the coordinate axes have changed because of the deformation.

The relations between the **x** and **X** axes can be described in terms of a *coordinate transformation*. Although this sounds suspiciously like the transformation of coordinate axes that we discussed in Chapter 3, it is quite different. The transformation of axes was just a way of looking at the *same thing* from two different points of view. The coordinate transformation here represents *two different states in time*, one in the present and another in the past. We could, for example, carry out a transformation of axes to take several different looks at the present state, but it would not help us to understand the relationship between the present and past states because the two are different.

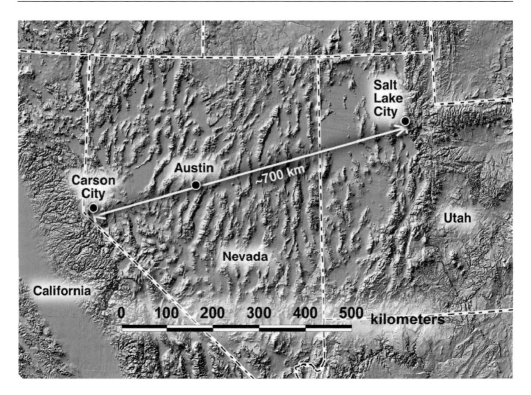

Figure 7.1 Map of the western United States showing the line along which we are interested in the one-dimensional strain. The line between Carson City and Salt Lake City is the **x** axis discussed in the text.

Figure 7.2 Two one-dimensional coordinate systems that describe the distances between the cities shown in Figure 7.1 before (b) and after (a) the Cenozoic extension in the western United States. You can think of the two axes as a hypothetical present-day road map (a) compared to a mid-Tertiary one (b). In both (a) and (b) the tick marks occur every 100 km.

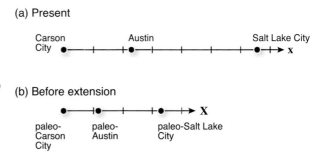

In the case of our example, the coordinate transformation is inhomogeneous because the change is not constant but depends on position. Salt Lake City has clearly been moved a greater distance from Carson City than the town of Austin has; the magnitude of the change depends on the position of the point of interest. We can write equations that express the position in one frame of reference as a function of the position in the other reference frame. For the case above,

$$\mathbf{x} = 2\mathbf{X} \qquad (7.2a)$$

or

$$\mathbf{X} = 0.5\mathbf{x} \qquad (7.2b)$$

7.2 Deformation and displacement gradients

Note that, even in this one-dimensional case, we are treating the positions of the towns as position vectors, rather than scalars. The first equation (7.2a) gives a point's position in present-day coordinates as a function of its position prior to the start of the extension. Salt Lake City started out at ~350 km from Carson City and, by Equation 7.2a, it should now be ~700 km away. The second equation (7.2b) yields the position of a point in the past, given its present coordinates. This is the equation we want in order to solve the question: "How far was Austin from Carson City before the extension began?" Substituting 240 km (i.e., the present distance between Carson City and Austin) for **x** in Equation 7.2b, we calculate that "paleo-Austin" was 120 km from "paleo-Carson City" before extension (Fig. 7.2b). In more precise terms, Equation 7.2a is a *Green transformation* (new in terms of old), whereas Equation 7.2b is a *Cauchy transformation* (old in terms of new).

Although the change in position is not constant, the *ratio* or *gradient* of the change is constant (in this oversimplified example). It is nothing more than the slopes in the above simple equations, which apply not only to Salt Lake and Austin, but every other point in between. We can write these gradients as

$$\frac{\Delta \mathbf{x}}{\Delta \mathbf{X}} = \lim \frac{\partial \mathbf{x}}{\partial \mathbf{X}} = 2 \tag{7.3a}$$

$$\frac{\Delta \mathbf{X}}{\Delta \mathbf{x}} = \lim \frac{\partial \mathbf{X}}{\partial \mathbf{x}} = 0.5 \tag{7.3b}$$

These ratios are known as the *deformation gradients* and they are homogeneous, unlike the coordinate transformations. We use ∂ to indicate *partial derivatives* in the above equations because, as we'll see below, the deformation gradients can be functions of positions along each of the three axes of the coordinate system. "Deformation gradient" is, in fact, just a fancy name for something with which we are already familiar. $\Delta \mathbf{x}$ gives the present, or final, length between any two points along the line (Salt Lake and Austin, for example); the $\Delta \mathbf{X}$ is the initial length between the same two points. Thus, the ratios in Equations 7.3 are nothing more than the *stretches* referred to either the initial or the final conditions as in Equation 7.1b:

$$S = \frac{\text{final length}, l_f}{\text{initial length}, l_i} = \frac{\Delta \mathbf{x}}{\Delta \mathbf{X}} = \frac{(\mathbf{x}_{\text{Salt Lake}} - \mathbf{x}_{\text{Austin}})}{(\mathbf{X}_{\text{Salt Lake}} - \mathbf{X}_{\text{Austin}})} = \frac{(670 - 240)\text{km}}{(335 - 120)\text{km}} = \frac{430}{215} = 2 \tag{7.4a}$$

$$\bar{S} = \frac{\text{initial length}, l_i}{\text{final length}, l_f} = \frac{\Delta \mathbf{X}}{\Delta \mathbf{x}} = \frac{215\,\text{km}}{430\,\text{km}} = 0.5 \tag{7.4b}$$

For translation alone, all points along the line move by the same amount and the coordinate transformations would be homogeneous. In this case, $\Delta \mathbf{x} = \Delta \mathbf{X}$, and the deformation gradients would be equal to one. Likewise, the stretches, S and $\bar{S}$, will also be one.

Another way to look at this problem is to consider the displacement vectors that connect the initial and final positions of points along the line (Fig. 7.3). Continuing with the Basin and Range example, we can write

$$\begin{aligned} \mathbf{u}_{\text{Salt Lake}} &= \mathbf{x}_{\text{Salt Lake}} - \mathbf{X}_{\text{Salt Lake}} = 670\,\text{km} - 335\,\text{km} = 335\,\text{km} \\ \mathbf{u}_{\text{Austin}} &= \mathbf{x}_{\text{Austin}} - \mathbf{X}_{\text{Austin}} = 240\,\text{km} - 120\,\text{km} = 120\,\text{km} \end{aligned} \tag{7.5}$$

In both cases, the vector, **u**, is equal to the initial position of the city and it is also equal to half the final position of the city. That is, the size of the displacement vector depends on its position, so we can write

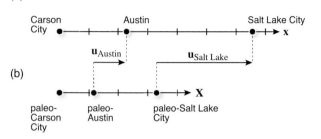

Figure 7.3 Same coordinate systems as in the previous figure, but now emphasizing the displacement vectors that connect initial and final positions of the cities.

$$\mathbf{u} = \mathbf{X} \quad (7.6a)$$
$$\mathbf{u} = 0.5\mathbf{x} \quad (7.6b)$$

The first equation (7.6a) is known as *Lagrange displacement* (in terms of old) and the second (7.6b) *Euler displacement* (in terms of new). Like the coordinate transformation, the displacement field is inhomogeneous because all vectors are not the same length. However, the change in displacements with position, or *displacement gradient*, is constant and homogeneous (i.e., because the slopes in Equations 7.6 are constant). Thus, we can write

$$\frac{\Delta \mathbf{u}}{\Delta \mathbf{X}} = \lim \frac{\partial \mathbf{u}}{\partial \mathbf{X}} = 1 \quad (7.7a)$$

$$\frac{\Delta \mathbf{u}}{\Delta \mathbf{x}} = \lim \frac{\partial \mathbf{u}}{\partial \mathbf{x}} = 0.5 \quad (7.7b)$$

Again, the partial derivatives are to prepare us for the future three-dimensional case.

The physical meaning of $\Delta \mathbf{u}$ becomes obvious when we expand it for the Basin and Range case that we have been considering throughout this section. $\Delta \mathbf{u}$ is just the difference between two vectors; in the case of our example, it is the difference between $\mathbf{u}_{\text{Salt Lake}}$ and $\mathbf{u}_{\text{Austin}}$ (Fig. 7.3). To simplify the following equations, we'll just use S as the subscript for Salt Lake and A as the subscript for Austin. We know from Equations 7.5 that

$$\mathbf{u}_S = \mathbf{x}_S - \mathbf{X}_S \text{ and } \mathbf{u}_A = \mathbf{x}_A - \mathbf{X}_A$$

Therefore, we can write the equation for $\Delta \mathbf{u}$:

$$\Delta \mathbf{u} = \mathbf{u}_S - \mathbf{u}_A = (\mathbf{x}_S - \mathbf{X}_S) - (\mathbf{x}_A - \mathbf{X}_A)$$

This equation can now be rearranged in terms of the initial and final lengths between Salt Lake and Austin:

$$\Delta \mathbf{u} = (\mathbf{x}_S - \mathbf{x}_A) - (\mathbf{X}_S - \mathbf{X}_A) = l_f - l_i$$

Now, we can rewrite Equations 7.7:

$$e = \frac{\Delta \mathbf{u}}{\Delta \mathbf{X}} = \frac{(\mathbf{u}_S - \mathbf{u}_A)}{(\mathbf{X}_S - \mathbf{X}_A)} = \frac{l_f - l_i}{l_i} \quad (7.8a)$$

$$\bar{e} = \frac{\Delta \mathbf{u}}{\Delta \mathbf{x}} = \frac{(\mathbf{u}_S - \mathbf{u}_A)}{(\mathbf{x}_S - \mathbf{x}_A)} = \frac{l_f - l_i}{l_f} \quad (7.8b)$$

Thus, the displacement gradients are just the *extensions* referred to the initial state (e) and to the final state ($\bar{e}$). So far we have only determined the same simple equations that everyone learns in their initiation to strain in a first course in structural geology. Equations 7.4 and 7.8 are not quite right because they imply that a scalar quantity – S in the case of 7.4 and e in the

7.3 Displacement and deformation gradients in three dimensions

	Old coordinates	New coordinates
Coordinate transformation	Green $x_1 = \dfrac{\partial x_1}{\partial X_1} X_1$	Cauchy $X_1 = \dfrac{\partial X_1}{\partial x_1} x_1$
Displacements	Lagrangian $u_1 = \dfrac{\partial u_1}{\partial X_1} X_1$	Eulerian $u_1 = \dfrac{\partial u_1}{\partial x_1} x_1$

Table 7.1

case of 7.8 – is equivalent to the ratio of two vectors. This tells us something important: *The stretch and the extension are really scalar components of tensor quantities.* The difference here is that we have been more accurate about what the quantities of interest are and we have developed these ideas in such a way that extension to three dimensions will be straightforward.

If we just had translation alone, the displacement field would be homogeneous because all of the vectors, **u**, would be the same length and orientation. The displacement gradient would be zero because there is no change in **u** anywhere:

$$\frac{\partial \mathbf{u}}{\partial \mathbf{X}} = \frac{\partial \mathbf{u}}{\partial \mathbf{x}} = 0$$

In preparation for three dimensions, we summarize what we've learned so far in Table 7.1.

The subscript 1 is used to indicated that the positions, vectors, and gradients are along the X_1 or x_1 axis. We use partial derivatives because, in general, there will be three axes in our Cartesian coordinate system and the transformations and displacements will depend on all three.

7.3 DISPLACEMENT AND DEFORMATION GRADIENTS IN THREE DIMENSIONS

7.3.1 Displacement of a point

We have seen in the previous section that the deformation and displacement gradients in one dimension are identical to the stretch and the extension that all students learn about in their introductory structural geology classes. To analyze real-world deformation, however, we need to work in three dimensions, not one. Thus, we'll now leave our (overly) simple Basin and Range example behind and plunge into three dimensions (Fig. 7.4). The *deformation gradient* matrix is

$$\begin{bmatrix} x_1 \\ x_2 \\ x_3 \end{bmatrix} = \begin{bmatrix} \dfrac{\partial x_1}{\partial X_1} & \dfrac{\partial x_1}{\partial X_2} & \dfrac{\partial x_1}{\partial X_3} \\ \dfrac{\partial x_2}{\partial X_1} & \dfrac{\partial x_2}{\partial X_2} & \dfrac{\partial x_2}{\partial X_3} \\ \dfrac{\partial x_3}{\partial X_1} & \dfrac{\partial x_3}{\partial X_2} & \dfrac{\partial x_3}{\partial X_3} \end{bmatrix} \begin{bmatrix} X_1 \\ X_2 \\ X_3 \end{bmatrix} \quad (7.9)$$

You can see that it would be very tedious to write matrices like this out all the time. It is much easier to use the Einstein summation convention. For the same deformation gradient with respect to the initial state, we write

$$x_i = \frac{\partial x_i}{\partial X_j} X_j \quad (7.10a)$$

With respect to the final state, it is

$$X_i = \frac{\partial X_i}{\partial x_j} x_j \quad (7.10b)$$

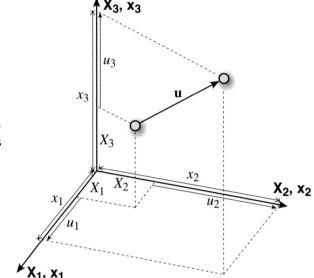

Figure 7.4 Displacement of a point by vector **u** in three dimensions. The axes **x** and **X** refer to the new and the old coordinate systems, respectively.

Likewise, the *displacement gradient* with respect to the initial state is

$$u_i = \frac{\partial u_i}{\partial X_j} X_j \qquad (7.11a)$$

and with respect to the final,

$$u_i = \frac{\partial u_i}{\partial x_j} x_j \qquad (7.11b)$$

Note that Equations 7.10 and 7.11, and Figure 7.4, describe the displacement of a point with coordinates $[X_1 \;\; X_2 \;\; X_3]$ to a position $[x_1 \;\; x_2 \;\; x_3]$. How does this case apply to our earlier example where there were two vectors, the displacement of Austin and the displacement Salt Lake (Fig. 7.3)? In Figure 7.4, there are, implicitly, two position vectors, one from the origin to the tail of the vector, and another from the origin to the head of the vector, **u**. But, while Equations 7.10 and 7.11 describe the displacement of a point, they don't really describe strain because we don't yet know how different points within the material move with respect to one another.

7.3.2 Difference between two displacement vectors

To address this shortcoming, we consider two displacement vectors, which for simplicity's sake are illustrated in two dimensions (Fig. 7.5). The situation shown in the blowup on the right side of Figure 7.5 is equivalent (in 2D) to that shown in Figure 7.4, but now the position vectors are shown explicitly. Vector PQ has coordinates $[\Delta X_1 \;\; \Delta X_2 \;\; \Delta X_3]$ and the distorted vector P'Q' has coordinates $[\Delta x_1 \;\; \Delta x_2 \;\; \Delta x_3]$. What we want to find is a relationship between the initial (or final) position of the vectors and the displacement vector, Δu_i, as shown in Figure 7.6.

7.3 Displacement and deformation gradients in three dimensions

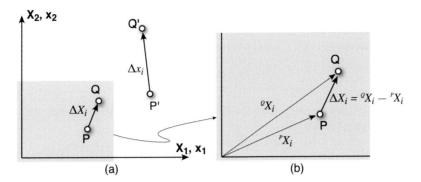

Figure 7.5 (a) Vector PQ is distorted to vector P′Q′ during deformation. (b) A blowup of the light gray box on the left, showing the physical meaning of ΔX_i, which is the difference between the position vector to point P and the position vector to point Q.

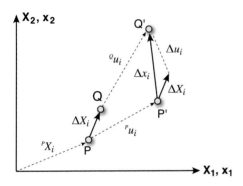

Figure 7.6 All of the vectors involved in describing the distortion of vector PQ to vector P′Q′.

You can see from Figure 7.6 that Δu_i is the *difference* between the two displacement vectors that connect PP′ ($^P u_i$) and QQ′ ($^Q u_i$). Alternatively, Δu_i is the difference between P′Q′ and PQ. We can write a simple vector addition equation that reflects this last statement:

$$P'Q' = \Delta x_i = \Delta X_i + \Delta u_i$$

Expanding Δu_i we get

$$\Delta u_i = {}^Q u_i - {}^P u_i$$

Substituting from Equation 7.11a:

$$\Delta u_i = {}^Q u_i - {}^P u_i = \frac{\partial u_i}{\partial X_j} {}^Q X_j - \frac{\partial u_i}{\partial X_j} {}^P X_j = \frac{\partial u_i}{\partial X_j} \left({}^Q X_j - {}^P X_j \right)$$

Since $\Delta X_i = {}^Q X_i - {}^P X_i$ we can write

$$\Delta u_i = \frac{\partial u_i}{\partial X_j} \Delta X_j \quad (i,j = 1, 2, 3)$$

Remember that we said that the displacement gradient is nothing more than the extension, so,

$$\frac{\partial u_i}{\partial X_j} = e_{ij} \text{ and } \Delta u_i = e_{ij} \Delta X_j \tag{7.12a}$$

Because Δu_i and ΔX_j are vectors, it follows that e_{ij} is a tensor, the *Lagrangian displacement gradient tensor* (referred to the initial state). Following the same steps as above, we can write the *Eulerian displacement gradient tensor* (referred to the final state):

$$\frac{\partial u_i}{\partial x_j} = \bar{e}_{ij} \text{ and } \Delta u_i = \bar{e}_{ij}\Delta x_j \qquad (7.12b)$$

To derive the equivalent expressions for the deformation gradient tensors, we can simply expand the equations for Δx_i:

$$\Delta x_i = {}^{Q'}x_i - {}^{P'}x_i = \frac{\partial x_i}{\partial X_j}{}^Q X_j - \frac{\partial x_i}{\partial X_j}{}^P X_j = \frac{\partial x_i}{\partial X_j}\left({}^Q X_j - {}^P X_j\right) = \frac{\partial x_i}{\partial X_j}\Delta X_j$$

As before, the deformation gradients are equivalent to the one-dimensional stretches, so

$$\frac{\partial x_i}{\partial X_j} = F_{ij} \text{ and } \Delta x_i = F_{ij}\Delta X_j \qquad (7.13a)$$

$$\frac{\partial X_i}{\partial x_j} = \bar{F}_{ij} \text{ and } \Delta X_i = \bar{F}_{ij}\Delta x_j \qquad (7.13b)$$

F_{ij} is the *Green deformation gradient tensor*, referred to the initial state, and $\bar{F}_{ij}$ is the *Cauchy deformation gradient tensor*, referred to the final state. All of the deformation and displacement gradient tensors as described here are valid for either small or large deformations. Having four different tensors to deal with is cumbersome at best, so it would be nice if there were some simplifying assumptions that would remove that complexity. In the next chapter, we'll explore just such a simplification.

7.4 GEOLOGICAL APPLICATION: GPS TRANSECTS

The Global Positioning System (GPS) has revolutionized the science of geodesy by making it possible to detect small movements of the Earth over very large distances. This technique has become the backbone of many studies in tectonics over the past decade and has enabled, for the first time, structural geologists to measure deformation "in real time." Many early GPS studies collected data in transects perpendicular to local structure.

A simple way to look at the strain between stations is to plot the displacement vectors of the stations against their positions along a transect. The slope in such a plot represents the difference in displacement of the two stations, Δu, over the difference in position of the two stations, ΔX (Fig. 7.7). From what we have just seen, this is nothing more than the displacement gradient, and from Equations 7.8 and 7.12 this ratio is, in one dimension, our old friend the extension:

$$\frac{\Delta u}{\Delta X} = \frac{du}{dX} = e \qquad (7.14)$$

To estimate this gradient, or slope, we commonly fit a straight line to the points (Fig. 7.7). Before proceeding with the solution to the geological problem, we need to address the general problem of fitting data to a straight line.

7.4.1 Least squares fit of data to a line

All graphics programs and spreadsheets now enable a user to fit a straight line to data, commonly called a *linear regression*. Undoubtedly, many professors have seen comically egregious misuse of this function, not only by their students, but also occasionally by their own colleagues! This is such a fundamental operation that, as scientists, we should know how

7.4 Geological application: GPS transects

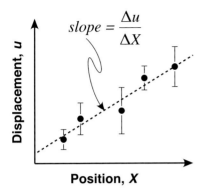

Figure 7.7 Plot of displacement versus position. The slope gives the extension, *e*, in the direction of the transect. Note the implicit sign convention of the displacements: positive displacements are those where the vector points in the positive direction of the position axis. If the transect were east–west, these vectors would point east; west-pointing vectors would have negative displacement values.

this actually works. Our discussion below gives only the briefest outline of least squares fitting and we highly recommend that you read the lucid treatment of this topic by Taylor (1997), the classic work of Bevington and Robinson (2003), or that found in any of a number of different statistics textbooks.

In the case shown in Figure 7.7, the equation for the straight-line fit will have the form

$$u = t + eX \tag{7.15}$$

where u is the displacement vector, t is the intercept along the displacement axis (i.e., the displacement at the position $X = 0$), the position along the transect is X and the slope of the line, also the extension as in Equation 7.14, is e. There are n GPS stations and the letter i designates a station i located at position X_i displaced by an amount u_i. Furthermore, we will assume that there are significant uncertainties, σ_i, in the displacement vector, u_i, but negligible uncertainties in the position of the station, X_i.

Let's explore, for a minute, what the uncertainty, σ_i, actually means. Suppose that you were able to make multiple measurements of the displacement vector, u_i, at station i. The measurements would, of course, not all be identical but would have some variation about the mean, or average, value. If that variation were well behaved, it would have a normal, or *Gaussian*, distribution – the familiar bell curve with a maximum centered at the mean. σ_i is known as the *standard deviation* of that normal distribution and its value is given by

$$\sigma = \sqrt{\frac{1}{N-1} \sum_{k=1}^{N} (u_k - \bar{u})^2} \tag{7.16}$$

where N is the number of measurements of displacement vector u and $\bar{u}$ is the average of all of the measurements of the displacement vector at station i. For simplicity's sake, we have omitted the subscript i from Equation 7.16. There is some discussion in the literature of whether or not to use N or $(N-1)$ in the denominator of Equation 7.16. Using $(N-1)$ makes sense because, in the limiting case where you have only one measurement, the standard deviation is undefined (divide by zero). However, we particularly like the attitude of Press *et al.* (1986), who wrote: "If the difference between N and $N-1$ ever matters to you, then you are probably up to no good anyway – e.g., trying to substantiate a questionable hypothesis with marginal data." Equation 7.16 gives the value of one standard deviation, where you have a 68% chance that the correct answer lies at 1σ or less from the mean. At 2σ or less from the mean, you are 95% confident, etc. The *variance*, another common measure, is equal to the square of the standard deviation.

We return, now, to the question of how to find the best-fit straight line for multiple stations. Without going into greater detail of normal distribution, suffice it to say that the difference between the observed value at a single station, u_i, and the value of u given by the best-fit straight line (Eq. 7.15) is

$$u_i - t - eX_i$$

and the statistic χ^2 is the sum of the squares of the above difference divided by the uncertainties, defined as

$$\chi^2 = \sum_{i=1}^{n} \frac{(u_i - t - eX_i)^2}{\sigma_i^2} \qquad (7.17)$$

This is why linear regression is known as least squares best fit. To find the best-fit line we must minimize χ^2, which means differentiating 7.17 with respect to t and e and setting it equal to zero:

$$\frac{\partial \chi^2}{\partial t} = -2 \sum_{i=1}^{n} \frac{u_i - t - eX_i}{\sigma_i^2} = 0 \qquad (7.18a)$$

and

$$\frac{\partial \chi^2}{\partial e} = -2 \sum_{i=1}^{n} \frac{X_i(u_i - t - eX_i)}{\sigma_i^2} = 0 \qquad (7.18b)$$

Equations 7.18 give us two equations for our two unknowns, t and e. Following Press *et al.* (1986), we define the following quantities:

$$S \equiv \sum_{i=1}^{n} \frac{1}{\sigma_i^2} \qquad S_X \equiv \sum_{i=1}^{n} \frac{X_i}{\sigma_i^2} \qquad S_u \equiv \sum_{i=1}^{n} \frac{u_i}{\sigma_i^2}$$
$$S_{XX} \equiv \sum_{i=1}^{n} \frac{X_i^2}{\sigma_i^2} \qquad S_{Xu} \equiv \sum_{i=1}^{n} \frac{X_i u_i}{\sigma_i^2} \qquad \Delta \equiv SS_{XX} - S_X^2 \qquad (7.19)$$

and with them can rewrite Equations 7.18 as

$$tS + eS_X = S_u \qquad (7.20a)$$

and

$$tS_X + eS_{XX} = S_{Xu} \qquad (7.20b)$$

We can now solve for t and e:

$$t = \frac{S_{XX} S_u - S_X S_{Xu}}{\Delta} \qquad (7.21a)$$

and

$$e = \frac{SS_{Xu} - S_X S_u}{\Delta} \qquad (7.21b)$$

Because we know the input uncertainties, σ_i, we can *propagate the errors* through to get the uncertainties on t and e. We will explore error propagation in a subsequent chapter; for now, we simply give the errors for the two parameters:

$$\sigma_t = \sqrt{\frac{S_{XX}}{\Delta}} \qquad (7.22a)$$

and

$$\sigma_e = \sqrt{\frac{S}{\Delta}} \qquad (7.22b)$$

7.4 Geological application: GPS transects

Finally, we introduce two additional parameters: The *covariance* of t and e is

$$\sigma_{te} = \frac{-S_X}{\Delta} \tag{7.23}$$

and the correlation coefficient, r_{te}, which varies between −1 and 1,

$$r_{te} = \frac{-S_X}{\sqrt{SS_{XX}}} \tag{7.24}$$

The correlation coefficient tells us how well u and X are correlated ($r_{te} \approx 1$), anticorrelated ($r_{te} \approx -1$), or uncorrelated ($r_{te} \approx 0$).

If you want to see how to code these from scratch, we highly recommend Chapters 14 and 15 of Press *et al.* (1986); for everyone interested in statistical treatment of data, the Introduction to Chapter 14 should be required reading! MATLAB® has built in functions for linear regression as well as a Figure menu for basic fitting (Tools → Basic Fitting). Suppose you have some data in vectors x and y. You can fit a straight line to the data just by typing:

```
p=polyfit (x,y,1);  %p(1) = slope and p(2) = intercept of line
R=corrcoef (x,y);   %R(1,2) = Correlation coefficient
```

In addition, the MATLAB Statistics Toolbox contains two functions, **regress** and **regstats**, to perform linear regression. **regstats** provides a complete statistics of the regression. Performing linear regression of data with errors in x and y, however, is not that simple. Fortunately, you can find functions to perform this task at the MATLAB Central File Exchange website, such as the function **york_fit** by Travis Wiens.

7.4.2 Strain (rate) in a GPS transect

Returning to our initial problem, how does the finding of strain in a GPS transect work in practice? Here, we will set up a real-world example and leave it to you to solve as part of the exercises. One of the earliest earthquakes to be captured by a modern GPS network was the 1995 M8.1 Antofagasta event (Klotz *et al.*, 1999). This earthquake occurred on the subducting plate boundary between the South American Plate and the oceanic Nazca Plate (Fig. 7.8). Because we are looking at coseismic deformation, this particular problem involves the calculation of strain, but in the more general case of GPS surveys capturing interseismic deformation, the data reported are displacements averaged over time or a velocity. Thus, analysis of interseismic GPS involves the determination of *strain rate*, rather than strain. The procedure described here is identical.

The first thing to notice about the Antofagasta data (Fig. 7.8) is that, rather than a single transect, there is a band of GPS stations. Second, the vectors are not oriented exactly east–west, but point towards the west-southwest. However, in our UTM-19 (or 19S) coordinate system, eastings (X_1) and northings (X_2) are positive. These facts lead to the first two steps in the analysis:

1. Determine the average or mean vector that characterizes, as best possible, the overall orientation of the vectors. To do this, you will use the mean vector calculation described in Chapter 2 (Section 2.4.1). Once you learn more about error propagation, you could redo this problem to calculate the uncertainties in the mean vector, but for right now, just add up all the vectors and find the mean direction.
2. Determine the two-dimensional transformation matrix a_{ij} needed for a new coordinate system where the X_1' axis is parallel to the mean vector direction.

3. Transform the east and north coordinates of the GPS vectors, and the east and north components of the errors into the new coordinate system using Equations 3.6 from Chapter 3.
4. Plot the u'_1 component of each displacement vector, and its error, against the X'_1 component of the station position.
5. Fit a straight line to approximately linear segments of the resulting curve using the relations or built-in functions from the previous sections.

One-dimensional plots like those you've just constructed are standard practice in articles describing GPS data and it is important to know how to read them. Once you have completed the Antofagasta earthquake exercise, however, you will see that there are artifacts that result from the fact that the strain varies in two dimensions because gradients in displacement exist in two dimensions, not just along a transect. Although we can tell from such a plot if there is shortening or extension between the two stations, we have no way of knowing how close these values are to the principal axes of infinitesimal strain, nor do we know anything about rotation. At the end of Chapter 8, we will see how to solve the more complete problem.

7.5 EXERCISES

In previous chapters, we have given you a lot of MATLAB code to carry out individual calculations. In this chapter, you can begin to put those pieces together to solve a very interesting geologic problem, the one described in the previous section.

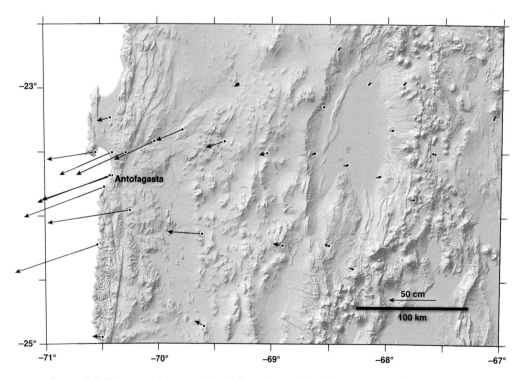

Figure 7.8 Shaded relief map of the Chilean Coastal Cordillera near Antofagasta, Chile, showing the station locations and coseismic GPS vectors of displacement during the 1995 Mw 8.1 earthquake. GPS data are from Klotz *et al.* (1999).

7.5 Exercises

1. The table below, modified from Klotz *et al.* (1999), lists the GPS data for locations shown in Figure 7.8. The units of the displacements and the errors are meters.

ID	Longitude	Latitude	U(E)	Error(E)	U(N)	Error(N)
julo	−70.546 0002	−23.526 0447	−0.527 000	0.002 600	−0.082 000	0.002 800
calc	−70.531 9985	−24.264 0430	−0.852 000	0.002 700	−0.283 000	0.002 800
cari	−70.499 0037	−24.947 0407	−0.149 000	0.002 700	0.003 000	0.002 800
caco	−70.471 9967	−23.766 0478	−0.832 000	0.003 200	−0.306 000	0.002 900
meji	−70.415 9964	−23.200 0390	−0.167 000	0.002 600	−0.039 000	0.002 800
unia	−70.419 9960	−23.702 0464	−0.756 000	0.002 600	−0.274 000	0.002 800
udan	−70.405 0006	−23.669 0402	−0.730 000	0.003 200	−0.246 000	0.001 800
antf	−70.401 0045	−23.544 0436	−0.562 000	0.010 000	−0.232 000	0.010 000
urib	−70.280 0027	−23.505 0416	−0.527 000	0.002 700	−0.232 000	0.002 800
live	−70.252 9978	−23.964 0425	−0.851 000	0.002 600	−0.133 000	0.002 800
mabl	−70.028 0003	−23.448 0389	−0.430 000	0.002 600	−0.199 000	0.002 800
baqu	−69.781 0030	−23.342 0390	−0.290 000	0.002 600	−0.122 000	0.002 800
minf	−69.606 0045	−24.105 0470	−0.382 000	0.002 800	0.022 000	0.002 800
coba	−69.589 0013	−24.824 0416	−0.130 000	0.002 700	0.068 000	0.002 800
loba	−69.416 0042	−23.448 0432	−0.219 000	0.002 900	−0.070 000	0.002 800
sigo	−69.297 0033	−22.926 0455	−0.078 000	0.002 700	−0.053 000	0.002 800
pael	−69.041 0009	−23.538 0448	−0.119 000	0.002 600	−0.028 000	0.002 800
esim	−68.898 0033	−24.226 0430	−0.135 000	0.002 600	0.027 000	0.002 800
cene	−68.631 9956	−23.551 0467	−0.074 000	0.002 700	−0.027 000	0.002 800
ceto	−68.546 0005	−23.177 0408	−0.049 000	0.002 600	−0.022 000	0.002 800
nrar	−68.493 9972	−24.254 0478	−0.086 000	0.010 000	0.024 000	0.010 000
pbar	−68.425 0025	−22.711 0477	−0.008 000	0.002 900	−0.015 000	0.002 800
peni	−68.345 9975	−23.640 0423	−0.049 000	0.004 000	−0.003 000	0.002 900
paso	−68.290 9993	−24.449 0472	−0.043 000	0.002 400	0.014 000	0.002 700
sanp	−68.164 9960	−22.966 0402	−0.015 000	0.002 800	−0.016 000	0.002 800
pein	−68.055 9972	−23.686 0468	−0.050 000	0.002 900	−0.010 000	0.002 800
toco	−67.949 9950	−23.285 0458	−0.030 000	0.002 600	−0.002 000	0.002 800
ctoc	−67.854 9994	−22.928 0390	−0.010 000	0.002 600	−0.007 000	0.002 800
cmin	−67.757 9993	−23.889 0396	−0.036 000	0.002 600	−0.006 000	0.002 800
saca	−67.603 0034	−23.542 0476	−0.014 000	0.002 600	0.006 000	0.002 800
paja	−67.072 9981	−23.225 0393	−0.006 000	0.002 600	−0.007 000	0.002 800

   a. This region is in UTM zone 19 south. Convert all of the station locations to eastings and northings so you can carry out the strain calculations by having both position and displacement in meters.
   b. Following the steps laid out in Section 7.4.2, calculate the coseismic strain during the 1995 Antofagasta earthquake. You may need to calculate the strain for more than one segment of the curve if the curve is not linear.
   c. Is the slope of your graph positive or negative and what does the sign indicate about the nature of the strain?
   d. Provide a plausible explanation for any points on your graph that do not seem to fit the general trend on your plot.
   e. You can define the station positions either by their position prior to, or after, the displacement. How does your choice of reference frame affect your calculation?
2. The town of Wendover, on the Utah–Nevada border, is about 510 km from Carson City, along the line between Carson City and Salt Lake City (it is actually about 50 km north of the line, but we'll assume that it lies along the line for the purposes of this problem). Use the relations developed in Section 7.2 to determine where Wendover was prior to Basin and Range extension. What assumptions have you used in this calculation?

Figure 7.9 Simplified cross section of the Viking Graben in the North Sea, after Fjeldskaar *et al.* (2004). The dark gray unit is the pre-rift basement. Syn- and post-rift strata are shown in light gray. The section is for use with Exercise 3.

3. Figure 7.9 shows a cross section from the Viking Graben in the North Sea (modified from Fjeldskaar *et al.*, 2004). The section is drawn with no vertical exaggeration and the units on the scale are in kilometers. The dark gray is the basement and the light gray is the graben and post-graben fill. Assuming that the top of the basement was originally horizontal, determine the Green and Cauchy deformation gradients, and the Lagrange and Euler displacement gradients. Note that the Green deformation gradient is known in basin modeling circles as the stretching, or β, factor (McKenzie, 1978).
4. Derive Equations 7.21 from Equations 7.18. Show your intermediate steps.

CHAPTER

EIGHT

Infinitesimal strain

8.1 SMALLER IS SIMPLER

In the last chapter, we sought a simplifying assumption to reduce the number of different tensors that we have to deal with. It turns out that, if we only deal with very small changes, we can cut in half the number of tensors that we've introduced so far. The same simplification has a number of other benefits as well.

Consider the simple deformation shown in Figure 8.1. If $\Delta X_1 = 1.0$ and $\Delta x_1 = 1.001$, then the displacement gradient is

$$\frac{\partial u_1}{\partial X_1} = \frac{0.001}{1.0} = 0.001\,000 \quad \text{and} \quad \frac{\partial u_1}{\partial x_1} = \frac{0.001}{1.001} = 0.000\,999$$

Thus, when strains are small, $\frac{\partial u_1}{\partial x_1} \approx \frac{\partial u_1}{\partial X_1}$ and the difference between the displacement gradients in the initial and final states is not important. Small strains are called *infinitesimal strains*. Small strains are important in a number of fields in earth sciences, perhaps most notably in geophysics.

8.1.1 The components of the displacement gradient tensor

Though we saw in the previous chapter that, in one dimension, the displacement gradient tensor, **e**, is equivalent to the linear extension that one learns in introductory structural geology, we are left with the nagging suspicion that the nine components of **e** in three dimensions are somewhat more complicated. The infinitesimal strain assumption allows us some further insight, in particular, into the meanings of the six off-diagonal components of the tensor.

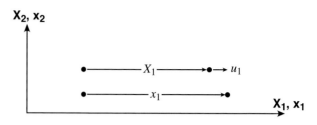

Figure 8.1 Simple illustration of a small extension parallel to one of the axes of the coordinate system.

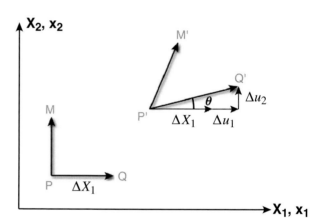

Figure 8.2 The special case of the deformation of two vectors that start out parallel to the axes of the coordinate system.

Consider the special case shown in Figure 8.2. Because PQ is perpendicular to the X_2 axis, $\Delta X_2 = 0$. Expanding Equation 7.12 in two dimensions, we get

$$\Delta u_1 = e_{11}\Delta X_1 + e_{12}\Delta X_2 = e_{11}\Delta X_1 + 0 = e_{11}\Delta X_1$$
$$\Delta u_2 = e_{21}\Delta X_1 + e_{22}\Delta X_2 = e_{21}\Delta X_1 + 0 = e_{21}\Delta X_1$$
(8.1)

From the first equation, you can see that

$$e_{11} = \frac{\Delta u_1}{\Delta X_1}$$

Because $\Delta u_1 = \Delta x_1 - \Delta X_1$, this equation says that e_{11} is equal to the final length minus the initial length, divided by the initial length. In other words, e_{11} is just the extension along the X_1 axis.

8.1.2 Significance of the off-diagonal components, e_{21} and e_{12}

From the geometry in Figure 8.2, we can see that

$$\tan\theta = \frac{\Delta u_2}{\Delta X_1 + \Delta u_1}$$

Remembering our assumption of *infinitesimal strain*, we see that $\Delta u_1 \ll \Delta X_1$ and therefore

$$\tan\theta \approx \frac{\Delta u_2}{\Delta X_1}$$

Furthermore, for small angles, the tangent of an angle is approximately equal to the angle itself, measured in radians, so $\tan\theta \approx \theta$. Thus,

8.1 Smaller is simpler

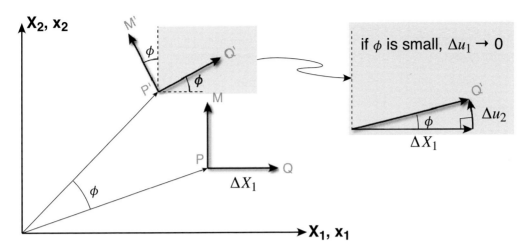

Figure 8.3 A special case where two originally orthogonal vectors, **PQ** and **PM**, are rotated about the origin by a constant angle, ϕ. The resulting vectors **P'Q'** and **P'M'** are still perpendicular to each other but each makes an angle of ϕ with the axes of the coordinate system, **X**.

$$\theta \approx \frac{\Delta u_2}{\Delta X_1}$$

and, again because of infinitesimal assumption, we can write

$$\theta \approx \frac{\Delta u_2}{\Delta X_1} = \frac{\partial u_2}{\partial X_1} = e_{21}$$

Thus, e_{21} measures the *counterclockwise rotation* of the vector PQ from the X_1 axis towards the X_2 axis; by the same reasoning, e_{12} measures the clockwise rotation of PM from the X_2 axis towards the X_1 axis.

8.1.3 Non-equivalence of e_{21} and e_{12}

To further understand the significance of the off-diagonal components, let's see what happens to the displacement gradient tensor when we have just rotation but no strain (Fig. 8.3). With the assumption of small rotation angles,

$$e_{11} = \frac{\Delta u_1}{\Delta X_1} = 0 \quad \text{and} \quad e_{21} = \frac{\Delta u_2}{\Delta X_1} = \tan \phi \approx \phi$$

Likewise, e_{12} will be approximately equal to $-\phi$ because it is a rotation of ΔX_1 towards X_2 (counterclockwise), whereas we just saw that e_{21} is a clockwise rotation of ΔX_2 towards X_1. So, for the case of pure rotation with no strain, the displacement gradient tensor in two dimensions has the form:

$$e_{ij} = \begin{bmatrix} 0 & -\phi \\ \phi & 0 \end{bmatrix}$$

Clearly, e_{21} does not necessarily equal e_{12}, and therefore the displacement gradient tensor, e_{ij}, is an asymmetric tensor that represents both strain and rotation.

8.1.4 Additive decomposition of the displacement gradient tensor

Any asymmetric tensor can be expressed as the sum of a symmetric tensor and an antisymmetric tensor, so in the case of the displacement gradient tensor:

$$e_{ij} = \varepsilon_{ij} + \omega_{ij} \tag{8.2}$$

where

$$\varepsilon_{ij} = \frac{1}{2}(e_{ij} + e_{ji}) \quad \text{and} \quad \omega_{ij} = \frac{1}{2}(e_{ij} - e_{ji})$$

You can easily prove to yourself that this is true by substituting:

$$\varepsilon_{ij} + \omega_{ij} = \left(\frac{e_{ij}}{2} + \frac{e_{ji}}{2} + \frac{e_{ij}}{2} - \frac{e_{ji}}{2}\right) = \left(\frac{e_{ij}}{2} + \frac{e_{ij}}{2}\right) = e_{ij}$$

We call ε_{ij} the *infinitesimal strain tensor*; it is a symmetric tensor. ω_{ij} is the *rotation tensor*; it is an antisymmetric tensor. When we do finite strain, you will see precisely what assumptions this entails. We will explore the significance of θ and ϕ in great detail in Chapter 10.

8.2 INFINITESIMAL STRAIN IN THREE DIMENSIONS

Obviously, this discussion carries over to three dimensions. For the strain tensor, we can write

$$\varepsilon_{ij} = \frac{1}{2}(e_{ij} + e_{ji}) = \begin{bmatrix} e_{11} & \frac{(e_{12} + e_{21})}{2} & \frac{(e_{13} + e_{31})}{2} \\ \frac{(e_{21} + e_{12})}{2} & e_{22} & \frac{(e_{23} + e_{32})}{2} \\ \frac{(e_{31} + e_{13})}{2} & \frac{(e_{32} + e_{23})}{2} & e_{33} \end{bmatrix}$$

and for the rotation tensor

$$\omega_{ij} = \frac{1}{2}(e_{ij} - e_{ji}) = \begin{bmatrix} 0 & \frac{(e_{12} - e_{21})}{2} & \frac{(e_{13} - e_{31})}{2} \\ \frac{(e_{21} - e_{12})}{2} & 0 & \frac{(e_{23} - e_{32})}{2} \\ \frac{(e_{31} - e_{13})}{2} & \frac{(e_{32} - e_{23})}{2} & 0 \end{bmatrix}$$

When it's written out in matrix form you can clearly see that ε_{ij} is symmetric and has six independent components. ω_{ij} is antisymmetric and has only three independent components.

The meanings of the e_{ij} terms are as follows:

- e_{11}, e_{22}, and e_{33} – extensions parallel to the axes of the reference system
- e_{12} – rotation of a line parallel to the 2 axis towards the 1 axis (about the 3 axis)
- e_{13} – rotation of a line parallel to the 3 axis towards the 1 axis (about the 2 axis; Fig. 8.4), etc.

8.2.1 Rotation axis from the antisymmetric tensor, ω

An antisymmetric tensor (e.g., ω_{jk}, above) is sometimes also known as an axial vector. To get the Cartesian coordinates, r_i, of that vector,

$$r_i = -\frac{b_{ijk}\omega_{jk}}{2} \tag{8.3}$$

8.2 Infinitesimal strain in three dimensions

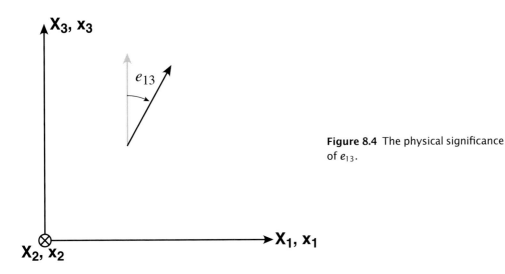

Figure 8.4 The physical significance of e_{13}.

where b_{ijk} is a permutation symbol which is equal to +1 if the suffixes are cyclic (e.g., 1-2-3), -1 if the suffixes are acyclic (e.g., 1-3-2), and 0 if any two suffixes are repeated. The three components of **r**, which give the orientation of the rotation axis, are

$$r_1 = \frac{-(\omega_{23} - \omega_{32})}{2} \quad r_2 = \frac{-(-\omega_{13} + \omega_{31})}{2} \quad \text{and} \quad r_3 = \frac{-(\omega_{12} - \omega_{21})}{2} \tag{8.4}$$

The amount of rotation in *radians* is just the length of the vector, **r**:

$$|\mathbf{r}| = r = \sqrt{r^2_1 + r^2_2 + r^2_3} \tag{8.5}$$

8.2.2 Homogeneous strain

If the deformation is the same throughout the region, then the displacements are not a function of position. We can express this condition as:

- e_{ij}'s are all constant, and
- $e_{ij} \neq f(X_i)$.

From Equation 7.12 we have $\Delta u_i = e_{ij}\Delta X_j$, which in the limit becomes

$$du_i = e_{ij} dX_j \tag{8.6}$$

Integrating both sides of Equation 8.6,

$$\int du_i = \int e_{ij} dX_j \tag{8.7}$$
$$u_i = t_i + e_{ij} X_j$$

where the constant of integration is the displacement of the origin. Note the similarity of this equation to Equation 7.15 in the previous chapter.

With Equation 8.7, there is a more elegant way to prove that rotation is antisymmetrical. We set up the problem with the rotation axis at the origin (Fig. 8.5). Because X_i is perpendicular to u_i, their dot product should equal zero: $u_i \cdot X_i = 0$. But,

$$u_i = e_{ij} X_j$$

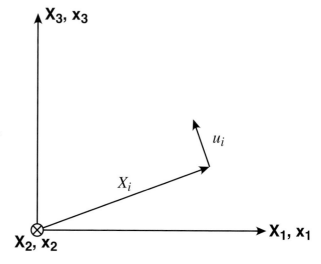

Figure 8.5 Illustration of the simple case of pure rotation of a line about the origin of the coordinate system with no strain.

so from a simple substitution of the last equation into the previous one,

$$e_{ij}X_iX_j = 0$$

The only way that this equation can be correct is if

- $e_{ij} = 0$ when $i = j$, and
- $e_{ij} = -e_{ji}$ when $i \neq j$.

Notice that the Equation 8.7,

$$u_i = t_i + e_{ij}X_j$$

represents three linear equations. It follows that

- straight lines remain straight, and
- parallel lines remain parallel.

The equation may be further broken down into

$$u_i = t_i + \varepsilon_{ij}X_j + \omega_{ij}X_j$$

Thus, in one equation we have the complete expression of deformation as a

$$\text{translation} + \text{strain} + \text{rotation}$$

Commonly in geology, we can't measure the translation or the rotation, so we just look at the displacement of points relative to other points within the same body. For example, we can measure the aspect ratio of a deformed oolite, but we have no way of knowing how far it moved or how much it rotated (obviously, this is not the case for all features, but it is for a majority). We write this equation as

$$\bar{u}_i = \varepsilon_{ij}X_j$$

8.3 TENSOR SHEAR STRAIN VS. ENGINEERING SHEAR STRAIN

The *angular shear*, ψ, is the change in angle of two lines that were originally perpendicular to each other (Fig. 8.6a). The angular shear is related to the *shear strain*, γ, by

$$\gamma = \tan\psi$$

8.4 Strain invariants

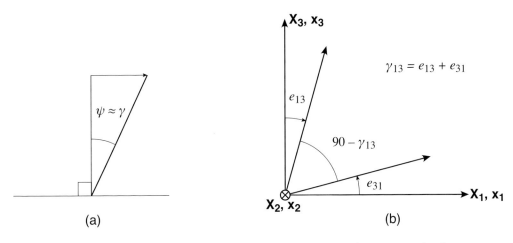

Figure 8.6 (a) Definition of engineering shear strain as the change in angle of two initially perpendicular lines. (b) The relationship between the tensor shear strain and the engineering shear strain.

As we assumed before, when strains are very small, the tangent of an angle is equal to the angle itself (in radians) so we can write

$$\gamma \approx \psi$$

This shear strain is known as the *engineering shear strain*, whereas the shear strains we derived above – e_{12}, e_{21}, e_{23}, e_{32}, e_{13}, and e_{31} – are known as the *tensor shear strains*. In Figure 8.6b, you can see that e_{13} and e_{31} are both positive, because both of the vectors are positive, even though the rotations implied by the two are opposite in sign. In Means (1976), you will see the infinitesimal strain tensor written as

$$\begin{bmatrix} e_{11} & e_{12} & e_{13} \\ e_{21} & e_{22} & e_{23} \\ e_{31} & e_{32} & e_{33} \end{bmatrix} = \begin{bmatrix} \varepsilon_{11} & \frac{\gamma_{12}}{2} & \frac{\gamma_{13}}{2} \\ \frac{\gamma_{21}}{2} & \varepsilon_{22} & \frac{\gamma_{23}}{2} \\ \frac{\gamma_{31}}{2} & \frac{\gamma_{32}}{2} & \varepsilon_{33} \end{bmatrix} \quad (8.8)$$

Means defines the off-diagonal components of the displacement gradient tensor such as e_{12} by

$$e_{12} = \frac{1}{2}\left(\frac{\partial u_1}{\partial X_2} + \frac{\partial u_2}{\partial X_1}\right)$$

In this book, we have used the following definitions:

$$e_{12} = \frac{\partial u_1}{\partial X_2} \quad \text{and} \quad \varepsilon_{12} = \frac{1}{2}\left(\frac{\partial u_1}{\partial X_2} + \frac{\partial u_2}{\partial X_1}\right)$$

By the way, you should note that $\begin{pmatrix} \varepsilon_{11} & \gamma_{12} & \gamma_{13} \\ \gamma_{21} & \varepsilon_{22} & \gamma_{23} \\ \gamma_{31} & \gamma_{32} & \varepsilon_{33} \end{pmatrix}$ is not a tensor!

8.4 STRAIN INVARIANTS

In Chapter 5, we showed that any tensor quantity has invariants, combinations of the components that do not change, regardless of the coordinate system you choose. In the case of strain, the first invariant,

$$\varepsilon_1 + \varepsilon_2 + \varepsilon_3 = \varepsilon_{11} + \varepsilon_{22} + \varepsilon_{33} \quad (8.9)$$

8.5 STRAIN QUADRIC AND STRAIN ELLIPSOID

Like any other tensor, the strain tensor has a quadric surface whose axes are given by $\varepsilon_1^{-0.5}$ and $\varepsilon_3^{-0.5}$; it is the vertically oriented ellipse (Fig. 8.7). The normal to the quadric surface at the point where the vector **X** intersects it gives the orientation of the displacement of the end of the vector to its new position. You can see (Fig. 8.7) that the elongation of the vector, ε, is just the component of **u** in the **X** direction. Therefore:

$$\varepsilon = \mathbf{u} \cdot \mathbf{X} = \varepsilon_{ij} X_i X_j \tag{8.10}$$

where the X's are the direction cosines of the original unit-length line. Expanding this equation we get

$$\varepsilon = \varepsilon_{11} X_1 X_1 + \varepsilon_{12} X_1 X_2 + \varepsilon_{13} X_1 X_3$$
$$+ \varepsilon_{21} X_2 X_1 + \varepsilon_{22} X_2 X_2 + \varepsilon_{23} X_2 X_3$$
$$+ \varepsilon_{31} X_3 X_1 + \varepsilon_{32} X_3 X_2 + \varepsilon_{33} X_3 X_3$$

Combining terms, we get

$$\varepsilon = \varepsilon_{11} X_1^2 + \varepsilon_{22} X_2^2 + \varepsilon_{33} X_3^2 + X_1 X_2 (\varepsilon_{12} + \varepsilon_{21}) + X_1 X_3 (\varepsilon_{13} + \varepsilon_{31}) + X_2 X_3 (\varepsilon_{23} + \varepsilon_{32})$$

You can see from Figure 8.7 that

$$x_1 = X_1(1 + \varepsilon_1) \quad x_2 = X_2(1 + \varepsilon_2) \quad \text{and} \quad x_3 = X_3(1 + \varepsilon_3)$$

Substituting these values into the equation for a sphere

$$X_1^2 + X_2^2 + X_3^2 = 1$$

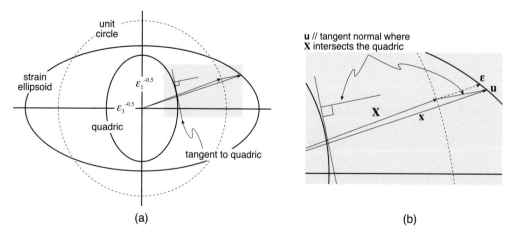

Figure 8.7 (a) The infinitesimal strain quadric and strain ellipse. (b) An expanded detail for the vectors **X** and **u** in (a). The normal to the quadric surface gives the orientation of the vector **u**, which indicates the displacement of the end of the vector **X**; it does not give the orientation of the new position of vector **x** directly.

8.6 Mohr circle for infinitesimal strain

we get

$$\frac{x_1^2}{(1+\varepsilon_1)^2} + \frac{x_2^2}{(1+\varepsilon_2)^2} + \frac{x_3^2}{(1+\varepsilon_3)^2} = 1 \tag{8.11a}$$

This is the equation for the *infinitesimal strain ellipse*. Using the identities in Equations 7.1, we can also write the equation for the strain ellipse in terms of the *principal stretches*:

$$\frac{x_1^2}{S_1^2} + \frac{x_2^2}{S_2^2} + \frac{x_3^2}{S_3^2} = 1 \tag{8.11b}$$

or the principal quadratic elongations:

$$\frac{x_1^2}{\lambda_1} + \frac{x_2^2}{\lambda_2} + \frac{x_3^2}{\lambda_3} = 1 \tag{8.11c}$$

8.6 MOHR CIRCLE FOR INFINITESIMAL STRAIN

As shown in Chapter 5, any tensor transformation can, in two dimensions, be represented by a Mohr circle construction (Fig. 8.8). For infinitesimal strain, we start with the strain tensor, ε_{ij}:

$$\varepsilon_{ij} = \begin{bmatrix} \varepsilon_1 & 0 & 0 \\ 0 & \varepsilon_2 & 0 \\ 0 & 0 & \varepsilon_3 \end{bmatrix}$$

add a transformation matrix:

$$a_{ij} = \begin{pmatrix} \cos\theta & 0 & \sin\theta \\ 0 & 1 & 0 \\ -\sin\theta & 0 & \cos\theta \end{pmatrix}$$

The tensor transformation equation is

$$\varepsilon'_{ij} = a_{ik}a_{jl}\varepsilon_{kl} \tag{8.12}$$

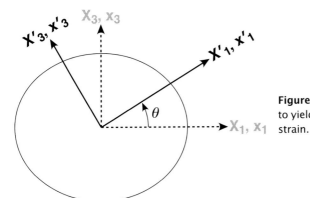

Figure 8.8 The coordinate transformation to yield the Mohr circle for infinitesimal strain.

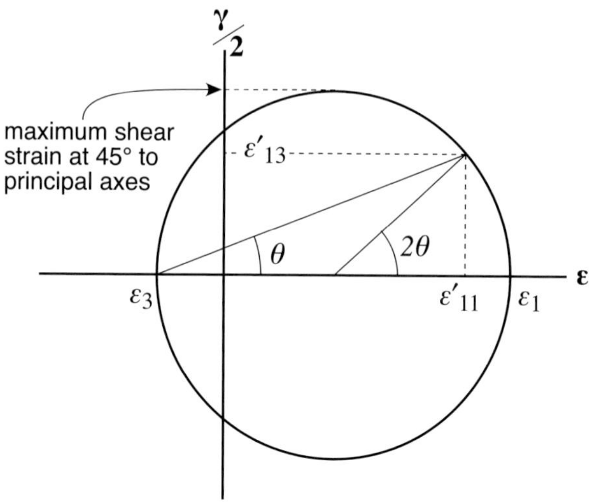

Figure 8.9 The Mohr circle for infinitesimal strain, showing the graphical calculation for a rotation of the coordinate system by the angle θ.

which gives us the new form of the strain tensor:

$$\varepsilon'_{ij} = \begin{bmatrix} \varepsilon'_{11} & 0 & \varepsilon'_{13} \\ 0 & \varepsilon_2 & 0 \\ \varepsilon'_{31} & 0 & \varepsilon'_{33} \end{bmatrix} = \begin{bmatrix} (\varepsilon_1 \cos^2\theta + \varepsilon_3 \sin^2\theta) & 0 & ((\varepsilon_3 - \varepsilon_1)\cos\theta\sin\theta) \\ 0 & \varepsilon_2 & 0 \\ ((\varepsilon_1 - \varepsilon_3)\cos\theta\sin\theta) & 0 & (\varepsilon_1 \sin^2\theta + \varepsilon_3 \cos^2\theta) \end{bmatrix} \quad (8.13)$$

Upon rearranging, we get

$$\varepsilon'_{11} = \frac{(\varepsilon_1 + \varepsilon_3)}{2} + \frac{(\varepsilon_1 - \varepsilon_3)}{2}\cos 2\theta$$

$$\varepsilon'_{13} = \frac{\gamma}{2} = \frac{(\varepsilon_1 - \varepsilon_3)}{2}\sin 2\theta \quad (8.14)$$

Equations 8.14 give the familiar Mohr circle (Fig. 8.9). Probably the most important thing illustrated by Figure 8.9 is that the two planes of maximum shear strain are oriented at +45° and 45° to the principal axes, ε_1 and ε_3. Turning this around, for infinitesimal strain in a shear zone, the shortening and extension directions are always at 45° to the shear zone boundary. This forms the basis for both fault slip and microstructure methods. For example, foliation at the edge of a mylonite zone, P and T axes of earthquakes, and the new tips of sigmoidal gash fractures are all oriented at 45° to the shear zone (Fig. 8.10). We will return to this in the geological problems section.

8.7 EXAMPLE OF CALCULATIONS

Problem

Given the following displacement gradient tensor, calculate ε_{ij}, ω_{ij}, and the magnitudes and orientations of the principal axes:

$$e_{ij} = \begin{bmatrix} 10 & 4 & -2 \\ -4 & 3 & 0 \\ 6 & 0 & 4 \end{bmatrix}$$

8.7 Example of calculations

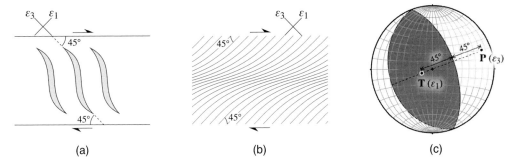

Figure 8.10 Three geological situations that illustrate the principle that the maximum infinitesimal shear strain planes are oriented at 45° to the principal axes of infinitesimal strain: (a) sigmoidal veins ("tension" gashes), (b) heterogeneous ductile shear zone in granitoid rocks, and (c) **P** and **T** axes of earthquakes.

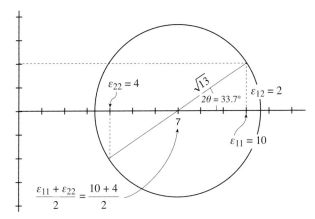

Figure 8.11 Mohr circle solution to the problem described in the text.

Solution

The strain and rotation tensors are easy to calculate:

$$\varepsilon_{ij} = \frac{1}{2}(e_{ij} + e_{ji}) = \begin{bmatrix} 10 & 0 & 2 \\ 0 & 3 & 0 \\ 2 & 0 & 4 \end{bmatrix} \quad \text{and} \quad \omega_{ij} = \frac{1}{2}(e_{ij} - e_{ji}) = \begin{bmatrix} 0 & 4 & -4 \\ -4 & 0 & 0 \\ 4 & 0 & 0 \end{bmatrix}$$

There are several ways to calculate the orientations and magnitudes of the principal axes. For example, we could solve the eigenvalue problem that we discussed when talking about generic tensors. In this case, however, there is an easier way.

Note the position of the zeros in the strain matrix. They indicate that the second, X_2, axis is already parallel to one of the principal axes. Thus, we can solve this problem graphically, using the Mohr circle construction (Fig. 8.11). All we need to do is rotate the coordinate system about the X_2 axis.

From the Pythagorean theorem, the radius of the circle is $\sqrt{(10-7)^2 + 2^2} = \sqrt{13}$, therefore

$$\varepsilon_1 = 7 + \sqrt{13} \quad \varepsilon_2 = 7 - \sqrt{13} \quad \text{and} \quad \varepsilon_3 = 3$$

Notice that $\varepsilon_1 + \varepsilon_2 + \varepsilon_3 = \varepsilon_{11} + \varepsilon_{22} + \varepsilon_{33}$ (Eq. 8.9). ω_{ij} is an antisymmetric tensor or an axial vector. The amount of rotation in radians is (Eqs. 8.4 and 8.5)

$$|\mathbf{r}| = r = \sqrt{\left(0^2 + (-4)^2 + (-4)^2\right)} = 5.6568 \text{ radians}$$

The MATLAB® function **InfStrain**, below, computes the strain and rotation tensors, principal strains, components of rotation, rotation magnitude, and rotation axis orientation from an input displacement gradient tensor. To solve the example above, just type in MATLAB:

```
e = [10 4 -2;-4 3 0;6 0 4]; %Displacement gradient tensor
[eps,ome,pstrains,rotc,rot] = InfStrain(e); %Solve for strain

function [eps,ome,pstrains,rotc,rot] = InfStrain(e)
%InfStrain computes infinitesimal strain from an input displacement
%gradient tensor
%
% USE: [eps,ome,pstrains,rotc,rot] = InfStrain(e)
%
% e = 3 x 3 displacement gradient tensor
% eps = 3 x 3 strain tensor
% ome = 3 x 3 rotation tensor
% pstrains = 3 x 3 matrix with magnitude (column 1), trend (column 2) and
%            plunge (column 3) of maximum (row 1), intermediate (row 2),
%            and minimum (row 3) principal strains
% rotc = 1 x 3 vector with rotation components
% rot = 1 x 3 vector with rotation magnitude and trend and plunge of
%            rotation axis
%
% NOTE: Output trends and plunges of principal strains and rotation axes
% are in radians
%
% InfStrain uses function CartToSph and ZeroTwoPi

%Initialize variables
eps = zeros(3,3);
ome = zeros(3,3);
pstrains = zeros(3,3);
rotc = zeros(1,3);
rot = zeros(1,3);

%Compute strain and rotation tensors (Eq. 8.2)
for i = 1:3
    for j = 1:3
        eps(i,j)= 0.5*(e(i,j)+e(j,i));
        ome(i,j)= 0.5*(e(i,j)-e(j,i));
    end
end
%Compute principal strains and orientations. Here we use the MATLAB
%function eig. D is a diagonal matrix of eigenvalues (i.e. principal
%strains), and V is a full matrix whose columns are the corresponding
%eigenvectors (i.e. principal strain directions)
```

```
[V,D] = eig(eps);
%Maximum principal strain
pstrains(1,1) = D(3,3);
[pstrains(1,2),pstrains(1,3)] = CartToSph(V(1,3),V(2,3),V(3,3));
%Intermediate principal strain
pstrains(2,1) = D(2,2);
[pstrains(2,2),pstrains(2,3)] = CartToSph(V(1,2),V(2,2),V(3,2));
%Minimum principal strain
pstrains(3,1) = D(1,1);
[pstrains(3,2),pstrains(3,3)] = CartToSph(V(1,1),V(2,1),V(3,1));

%Calculate rotation components (Eq. 8.4)
rotc(1)=(ome(2,3)-ome(3,2))*-0.5;
rotc(2)=(-ome(1,3)+ome(3,1))*-0.5;
rotc(3)=(ome(1,2)-ome(2,1))*-0.5;

%Compute rotation magnitude (Eq. 8.5)
rot(1) = sqrt(rotc(1)^2+rotc(2)^2+rotc(3)^2);
%Compute trend and plunge of rotation axis
[rot(2),rot(3)] = CartToSph(rotc(1)/rot(1),rotc(2)/rot(1),rotc(3)/rot(1));
%If plunge is negative
if rot(3) < 0.0
    rot(2) = ZeroTwoPi(rot(2)+pi);
    rot(3) = -rot(3);
    rot(1) = -rot(1);
end
end
```

8.8 GEOLOGICAL APPLICATIONS OF INFINITESIMAL STRAIN

The applications of infinitesimal strain are virtually unlimited, especially in geophysics where the changes observed are very small relative to the distances over which they occur. For example, even a very large earthquake may have less than 10 m of slip on a fault plane whose dimensions span many tens to hundreds of kilometers.

8.8.1 Fault-slip and earthquake data

Analysis of strain for small faults and earthquakes is essentially the same because both represent small deformations in large regions. Additionally, both represent essentially plane strain deformation because there is no change in the direction perpendicular to the slip vector, $\Delta \mathbf{u}$ (Fig. 8.12). Our derivation follows Molnar (1983). Initially, we choose a coordinate system so that X_2 is parallel to the strike of the fault plane and perpendicular to the slickensides on the fault. Later on, we will show the more general case where neither the slickensides nor the fault plane are parallel, or orthogonal, to the axes. In this specialized case, θ is the angle between the fault surface and the vertical axis, that is, 90 - dip of the fault.

Derivation of the displacement gradient tensor

We have already derived the displacement gradient tensor:
$$\Delta u_i = e_{ij} \Delta X_j \quad \text{where} \quad e_{ij} = \frac{\partial u_i}{\partial X_j}$$

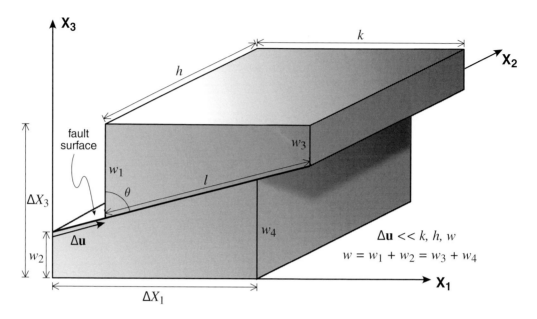

Figure 8.12 Block diagram illustrating the coordinate system used in the calculation of strain and rotation from earthquakes and/or small faults.

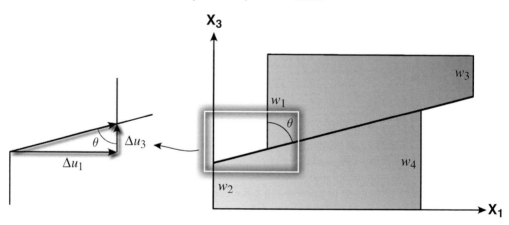

Figure 8.13 View of the faulted block parallel to the X_2 axis edge-on to the fault plane. Detail at left shows the angular relations of the components of the slip vector.

The components of $\Delta \mathbf{u}$ are easily determined from the trigonometry of the block (Fig. 8.13):

$$\Delta u_1 = \Delta \mathbf{u} \sin \theta \quad \text{and} \quad \Delta u_3 = \Delta \mathbf{u} \cos \theta \tag{8.15}$$

Likewise, the length in the ΔX_3 is simple because the fault does not cut the top or bottom of the block (i.e., the sides of the block that are perpendicular to the X_3 axis):

$$\Delta X_3 = w = (w_1 + w_2) = (w_3 + w_4) \tag{8.16}$$

Therefore, the extension parallel to the X_3 axis, e_{33}, is

$$e_{33} = \frac{\Delta u_3}{\Delta X_3} = \frac{\Delta \mathbf{u} \cos \theta}{w} \tag{8.17}$$

8.8 Geological applications of infinitesimal strain

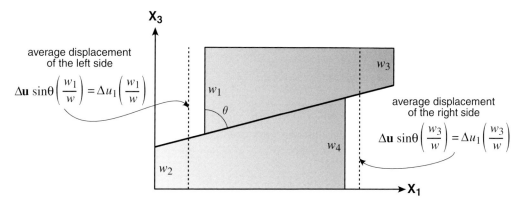

Figure 8.14 Illustration of the average displacements of the sides of the block perpendicular to X_1 as a function of the hanging wall length to the total length.

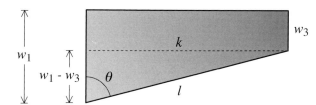

Figure 8.15 Relationships between l, k, and w.

and the rotation towards X_1 of a line originally parallel to X_3, the off-diagonal component e_{13}, is

$$e_{13} = \frac{\Delta u_1}{\Delta X_3} = \frac{\Delta \mathbf{u} \sin \theta}{w} \tag{8.18}$$

The calculation ΔX_1 is more complicated because the fault has offset the sides of the block that are perpendicular to X_1. The *average* displacement of those sides of the block is a function of the ratio of initial length of the side, w for both left and right sides, to the length of the side in the hanging wall only (in our footwall fixed reference frame). So the average displacement of the left side of the block is w_1/w and of the right side is w_3/w (Fig. 8.14). With this insight, we are now ready to calculate the e_{11} component of the displacement gradient tensor:

$$e_{11} = \frac{\text{change in length}}{\text{initial length}} = \frac{\left(\frac{\Delta u_1 w_3}{w} - \frac{\Delta u_1 w_1}{w}\right)}{k} = \frac{\Delta u_1 \left(\frac{w_3 - w_1}{w}\right)}{k} = \frac{\Delta u_1}{\left(\frac{kw}{(w_3 - w_1)}\right)}$$

From the previous equation, and the relations depicted in Figure 8.15, we can see that

$$w_3 - w_1 = -l \cos \theta$$

so ΔX_1 is

$$\Delta X_1 = \frac{wk}{(w_3 - w_1)} = \frac{wk}{-l \cos \theta} = \frac{w(l \sin \theta)}{-l \cos \theta} = -\frac{w \sin \theta}{\cos \theta}$$

Thus we can write for e_{11}:

$$e_{11} = \frac{-\Delta \mathbf{u} \, l \sin \theta \cos \theta}{wk} = -\frac{\Delta \mathbf{u} \cos \theta}{w} \tag{8.19}$$

The rotation towards X_3 of a line originally parallel to X_1, the off-diagonal component e_{31}, in terms of the slip is

$$e_{31} = \frac{\Delta u_3}{\Delta X_1} = \frac{-\Delta u \, l \cos^2 \theta}{wk} = -\frac{\Delta u \cos^2 \theta}{w \sin \theta} \qquad (8.20)$$

The concept of seismic and geometric moment

We can further simplify the equations that have been derived so far by borrowing a concept from geophysics. Seismologists commonly use a scalar parameter known as the *seismic moment*:

$$M_o = \mu A \Delta u \qquad (8.21)$$

where μ is the shear modulus, A is the fault surface area, and Δu is the average slip. For the purposes of fault-slip data analysis, we can omit the shear modulus (which has units of stress) from the above equation because we are only interested in the strain; we are left with the *geometric moment*:

$$M_g = A \Delta u \qquad (8.22)$$

For the faulted block in Figure 8.12, the geometric moment would be

$$M_g = lh \Delta u \qquad (8.23)$$

We can rearrange and simplify this equation by solving for h:

$$h = \frac{V}{kw} = \frac{V}{(l \sin \theta) w} \qquad (8.24)$$

where V is the volume of the region being deformed and the other variables are as shown in Figure 8.12. Substituting Equation 8.24 into 8.23, the geometric moment can be written

$$M_g = \frac{V \Delta u}{w \sin \theta} \qquad (8.25)$$

Finally, the geometric moment divided by the volume gives us a quantity that shows up repeatedly in the equations that we derived for the displacement gradient tensor:

$$\frac{M_g}{V} = \frac{\Delta u}{w \sin \theta} \qquad (8.26)$$

Substituting Equation 8.26 into Equations 8.17, 8.18, 8.19, and 8.20, and writing the result out in matrix format, we get our final expression for the displacement gradient tensor in two dimensions:

$$e_{ij} = \frac{M_g}{V} \begin{bmatrix} -\sin \theta \cos \theta & \sin^2 \theta \\ -\cos^2 \theta & \sin \theta \cos \theta \end{bmatrix} \qquad (8.27)$$

Displacement gradient tensor in terms of fault orientation

Equation 8.27 shows that the displacement gradient tensor is composed of a scalar quantity – the geometric moment divided by the volume of the region – times a tensor that is composed of nothing more than trigonometric functions of the fault plane orientation, θ. By exploring this tensor a bit more, we can easily see how to extend it into three dimensions.

From the geometry in Figure 8.16, you can see that the complete orientation of the fault-slip system can be defined by two unit vectors, one parallel to the slip direction, $\hat{u}$, and the other the

8.8 Geological applications of infinitesimal strain

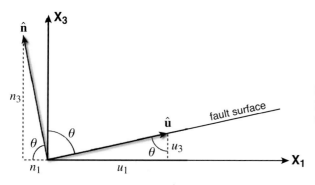

Figure 8.16 Edge-on view of the fault plane, showing the geometry of the unit normal and slip vectors.

pole, or normal, to the fault plane, $\hat{n}$. Because these are unit vectors, they can be written in terms of the angles that they make with the coordinate system:

$$\hat{u} = [\sin\theta \quad \cos\theta] \quad \text{and} \quad \hat{n} = [-\cos\theta \quad \sin\theta] \tag{8.28}$$

The dyad product (Eqs. 4.19, 4.20, 5.7) of $\hat{u}$ and $\hat{n}$ in two dimensions is

$$\hat{u} \otimes \hat{n} = \begin{bmatrix} \sin\theta \\ \cos\theta \end{bmatrix} [-\cos\theta \quad \sin\theta] = \begin{bmatrix} -\sin\theta\cos\theta & \sin^2\theta \\ -\cos^2\theta & \sin\theta\cos\theta \end{bmatrix} \tag{8.29}$$

This is clearly the same matrix as in Equation 8.27, so we can now rewrite that expression for the displacement gradient tensor as

$$e_{ij} = \left(\frac{M_g}{V}\right) \hat{u} \otimes \hat{n} = \left(\frac{M_g}{V}\right) u_i n_j \tag{8.30}$$

where u_i and n_j are the direction cosines of the unit vector parallel to the displacement vector and the unit normal vector of the upward pointing pole, respectively. Equation 8.30 is general for any coordinate system, not just the special case that we started out with.

Summing multiple faults, additive decomposition, principal axes

Where the volume of rock has multiple faults (or earthquakes), because we are dealing with infinitesimal strain, the individual faults and their moments can be summed and divided by the total volume:

$$(e_{ij})_{total} = \frac{\sum\limits^{n\,\text{faults}} (M_g u_i n_j)}{V} \tag{8.31}$$

Recall that the displacement gradient tensor, e_{ij}, is asymmetric. We can additively decompose it to yield the symmetric infinitesimal strain tensor and an antisymmetric rotation tensor:

$$e_{ij} = \varepsilon_{ij} + \omega_{ij} = \frac{M_g(u_i n_j + u_j n_i)}{2V} + \frac{M_g(u_i n_j - u_j n_i)}{2V} \tag{8.32}$$

Because $M_g/2V$ is a scalar, the orientations of the principal axes of ε_{ij} are identical to the principal axes of $(u_i n_j + u_j n_i)$ which, for a single fault, is a function of only the fault plane and slip system orientation. Those principal axes, which you can calculate either by an eigenvalue problem or more simply with the Mohr circle for infinitesimal strain (Fig. 8.9), lie in the plane of the pole and the slip vector (known in faulting analysis literature as the *movement plane*) at 45° to the pole and the fault plane.

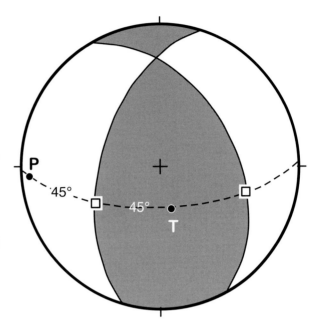

Figure 8.17 A typical earthquake focal mechanism solution. The two nodal planes are potential slip surfaces with potential slip vectors shown as white boxes. Note that the slip vector on one plane is also the pole to the other plane. The nodal planes define a tension quadrant in gray bisected by the **T** axis and a pressure quadrant in white bisected by the **P** axis. The movement plane is shown as a dashed line.

Earthquake seismologists commonly depict earthquake data as focal mechanism solutions with **P** (pressure) and **T** (tension) axes bisecting the appropriate quadrants (Fig. 8.17). Despite the stress terminology used to name them, we can see from the above analysis that the **P** and **T** axes are in fact the principal axes of infinitesimal strain (strictly speaking, the eigenvectors that are unit vectors parallel to the principal axes). Calculating these axes requires knowing nothing more than the pole to the plane and the slip vector (for which one needs to know both the direction and the sense of slip).

Some further remarks about fault slip and earthquake analyses

Molnar (1983) referred to the quantity $M_g u_i n_j$ in Equation 8.31 as the "asymmetric moment tensor" to distinguish it from the more familiar seismic moment tensor described by Kostrov (1974). Kostrov's moment tensor is symmetric because it was derived specifically for the case of earthquakes, where one commonly does not know which nodal plane is the true slip surface; it is identical to the symmetric part of Molnar's tensor (i.e., ε_{ij}, the symmetric part of Eq. 8.32). Jackson and McKenzie (1988) have questioned whether or not it is ever possible to determine the antisymmetric part of the Molnar's tensor for either earthquakes or faults. It is a question of frame of reference. Molnar's analysis assumes that the reference frame is fixed to the footwall, but usually in geology we don't know whether the footwall and fault plane are fixed or whether both rotate during the deformation, domino style (Fig. 8.18).

Throughout this analysis, we have kept the scalar terms M_g/V separate from the orientation terms $u_i n_j$ for a very practical reason. Particularly for the field structural geologist, the scalar terms are difficult to determine with any degree of accuracy. In any practical situation, because of the two-dimensional nature of most outcrops, it is virtually impossible to measure the fault surface area directly and one has little idea whether the displacement observed in the field is anything close to the "average" displacement. Likewise, there is quite a lot of ambiguity surrounding the choice of the volume of a region, even in the case of earthquakes. One can

8.8 Geological applications of infinitesimal strain 153

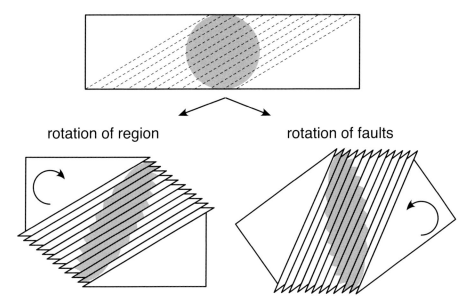

Figure 8.18 Ambiguity of the rotation determined from the antisymmetric part of the displacement gradient tensor in Equation 8.32.

estimate these parameters via a variety of fractal scaling relations (e.g., Marrett and Allmendinger, 1990), but these are also subject to order of magnitude uncertainty.

In contrast, calculation of **P** and **T** axes (i.e., infinitesimal strain axes orientations) from the orientation terms is robust and, assuming good outcrop, relatively free from large uncertainty (Marrett and Allmendinger, 1990). In the field, the most uncertain measurement is the determination of sense of slip. The only time a **P** and **T** axes analysis is likely to fail is when the largest fault in the region studied has a particularly different kinematics than the rest of the faults measured. Being able to calculate the geometric moment for that fault would allow one to correct for this error.

Finally, in as much as this is a chapter on infinitesimal strain, the importance of Equation 8.31 should be emphasized: Matrix addition is commutative. That means that with small faults, we can add them together in whatever order we want. In Chapter 9, we will see that when faults, and strains, become large we can no longer add the faults together in whatever order; for large faults, we have to know the *order* of formation to calculate strain correctly.

The MATLAB function **PTAxes**, below, computes the **P** and **T** axes from the orientation of several fault planes and their slip vectors. It also plots the solution in an equal area stereonet.

```
function [P,T] = PTAxes(fault,slip)
%PTAxes computes the P and T axes from the orientation of several fault
%planes and their slip vectors. Results are plotted in an equal area
%stereonet
%
%   USE: [P,T] = PTAxes(fault,slip)
%
%   fault = nfaults x 2 vector with strikes and dips of faults
%   slip = nfaults x 2 vector with trends and plunges of slip vectors
%   P = nfaults x 2 vector with trends and plunges of the P axes
```

```
%       T = nfaults x 2 vector with trends and plunges of the T axes
%
%       NOTE: Input/Output angles are in radians
%
%       PTAxes uses functions SphToCart, CartToSph, Stereonet, GreatCircle and
%       StCoordLine

%Initialize some vectors
n = zeros(1,3);
u = zeros(1,3);
eps = zeros(3,3);
P = zeros(size(fault,1),2);
T = zeros(size(fault,1),2);

%   For all faults
for i = 1:size(fault,1)
    %   Direction cosines of pole to fault and slip vector
    [n(1),n(2),n(3)] = SphToCart(fault(i,1),fault(i,2),1);
    [u(1),u(2),u(3)] = SphToCart(slip(i,1),slip(i,2),0);
    %   Compute u(i)*n(j) + u(j)*n(i)  (Eq. 8.32)
    for j = 1:3
        for k = 1:3
            eps(j,k)=(u(j)*n(k)+u(k)*n(j));
        end
    end
    %   Compute orientations of principal axes of strain. Here we use the
    %   MATLAB function eig
    [V,D] = eig(eps);
    %   P orientation [P(i,1),
    P(i,2)] = CartToSph(V(1,3),V(2,3),V(3,3));
    %   T orientation
    [T(i,1),T(i,2)] = CartToSph(V(1,1),V(2,1),V(3,1));
end

%   Plot stereonet
Stereonet(0,90*pi/180,10*pi/180,1);
hold on;
%   Plot other elements
for i = 1:size(fault,1)
    %   Plot fault
    [path] = GreatCircle(fault(i,1),fault(i,2),1);
    plot(path(:,1),path(:,2),'r');
    %   Plot Slip vector (red square)
    [xp,yp] = StCoordLine(slip(i,1),slip(i,2),1);
    plot(xp,yp,'rs');
    %   Plot P axis (black, filled circle)
    [xp,yp] = StCoordLine(P(i,1),P(i,2),1);
    plot(xp,yp,'ko','MarkerFaceColor','k');
```

8.8 Geological applications of infinitesimal strain

```
    % Plot T axis (black circle)
    [xp,yp] = StCoordLine(T(i,1),T(i,2),1);
    plot(xp,yp,'ko');
end

% Release plot
hold off;
end
```

8.8.2 Displacement fields and two-dimensional strain from GPS data

Because the changes in distance between GPS stations with time are extremely small (tens of millimeters) relative to the distance between stations (tens of kilometers), the strains measured by GPS are truly infinitesimal. The two-dimensional problem is shown in Figure 8.19.

We know how to solve this problem:

$$u_i = t_i + e_{ij}X_j \quad \text{where} \quad \frac{\partial u_i}{\partial X_j} = e_{ij} \qquad (8.33)$$

and

$$x_i = q_i + F_{ij}X_j \quad \text{where} \quad \frac{\partial x_i}{\partial X_j} = F_{ij}$$

and t_i is translation of a point at the origin of the coordinate system. From either of these equations, you can see that there are six unknowns: the two components of the translation vector $((t_1, t_2))$, and the four components of the displacement gradient tensor or deformation gradient tensor (e_{11}, e_{12}, e_{21}, e_{22} or F_{11}, F_{12}, F_{21}, F_{22}). Each station furnishes two equations. Therefore one needs a minimum of three non-colinear GPS stations to determine the two-dimensional strain ellipse. In three dimensions there are twelve unknowns, and each station furnishes three equations. Therefore one needs a minimum of four non-coplanar stations to determine the three-dimensional strain ellipsoid.

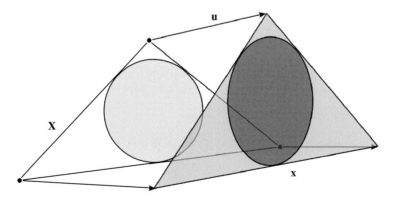

Figure 8.19 Displacements at three stations. The triangle described by the three stations prior to the deformation, represented by **X**, has a circle inscribed in it. Upon deformation, the three stations are displaced by three non-parallel vectors of unequal length to their new positions, **x**. The inscribed circle becomes the strain ellipse.

To solve this system of linear equations using standard linear algebra methods, we need to recast the equations into three matrices, two of which contain only known quantities and one that contains just the unknown quantities. The equations below will do the trick in two dimensions, as you can prove to yourself by a standard matrix multiplication:

$$\begin{bmatrix} {}^1u_1 \\ {}^1u_2 \\ {}^2u_1 \\ {}^2u_2 \\ \vdots \\ \vdots \\ {}^nu_1 \\ {}^nu_2 \end{bmatrix} = \begin{bmatrix} 1 & 0 & {}^1X_1 & {}^1X_2 & 0 & 0 \\ 0 & 1 & 0 & 0 & {}^1X_1 & {}^1X_2 \\ 1 & 0 & {}^2X_1 & {}^2X_2 & 0 & 0 \\ 0 & 1 & 0 & 0 & {}^2X_1 & {}^2X_2 \\ \vdots & \vdots & \vdots & \vdots & \vdots & \vdots \\ \vdots & \vdots & \vdots & \vdots & \vdots & \vdots \\ 1 & 0 & {}^nX_1 & {}^nX_2 & 0 & 0 \\ 0 & 1 & 0 & 0 & {}^nX_1 & {}^nX_2 \end{bmatrix} \begin{bmatrix} t_1 \\ t_2 \\ e_{11} \\ e_{12} \\ e_{21} \\ e_{22} \end{bmatrix} \quad (8.34)$$

Similarly, we could solve for the deformation gradients rather than the displacement gradients:

$$\begin{bmatrix} {}^1x_1 \\ {}^1x_2 \\ {}^2x_1 \\ {}^2x_2 \\ \vdots \\ \vdots \\ {}^nx_1 \\ {}^nx_2 \end{bmatrix} = \begin{bmatrix} 1 & 0 & {}^1X_1 & {}^1X_2 & 0 & 0 \\ 0 & 1 & 0 & 0 & {}^1X_1 & {}^1X_2 \\ 1 & 0 & {}^2X_1 & {}^2X_2 & 0 & 0 \\ 0 & 1 & 0 & 0 & {}^2X_1 & {}^2X_2 \\ \vdots & \vdots & \vdots & \vdots & \vdots & \vdots \\ \vdots & \vdots & \vdots & \vdots & \vdots & \vdots \\ 1 & 0 & {}^nX_1 & {}^nX_2 & 0 & 0 \\ 0 & 1 & 0 & 0 & {}^nX_1 & {}^nX_2 \end{bmatrix} \begin{bmatrix} q_1 \\ q_2 \\ F_{11} \\ F_{12} \\ F_{21} \\ F_{22} \end{bmatrix} \quad (8.35)$$

Of course, we only need to solve one of the above two systems of equations because the displacement and deformation gradient tensors are simply related by the identity matrix:

$$\mathbf{e} = \mathbf{F} - \mathbf{I} \quad \text{or} \quad e_{ij} = F_{ij} - \delta_{ij}$$

Notice that the above Equations 8.34 and 8.35 are written not for three equations but for n equations. With more than three equations, the system is over-constrained, that is, there are more equations than unknowns. In such a case, we can actually use the extra information to assess the uncertainties in the assumption that strain in the region encompassed by the GPS stations is homogeneous.

The solution to Equations 8.34 or 8.35 is a classic application of inverse theory (see Menke, 1984). These equations are in the form of Equation 4.29, which is repeated here:

$$\mathbf{y} = \mathbf{Mx} \quad (8.36)$$

To solve for $\mathbf{x}$, we multiply $\mathbf{y}$ by the inverse of matrix $\mathbf{M}$, that is, $\mathbf{M}^{-1}$:

$$\mathbf{x} = \mathbf{M}^{-1}\mathbf{y} \quad (8.37)$$

In the case of Equations 8.34 and 8.35, the large matrix with six columns and the number of rows equal to twice the stations used (a minimum of six rows), commonly called the *design matrix*, is equivalent to $\mathbf{M}$ in Equation 8.36. It is this matrix that we must calculate the inverse of to solve this problem. As described in Chapter 4, determining the inverse of even a 3×3 matrix is tedious; the minimum size of our design matrix is 6×6!

8.8 Geological applications of infinitesimal strain

For perfectly constrained cases of just three GPS stations, one may use a procedure known as LU decomposition. For the over-constrained situation, the matrix is no longer square and cannot be inverted directly but a least squares best fit may be made. We highly recommend reading the relevant sections in Press *et al.* (Chapters 2 and 15, Press *et al.*, 1986). Menke (1984) gives the basic least squares solution to Equation 8.37 as

$$\mathbf{x} = \left[\mathbf{M}^T\mathbf{M}\right]^{-1}\mathbf{M}^T\mathbf{y} \qquad (8.38)$$

In the context of the GPS problem, **M** is the large matrix of 1's, 0's, and position vectors, **X**. All of the displacement vectors are held in **y**, and the unknowns (t_1, t_2, e_{11}, e_{12}, e_{21}, e_{22}) are in **x**. One could imagine using Equations 8.34 and 8.38 to calculate a single best-fit displacement gradient tensor to all of the stations in a GPS network, but the likelihood of that producing a meaningful result is small.

There are several potential strategies for calculating a more insightful result that demonstrates how the gradients, **e**, vary across a region. One can, for example, construct a network of triangles, know as a *Delaunay triangulation*, from the GPS stations. In this approach, each triangle provides the minimum number of stations necessary to calculate a deformation gradient in that triangle, but the triangles are all of different shapes and sizes, providing a very irregular view of the deformation (Fig. 8.20). Alternatively, one can establish a regular grid over a region and calculate the deformation based on the *n* stations nearest to a grid node, where $n \geq 3$. This is an improvement over the triangles method, but is still subject to artifacts produced by the irregular spacing of stations in a typical GPS network. It is difficult to know, in these cases, whether a particular pattern is due to heterogeneous strain or heterogeneous station spacing.

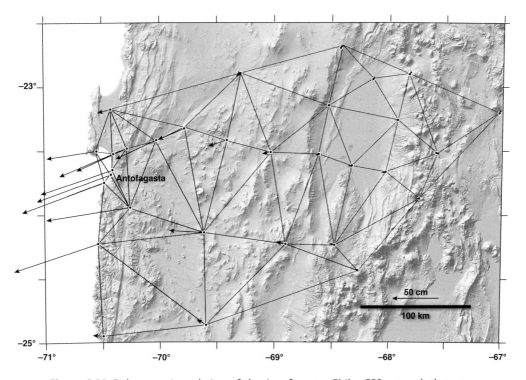

Figure 8.20 Delaunay triangulation of the Antofagasta, Chile, GPS network that was described in Chapter 7.

A third alternative exists. As before, one establishes a regular grid over the region of interest, but at each node in the grid, one uses *all* of the stations in the network, weighting the contribution of each station according to its distance from the node. This method is called a *weighted* least squares approach (Allmendinger *et al.*, 2009; Menke, 1984; Shen *et al.*, 1996). The basic form of the weighted least squares solution is

$$\mathbf{x} = \left[\mathbf{M}^T\mathbf{W}\mathbf{M}\right]^{-1}\mathbf{M}^T\mathbf{W}\mathbf{y}$$

where **W** is the diagonalized matrix of weighting values, W, given by

$$W = \exp\left[\frac{-d^2}{2\alpha^2}\right]$$

The parameter d is the distance of any particular station from the grid node and α is a distance weighting constant that specifies how the effect of a particular station decays with distance. A larger value of α produces greater smoothing, damping out local variations.

This raises an extremely important question with respect to strain: What is the proper length scale at which to calculate strain? It may come as a surprise that there is no single correct answer to this question, especially where strain is heterogeneous and discontinuous as in any study of strain over large areas at the surface of the Earth (Allmendinger *et al.*, 2009). In part, the answer to this question, regardless of whether one is interested in infinitesimal strain (this chapter) or finite strain (the next chapter), depends on the problem in which you are interested. In a thrust belt (Fig. 8.21), for example, the strain at the scale of the entire belt is entirely different than the strain within a single bed; there is no one correct strain measure.

Once we have found the displacement gradient tensor at a particular point, we may separate it into the symmetric infinitesimal strain tensor, ε_{ij}, and the antisymmetric rotation tensor, ω_{ij}, by Equation 8.2. The eigenvalues and eigenvectors of the infinitesimal strain tensor will give us the principal strains and the antisymmetric part will give us the rotation axis from Equation 8.3. This is quite a lot more than we could learn from the one-dimensional plot alone! Note that the only part of this problem that relies on infinitesimal strain assumptions is this final additive decomposition into symmetric and antisymmetric tensors. Everything else could equally well be carried out for finite strain.

The MATLAB function **GridStrain**, below, computes the two-dimensional infinitesimal strain of a displacement network using Delaunay triangulation (k = 0), nearest neighbor (k = 1), or the distance weighted method (k = 2). After the computation, the function plots the grid colored by the parameter chosen in variable **plotpar**. **Gridstrain** uses the MATLAB built-in function **lscov** to solve the simple or weighted least squares problem of Equation 8.38. In a way, **GridStrain** is a

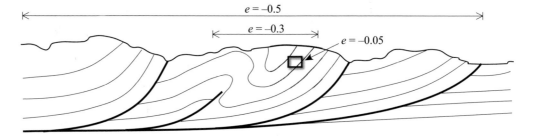

Figure 8.21 Cartoon cross section of a hypothetical thrust belt showing three different, valid measures of horizontal extension. In this, the extension in hand sample is completely different than that in a train of fold and that for the entire thrust belt. The length scale for measuring strain depends on the problem in which one is interested.

8.8 Geological applications of infinitesimal strain

miniature version of our Macintosh program SSPX (Cardozo and Allmendinger, 2009). You will get the chance to try **GridStrain** in the exercises section.

```
function [cent,eps,ome,pstrains,rotc] = GridStrain (pos,disp,k,par,plotpar)
%GridStrain computes the infinitesimal strain of a network of stations with
%displacements in x (east) and y (north). Strain in z is assumed to be zero
%
%   USE: [cent,eps,ome,pstrains,rotc] = GridStrain(pos,disp,k,par,plotpar)
%
%   pos = nstations x 2 matrix with x (east) and y (north) positions
%       of stations
%   disp = nstations x 2 matrix with x (east) and y (north) displacements
%        of stations
%   k = Type of computation: Delaunay (k = 0), nearest neighbor (k = 1), or
%       distance weighted (k = 2).
%   par = Parameters for nearest neighbor or distance weighted computation.
%         If Delaunay (k = 0), enter a scalar corresponding to the minimum
%         internal angle of a triangle valid for computation.
%         If nearest neighbor (k = 1), input a 1 x 3 vector with grid
%         spacing, number of nearest neighbors, and maximum distance
%         to neighbors.
%         If distance weighted (k = 2), input a 1 x 2 vector with grid
%         spacing and distance weighting factor alpha
%   plotpar = Parameter to color the cells: Maximum elongation
%             (plotpar = 0), minimum elongation (plotpar = 1),
%             rotation (plotpar = 2), or dilatation (plotpar = 3)
%   cent = ncells x 2 matrix with x and y positions of cells centroids
%   eps = 3 x 3 x ncells array with strain tensors of the cells
%   ome = 3 x 3 x ncells array with rotation tensors of the cells
%   pstrains = 3 x 3 x ncells array with magnitude and orientation of
%             principal strains of the cells
%   rotc = ncells x 3 matrix with rotation components of cells
%
%   NOTE: Input/Output angles should be in radians. Output azimuths are
%         given with respect to north
%         pos, disp, grid spacing, max. distance to neighbors, and alpha
%         should be in the same units of length
%
%   GridStrain uses function InfStrain

%   If Delaunay
if k == 0
    %   Indexes of triangle vertices: Use MATLAB built-in function delaunay
    inds = delaunay(pos(:,1),pos(:,2));
    %   Number of cells
    ncells = size(inds,1);
    %   number of stations per cell = 3
    nstat = 3;
    %   centers of cells
```

```
cent = zeros(ncells,2);
for i = 1:ncells
    %  Triangle vertices
    v1x = pos(inds(i,1),1); v2x = pos(inds (i,2),1); v3x = pos(inds
       (i,3),1);
    v1y = pos(inds (i,1),2); v2y = pos(inds (i,2),2); v3y = pos(inds
       (i,3),2);
    %  Center of cell
    cent(i,1)=(v1x + v2x + v3x)/3.0;
    cent(i,2)=(v1y + v2y + v3y)/3.0;
    %  Triangle internal angles
    s1 = sqrt((v3x-v2x)^2 + (v3y-v2y)^2);
    s2 = sqrt((v1x-v3x)^2 + (v1y-v3y)^2);
    s3 = sqrt((v2x-v1x)^2 + (v2y-v1y)^2);
    a1 = acos((v2x-v1x)*(v3x-v1x)/(s3*s2)+(v2y- v1y)*(v3y-v1y)/(s3*s2));
    a2 = acos((v3x-v2x)*(v1x-v2x)/(s1*s3)+(v3y- v2y)*(v1y-v2y)/(s1*s3));
    a3 = acos((v2x-v3x)*(v1x-v3x)/(s1*s2)+(v2y- v3y)*(v1y-v3y)/(s1*s2));
    %  If any of the internal angles is less than specified minimum,
    %  invalidate triangle
    if a1 < par || a2 < par || a3 < par
        inds(i,:) = zeros(1,3);
    end
end
%  Else if nearest neighbor or distance weighted
else
    %  Construct grid
    xmin = min(pos(:,1)); xmax = max(pos(:,1));
    ymin = min(pos(:,2)); ymax = max(pos(:,2));
    cellsx = ceil((xmax-xmin)/par(1));
    cellsy = ceil((ymax-ymin)/par(1));
    xgrid = xmin:par(1):(xmin+cellsx*par(1));
    ygrid = ymin:par(1):(ymin+cellsy*par(1));
    [XX,YY] = meshgrid(xgrid,ygrid);
    %  Number of cells
    ncells = cellsx * cellsy;
    %  Number of stations per cell. If nearest neighbor
    if k == 1
        nstat = par(2); %  Number of nearest neighbors
    %  If distance weighted
    elseif k == 2
        nstat = size(pos,1); %  All stations
    end
    %  centers of cells
    cent = zeros(ncells,2);
    count = 1;
    for i = 1:cellsy
        for j = 1:cellsx
            cent(count,1) = (XX(i,j)+XX(i,j+1))/2.0;
```

8.8 Geological applications of infinitesimal strain

```
                cent(count,2) = (YY(i,j)+YY(i+1,j))/2.0;
                count = count + 1;
            end
        end
    end
    %  Initialize indexes of stations for cells
    inds = zeros(ncells,nstat);
    %  Initialize weight factor matrix for distance weighted method
    wv = zeros(ncells,nstat*2);
    %  For all cells set inds and wv (if distance weighted method)
    for i = 1:ncells
        %  Initialize distances to nearest stations to -1.0
        dists = ones(1,nstat)*-1.0;
        %  For all stations
        for j = 1:size(pos,1)
            %  Distance from center of cell to station
            distx = cent(i,1) - pos(j,1);
            disty = cent(i,2) - pos(j,2);
            dist = sqrt(distx^2+disty^2);
            %  If nearest neighbor
            if k == 1
                %  If within the specified maximum distance to neighbors
                if dist <= par(3)
                [mind,mini] = min(dists);
                %  If number of neighbors are less than maximum
                if mind == -1.0
                    dists(mini) = dist;
                    inds(i,mini) = j;
                %  Else if maximum number of neighbors
                else
                    %  If current distance is lower than maximum distance
                    [maxd,maxi] = max(dists);
                    if dist < maxd
                        dists(maxi) = dist;
                        inds(i,maxi) = j;
                    end
                end
                end
            %  If distance weighted
            elseif k == 2
                inds(i,:) = 1:nstat; %  All stations
                %  weight factor
                weight = exp(-dist^2/(2.0*par(2)^2));
                wv(i,j*2-1) = weight;
                wv(i,j*2) = weight;
            end
        end
    end
end
```

```
%   Initialize arrays
y = zeros(nstat*2,1);M = zeros(nstat*2,6); e = zeros(3,3);
eps = zeros(3,3,ncells); ome = zeros (3,3,ncells);
pstrains = zeros(3,3,ncells); rotc = zeros(ncells,3);

%   For each cell
for i = 1:ncells
    %   If required minimum number of stations
    if min(inds(i,:)) > 0
    %   Fill displacements column vector y and design matrix M
    %   Use X1 = North, X2 = East
    for j = 1:nstat
        y(j*2-1) = disp(inds(i,j),2);
        y(j*2) = disp(inds(i,j),1);
        M(j*2-1,:) = [1 0 pos(inds(i,j),2) pos (inds(i,j),1) 0 0];
        M(j*2,:) = [0 1 0 0 pos(inds(i,j),2) pos (inds(i,j),1)];
    end
        %   Compute x (Eqs. 8.37 and 8.38): Use MATLAB function lscov
        %   If Delaunay or nearest neighbor
        if k == 0 || k == 1
            x = lscov(M,y);
        %   If distance weighted
        elseif k == 2
            x = lscov(M,y,wv(i,:));
        end
        %   Displacement gradient tensor
        for j = 1:2
            e(j,1) = x(j*2+1);
            e(j,2) = x(j*2+2);
        end
        %   Compute strain
        [eps(:,:,i),ome(:,:,i),pstrains(:,:,i),rotc (i,:)] = InfStrain(e);
    end
end

%   Variable to plot
%   If maximum principal strain
if plotpar == 0
    vp = pstrains(1,1,:);
    cbt = 'e1';
%   If minimum principal strain
elseif plotpar == 1
    vp = pstrains(3,1,:);
    cbt = 'e3';
%   If rotation: Since we are assuming plane strain, rotation = rotc(3)
elseif plotpar == 2
    vp = rotc(:,3)*180/pi;
    cbt = 'rot (deg)';
```

8.8 Geological applications of infinitesimal strain

```
%   If dilatation
elseif plotpar == 3
    vp = pstrains(1,1,:)+pstrains(2,1,:)+pstrains (3,1,:);
    cbt = 'dilat';
end

%  scale variable to plot so that is between 0 and 1
minvp = min(vp); maxvp = max(vp); rangvp = maxvp-minvp;
vps = (vp-minvp)/rangvp;

%  colormap
colormap(jet);

%  Plot cells
%  If Delaunay
if k == 0
    for i = 1:ncells
        %  If required minimum number of stations
            if min(inds(i,:)) > 0
                xp = [pos(inds(i,1),1);pos(inds (i,2),1);pos(inds(i,3),1)];
                yp = [pos(inds(i,1),2);pos(inds (i,2),2);pos(inds(i,3),2)];
                patch(xp,yp,vps(i),'EdgeColor','k');
            end
    end
end
%  If nearest neighbor or distance weighted
if k == 1 || k == 2
    count = 1;
    for i = 1:cellsy
        for j = 1:cellsx
            %  If required minimum number of stations
            if min(inds(count,:)) > 0
                xp = [XX(i,j) XX(i,j+1) XX(i+1,j+1) XX(i+1,j)];
                yp = [YY(i,j) YY(i,j+1) YY(i+1,j+1) YY(i+1,j)];
                patch(xp,yp,vps (count),'EdgeColor','k');
            end
            count = count + 1;
        end
    end
end

%  colorbar
ytick = [0 0.2 0.4 0.6 0.8 1.0];
cb = colorbar('Ytick',ytick,'YTickLabel',{num2str(minvp),...
     num2str(minvp+rangvp/5),num2str(minvp +2*rangvp/5),...
     num2str(minvp+3*rangvp/5),num2str(minvp +4*rangvp/5),num2str(maxvp)});
set(get(cb,'title'),'String',cbt);
```

```
%   Axes
axis equal;
xlabel('x'); ylabel('y');
end
```

8.9 EXERCISES

1. In Section 8.2.2, we asserted that "straight lines remain straight" and "parallel lines remain parallel" in any homogeneous deformation. Prove this to yourself by straining any two initially parallel line segments according to Equation 8.7.
2. Prove that Equation 8.7, for the case of a pure rotation of a line, requires that the principal diagonal of the displacement gradient tensor be all zero and that the off-diagonal elements above the principal diagonal equal the negative of those below.
3. Write Equation 8.29 in terms of the direction cosines of the unit normal and unit displacement vectors. How would your equation change if you use a north-east-down coordinate system?
4. Fifteen measurements of faults and their slickensides are given in the table below. Calculate the **P** and **T** axes of the individual faults and then calculate an unweighted moment tensor summation. The slickensides, of course, give the slip direction, but you will have to establish a sign convention in order to incorporate the sense of slip into the slip vector. Use function **PTAxes** or a spreadsheet program to solve this problem.

Fault Plane			Slickensides		Sense of slip
Strike	Dip	Direction	Trend	Plunge	
149.5	47.2	W	164.4	15.4	Left lateral
127.6	60	S	134.6	11.9	Left lateral
189.4	34.6	W	349.6	13.1	Left lateral
328	42.5	E	335.3	6.6	Left lateral
22.9	50.2	E	182	23.2	Thrust
108.8	31.1	S	169.2	27.7	Normal
184.6	39.8	W	317.1	31.6	Thrust
93.7	65	S	269.6	8.8	Right lateral
297.6	64.1	N	300.2	5.4	Right lateral
272.5	34.5	N	284.4	8	Right lateral
151.6	58.1	W	154.9	5.3	Left lateral
302.7	47	N	105.3	17.7	Right lateral
349.4	33.7	E	145.2	15.3	Thrust
90.9	71.1	S	96	14.6	Right lateral
189.7	36.6	W	247.9	32.3	Thrust

5. Using the same GPS data set that you were given for Exercise 1 in Chapter 7, calculate the Delaunay two-dimensional strain field for the coseismic displacements associated with the 1995 M 8.1 Antofagasta earthquake. (a) Plot the horizontal extension magnitudes and compare them to the answer you obtained using a one-dimensional transect in Chapter 7. (b) Plot the vertical axis rotations and explain the pattern that you see. (c) Repeat a and b but this time using the nearest neighbor method. (d) Repeat a and b using the distance weighted method. It is up to you to decide the parameters of the calculation in a to d. Discuss how these parameters, and specially α in d, affect your results. Hint: Create a text file with the east and north coordinates and displacements of the GPS stations. Read these in vectors **pos** and **disp** in MATLAB, and use them in the function **GridStrain** accordingly.

CHAPTER

NINE

Finite strain

9.1 INTRODUCTION

Because processes in the Earth work very slowly, the assumptions of infinitesimal strain work very well for deformation that happens on the scale of years to centuries. Thus geophysicists, who measure "real-time" deformation with seismometers, GPS satellites, and InSAR (interferometric radar), are content to stay in the realm of infinitesimal strain. Structural geologists, however, deal with deformation that accrues over millions of years or more. The assumptions of infinitesimal strain are commonly not appropriate for the large magnitude strains that result from deformation that accumulates over those long time frames. Thus, this chapter and the next will explore *finite strain*. In this chapter, we will look at finite strain simply as the difference between an initial and a final state; in the next chapter, we will see how strain accrues over time. Finite strain is considerably messier than infinitesimal strain, but we'll learn some interesting things along the way.

When deformations are large, we can no longer assume that the initial and final states are nearly identical:

$$dX_i \neq dx_i \quad \text{and} \quad \frac{\partial u_i}{\partial X_i} \neq \frac{\partial u_i}{\partial x_i}$$

and we have to go back to our four measures of deformation:

	Old coordinates	New coordinates
Coordinate transformation	$dx_i = \dfrac{\partial x_i}{\partial X_j} dX_j$ (Green)	$dX_i = \dfrac{\partial X_i}{\partial x_j} dx_j$ (Cauchy)
Displacements	$du_i = \dfrac{\partial u_i}{\partial X_j} dX_j$ (Lagrange)	$du_i = \dfrac{\partial u_i}{\partial x_j} dx_j$ (Euler)

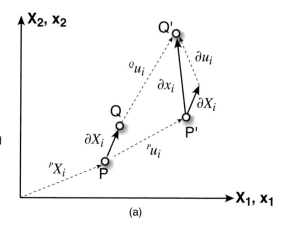

Figure 9.1 Distortion of a vector **PQ** to **P′Q′**. Diagram is the same as Figure 7.6 except that we now use derivatives rather than deltas. The axes **X** and **x** refer to the initial and final states, respectively.

9.2 DERIVATION OF THE LAGRANGIAN STRAIN TENSOR

To derive the basic equations of finite strain, we will return to Figure 7.6, but now with some minor relabeling (Fig. 9.1). As has been our habit throughout the book, we will do the basic derivations in two dimensions, which we will then generalize by use of the summation convention. From the diagram, you can see that

$$dx_i = dX_i + du_i \tag{9.1}$$

The Lagrangian displacement gradient tensor, e_{ij}, gives us the value of du_i:

$$du_i = \frac{\partial u_i}{\partial X_j} dX_j = e_{ij} dX_j \tag{9.2}$$

Combining Equations 9.1 and 9.2, we get an expression for the length of $P'Q'$, dx_i:

$$dx_i = dX_i + \frac{\partial u_i}{\partial X_j} dX_j = dX_i + e_{ij} dX_j \tag{9.3}$$

Now, let's look at the difference in the *squared lengths* of the two vectors:

$$|PQ|^2 = dX_i dX_i = dX_1^2 + dX_2^2 \quad \text{and} \quad |P'Q'|^2 = dx_i dx_i = dx_1^2 + dx_2^2$$

Expanding the expression for the deformed length, $|P'Q'|^2$, by substituting Equation 9.3 we get

$$|P'Q'|^2 = (dX_1 + e_{11} dX_1 + e_{12} dX_2)^2 + (dX_2 + e_{21} dX_1 + e_{22} dX_2)^2$$

With further expansion and rearranging of terms, we get

$$|P'Q'|^2 = [1 + 2e_{11} + e_{11}^2 + e_{21}^2] dX_1^2 + [1 + 2e_{22} + e_{22}^2 + e_{12}^2] dX_2^2 \\ + [e_{12} + e_{11} e_{12} + e_{21} + e_{21} e_{22}] 2 dX_1 dX_2$$

Let's simplify this equation by making the following substitutions:

$$E_{11} = e_{11} + \frac{1}{2}(e^2{}_{11} + e^2{}_{21}) \qquad E_{22} = e_{22} + \frac{1}{2}(e^2{}_{22} + e^2{}_{12})$$

$$E_{12} = \frac{1}{2}(e_{12} + e_{21}) + \frac{1}{2}(e_{11} e_{12} + e_{21} e_{22})$$

Now, we can write the difference in the lengths of the vector before and after deformation as

$$|P'Q'|^2 - |PQ|^2 = 2[E_{11} dX_1^2 + E_{22} dX_2^2 + E_{12} dX_1 dX_2]$$

9.4 Derivation of the Green deformation tensor

This equation, in general three-dimensional form using the Einstein summation convention, can be written

$$|P'Q'|^2 - |PQ|^2 = 2 dX_i E_{ij} dX_j$$

E_{ij} is the *Lagrangian finite strain tensor* and it has the general form

$$E_{ij} = \frac{1}{2}\left[\frac{\partial u_i}{\partial X_j} + \frac{\partial u_j}{\partial X_i} + \frac{\partial u_k}{\partial X_i}\frac{\partial u_k}{\partial X_j}\right] = \frac{1}{2}\left[e_{ij} + e_{ji} + e_{ki}e_{kj}\right] \tag{9.4}$$

where e_{ij} is the Lagrangian displacement gradient tensor that we first defined in Chapter 7. In infinitesimal strain, we assume that the last term in the equation is small enough to neglect:

$$\frac{\partial u_k}{\partial X_i}\frac{\partial u_k}{\partial X_j} = e_{ki}e_{kj} = 0$$

and we are left with

$$\varepsilon_{ij} = \frac{1}{2}(e_{ij} + e_{ji}) = \frac{1}{2}\left(\frac{\partial u_i}{\partial X_j} + \frac{\partial u_j}{\partial X_i}\right)$$

By way of example, the following shows you how to expand the Lagrangian strain tensor for $i, j = 1, 1$ and for $i, j = 1, 3$:

$$E_{11} = \frac{\partial u_1}{\partial X_1} + \frac{1}{2}\left[\left(\frac{\partial u_1}{\partial X_1}\right)^2 + \left(\frac{\partial u_2}{\partial X_1}\right)^2 + \left(\frac{\partial u_3}{\partial X_1}\right)^2\right]$$

and

$$E_{13} = \frac{1}{2}\left(\frac{\partial u_1}{\partial X_3} + \frac{\partial u_3}{\partial X_1}\right) + \frac{1}{2}\left[\frac{\partial^2 u_1}{\partial X_1 \partial X_3} + \frac{\partial^2 u_2}{\partial X_1 \partial X_3} + \frac{\partial^2 u_3}{\partial X_1 \partial X_3}\right]$$

9.3 EULERIAN FINITE STRAIN TENSOR

If we wish to reference our analysis of the deformation to the present deformed state, then we'll use the *Eulerian finite strain tensor*. Its form is quite similar to that just presented and we do not go through the derivation:

$$\bar{E}_{ij} = \frac{1}{2}\left[\frac{\partial u_i}{\partial x_j} + \frac{\partial u_j}{\partial x_i} - \frac{\partial u_k}{\partial x_i}\frac{\partial u_k}{\partial x_j}\right] \tag{9.5}$$

Notice the difference in the sign of the last term in Equations 9.4 and 9.5. Basically, you can see that, when we go from infinitesimal to finite strain, we are going from linear partial differential equations to non-linear partial differentials.

9.4 DERIVATION OF THE GREEN DEFORMATION TENSOR

In this case, we are just interested in finding the new squared length of the vector, $|P'Q'|^2$, in terms of the old length and the orientation of the vector. As before,

$$|P'Q'|^2 = dx_i\, dx_i = dx_1^2 + dx_3^2 \quad \text{and} \quad dx_i = \frac{\partial x_i}{\partial X_j} dX_j = F_{ij}\, dX_j$$

Substituting and expanding, we get

$$|P'Q'|^2 = (F_{11}\, dX_1 + F_{12}\, dX_2)^2 + (F_{21}\, dX_1 + F_{22}\, dX_2)^2$$

Expanding as before:

$$|P'Q'|^2 = \left[F_{11}^2 + F_{21}^2\right]dX_1^2 + \left[F_{22}^2 + F_{12}^2\right]dX_3^2 + [F_{11}F_{12} + F_{21}F_{22}]\,2dX_1\,dX_3$$

We can write this in simplified terms (and in three dimensions) as

$$|P'Q'|^2 = dX_i\,C_{ij}\,dX_j$$

This is the same equation as Equation 22.4 in Means (1976). C_{ij} is the *Green deformation tensor*, and it has the form

$$C_{ij} = \frac{\partial x_k}{\partial X_i}\frac{\partial x_k}{\partial X_j} = F_{ki}F_{kj} \tag{9.6}$$

Without derivation (which is very similar to what we have just suffered through), the *Cauchy deformation tensor* is

$$|PQ|^2 = dx_i\,\bar{C}_{ij}\,dx_j \quad \text{where} \quad \bar{C}_{ij} = \frac{\partial X_k}{\partial x_i}\frac{\partial X_k}{\partial x_j} \tag{9.7}$$

9.5 RELATIONS BETWEEN THE FINITE STRAIN AND DEFORMATION TENSORS

From the previous derivations, we have

$$|P'Q'|^2 - |PQ|^2 = 2dX_i\,E_{ij}\,dX_j \quad \text{and} \quad |P'Q'|^2 = dX_i\,C_{ij}\,dX_j$$

Therefore,

$$|PQ|^2 = dX_i\,dX_i = dX_i\,C_{ij}\,dX_j - 2dX_i\,E_{ij}\,dX_j$$

We can simplify this further using the substitution property of the Kronecker delta:

$$dX_i = \delta_{ij}\,dX_j$$

So,

$$dX_i\,\delta_{ij}\,dX_j = (C_{ij} - 2E_{ij})\,dX_i\,dX_j$$

The dX's cancel out and we have

$$E_{ij} = \frac{1}{2}(C_{ij} - \delta_{ij}) \quad \text{and} \quad C_{ij} = 2E_{ij} + \delta_{ij} \tag{9.8}$$

Likewise, we can write the relationship between the two tensors referred to the final state:

$$\bar{E}_{ij} = \frac{1}{2}(\delta_{ij} - \bar{C}_{ij}) \tag{9.9}$$

Thus, the strain tensors do not contain any more information than the deformation tensors and vice versa. If you carefully inspect all of these equations, you will see that they are all symmetric tensors. Thus, these tensors can all be represented by Mohr circles, they all have invariants, principal axes, etc.

From the above two equations, it is clear that:

- E_{ij} and C_{ij} have the same principal axes, and
- $\bar{E}_{ij}$ and $\bar{C}_{ij}$ have the same principal axes.

This is because, if the off-diagonal elements of one tensor are zero, the off-diagonal elements of the other have to be zero. Note, however, that E_{ij} and $\bar{E}_{ij}$ do not have the same principal axes. The difference in orientation of the principal axes between the initial and final state, as we will see, is defined as the *rotation tensor*, **R**.

9.6 RELATIONS TO THE DEFORMATION GRADIENT, F

Recall that the deformation gradient, **F**, is

$$d\mathbf{x} = \mathbf{F} \cdot d\mathbf{X} \quad \text{or} \quad dx_i = F_{ij}\,dX_j \quad \text{where} \quad F_{ij} = \frac{\partial x_i}{\partial X_j}. \tag{9.10}$$

We can also write

$$d\mathbf{x} = d\mathbf{X} \cdot \mathbf{F}^T \quad \text{or} \quad dx_i = dX_j F_{ji} \quad \text{where} \quad F_{ji} = \frac{\partial x_j}{\partial X_i} \tag{9.11}$$

The dot, $\cdot$, in the above equations represents *Gibbs dyadic notation*, which is different than standard matrix multiplication. Recall that the tensor, or dyad, product of two vectors results when you multiply a row vector times a column vector. In terms of matrix multiplication, while Equation 9.10 is equivalent to multiplying a 3×3 matrix times a 3×1 column vector, Equation 9.11 represents multiplying a 1×3 row vector times a 3×3 matrix. In the above case, we can expand 9.10 as

$$\begin{bmatrix} x_1 \\ x_2 \\ x_3 \end{bmatrix} = \begin{bmatrix} F_{11} & F_{12} & F_{13} \\ F_{21} & F_{22} & F_{23} \\ F_{31} & F_{32} & F_{33} \end{bmatrix} \begin{bmatrix} X_1 \\ X_2 \\ X_3 \end{bmatrix} \quad \text{or} \quad \begin{aligned} x_1 &= F_{11}X_1 + F_{12}X_2 + F_{13}X_3 \\ x_2 &= F_{21}X_1 + F_{22}X_2 + F_{23}X_3 \\ x_3 &= F_{31}X_1 + F_{32}X_2 + F_{33}X_3 \end{aligned}$$

Likewise, 9.11 is expanded as

$$\begin{bmatrix} x_1 & x_2 & x_3 \end{bmatrix} = \begin{bmatrix} X_1 & X_2 & X_3 \end{bmatrix} \begin{bmatrix} F_{11} & F_{21} & F_{31} \\ F_{12} & F_{22} & F_{32} \\ F_{13} & F_{23} & F_{33} \end{bmatrix}$$

or

$$\begin{aligned} x_1 &= X_1 F_{11} + X_2 F_{12} + X_3 F_{13} \\ x_2 &= X_1 F_{21} + X_2 F_{22} + X_3 F_{23} \\ x_3 &= X_1 F_{31} + X_2 F_{32} + X_3 F_{33} \end{aligned}$$

Clearly the two are the same.

We have just seen that

$$|P'Q'|^2 = (ds)^2 = d\mathbf{x}\,d\mathbf{x} = \left(d\mathbf{X} \cdot \mathbf{F}^T\right)(\mathbf{F} \cdot d\mathbf{X}) = d\mathbf{X}\left(\mathbf{F}^T \cdot \mathbf{F}\right)d\mathbf{X}$$

So

$$\mathbf{C} = \mathbf{F}^T \cdot \mathbf{F} \quad \text{or} \quad C_{ij} = \frac{\partial x_k}{\partial X_i}\frac{\partial x_k}{\partial X_j} \tag{9.12}$$

And without proof,

$$\bar{\mathbf{C}} = \mathbf{B}^{-1} = \left(\mathbf{F}^{-1}\right)^T \cdot \mathbf{F}^{-1} \quad \text{or} \quad \bar{C}_{km} = \frac{\partial X_j}{\partial x_k}\frac{\partial X_j}{\partial x_m}$$

B is yet another finite strain tensor called either the left Cauchy-Green tensor or Finger's tensor. By substituting into Equations 9.8 and 9.9, we can also derive:

$$\mathbf{E} = \frac{1}{2}\left(\mathbf{F}^T \cdot \mathbf{F} - \mathbf{1}\right)$$

and

$$\bar{\mathbf{E}} = \frac{1}{2}\left(\mathbf{1} - \left(\mathbf{F}^{-1}\right)^T \cdot \mathbf{F}^{-1}\right)$$

9.7 PRACTICAL MEASURES OF STRAIN

9.7.1 Stretch and quadratic elongation

Now, we want to relate our tensor description of finite strain to the well-known scalar measures such as the stretch, elongation, quadratic elongation, and angular shear. Let

$$dS = \left|\overrightarrow{PQ}\right| \quad \text{and} \quad ds = \left|\overrightarrow{P'Q'}\right|$$

where dS and ds represent the scalar magnitudes of the lines in the undeformed and deformed states, respectively. The square of the stretch, then, is just

$$S^2 = \frac{(ds)^2}{(dS)^2} = \frac{dX_i}{dS} C_{ij} \frac{dX_j}{dS} = \lambda$$

where λ is the quadratic elongation. Expanding this equation, we get

$$\begin{aligned}S^2 &= \frac{dX_1}{dS} C_{11} \frac{dX_1}{dS} + \frac{dX_1}{dS} C_{12} \frac{dX_2}{dS} + \frac{dX_1}{dS} C_{13} \frac{dX_3}{dS} \\ &+ \frac{dX_2}{dS} C_{21} \frac{dX_1}{dS} + \frac{dX_2}{dS} C_{22} \frac{dX_2}{dS} + \frac{dX_2}{dS} C_{23} \frac{dX_3}{dS} \\ &+ \frac{dX_3}{dS} C_{31} \frac{dX_1}{dS} + \frac{dX_3}{dS} C_{32} \frac{dX_2}{dS} + \frac{dX_3}{dS} C_{33} \frac{dX_3}{dS}\end{aligned} \quad (9.13)$$

For a vector parallel to the X_1 axis,

$$dX_1 = (1, \ 0, \ 0) \quad \Rightarrow \quad dX_1 = dS \quad dX_2 = 0 \quad \text{and} \quad dX_3 = 0$$

By substituting these values into the previous equation, 9.13, you can see that, if λ is parallel to a coordinate axis (the X_1 axis in this case), then

$$\lambda_{(1)} = C_{11} = 1 + 2E_{11} \quad (9.14)$$

Without going into the expansion, you can see that the same expression with respect to the final state is

$$\frac{1}{S^2} = \frac{1}{\lambda} = \frac{dx_i}{ds} \bar{C}_{ij} \frac{dx_j}{ds} \quad (9.15)$$

and the stretch parallel to the x_1 axis is

$$\frac{1}{S^2_{(1)}} = \frac{1}{\lambda_{(1)}} = \bar{C}_{11} = 1 - 2\bar{E}_{11} \quad (9.16)$$

9.7.2 Elongation

The elongation, e, is just the stretch minus 1. So,

$$e = \frac{ds - dS}{ds} = \left(\frac{dX_i}{dS} C_{ij} \frac{dX_j}{dS}\right)^{\frac{1}{2}} - 1 \quad (9.17)$$

and for an element parallel to the X_1 axis:

$$E_{(1)} = \sqrt{C_{11}} - 1 = \sqrt{1 + 2E_{11}} - 1$$

9.7 Practical measures of strain

Rearranging this last equation, we get

$$E_{11} = E_{(1)} + \frac{1}{2}E_{(1)}^2$$

In infinitesimal strain, we ignore the final, quadratic term of this equation.

9.7.3 Volume ratio

You can calculate the volume ratio as follows:

$$\frac{dv}{dV} = S_1 S_2 S_3 = \sqrt{C_1 C_2 C_3} = \sqrt{(1+2E_1)(1+2E_2)(1+2E_3)} = \sqrt{III_C} \qquad (9.18)$$

Where S_1, S_2, and S_3 are the principal stretches, and III_C is the third invariant of tensor C_{ij}.

9.7.4 Angle between two lines

In structural geology, the shear strain is defined as the tangent of the *change* in angle of two originally perpendicular lines (e.g., Fig. 8.6a). The key to solving this problem is to remember that the dot product of two vectors is related to the angle between them:

$$\cos \theta = \frac{u_i v_j}{|u||v|}$$

This problem is easier to do in matrix notation. First, we define the deformation gradient, **F**, as in Equation 9.10:

$$d\mathbf{x} = \mathbf{F} \cdot d\mathbf{X} = d\mathbf{X} \cdot \mathbf{F}^T \quad \text{where} \quad \mathbf{F} = F_{ij} = \frac{\partial x_i}{\partial X_j}$$

$$(ds)^2 = d\mathbf{x}\, d\mathbf{x} = d\mathbf{X} \cdot \mathbf{F}^T \cdot \mathbf{F} \cdot d\mathbf{X} = d\mathbf{X} \cdot \mathbf{C} \cdot d\mathbf{X} \quad \text{where} \quad \mathbf{C} = \mathbf{F}^T \cdot \mathbf{F}$$

$$\cos \theta_f = \frac{d\mathbf{x}_{(1)} \cdot d\mathbf{x}_{(2)}}{|d\mathbf{x}_{(1)}||d\mathbf{x}_{(2)}|} = \frac{(d\mathbf{X}_{(1)} \cdot \mathbf{F}^T)(\mathbf{F} \cdot d\mathbf{X}_{(2)})}{(d\mathbf{X}_{(1)} \mathbf{C} d\mathbf{X}_{(1)})^{\frac{1}{2}}(d\mathbf{X}_{(2)} \mathbf{C} d\mathbf{X}_{(2)})^{\frac{1}{2}}}$$

where θ_f is the angle between the two lines in the final state. If the lines were *unit vectors* in the material state, then

$$\cos \theta_f = \frac{d\hat{\mathbf{X}}_{(1)} \cdot \mathbf{C} \cdot d\hat{\mathbf{X}}_{(2)}}{S_{(1)} S_{(2)}} = \frac{dX_{(1)_i} C_{ij} dX_{(2)_j}}{S_{(1)} S_{(2)}}$$

Expanding this last equation:

$$\cos \theta_f = \frac{1}{S_{(1)} S_{(2)}} \left\{ \begin{array}{l} dX_{(1)_1} C_{11}\, dX_{(2)_1} + dX_{(1)_1} C_{12}\, dX_{(2)_2} + dX_{(1)_1} C_{13}\, dX_{(2)_3} \\ + dX_{(1)_2} C_{21}\, dX_{(2)_1} + dX_{(1)_2} C_{22}\, dX_{(2)_2} + dX_{(1)_2} C_{23}\, dX_{(2)_3} \\ + dX_{(1)_3} C_{31}\, dX_{(2)_1} + dX_{(1)_3} C_{32}\, dX_{(2)_2} + dX_{(1)_3} C_{33}\, dX_{(2)_3} \end{array} \right\}$$

The angle of the line in the initial state, θ_i, is just

$$\cos \theta_i = d\hat{\mathbf{X}}_{(1)} \cdot d\hat{\mathbf{X}}_{(2)} = \left\{ dX_{(1)_1} dX_{(2)_1} + dX_{(1)_2} dX_{(2)_2} + dX_{(1)_3} dX_{(2)_3} \right\}$$

To take a simple case, let's look at the change in the angle of two initially perpendicular lines that start out parallel to the X_1 and X_2 axes:

$$d\hat{\mathbf{X}}_{(1)} = (1,\ 0,\ 0) \quad \text{and} \quad d\hat{\mathbf{X}}_{(2)} = (0,\ 1,\ 0)$$

The initial angle between them is 90°, so

$$\cos \theta_i = \cos 90 = 0$$

The final angle is

$$\cos\theta_f = \frac{C_{12}}{S_{(1)}S_{(2)}} = \frac{C_{12}}{(C_{11}C_{22})^{\frac{1}{2}}} = \frac{2E_{12}}{[(1+2E_{11})(1+2E_{22})]^{\frac{1}{2}}} \qquad (9.19)$$

The MATLAB® function **FinStrain**, below, summarizes all the finite strain concepts we have discussed so far. **FinStrain** computes the Lagrangian (frame = 0) or Eulerian (frame = 1) strain tensor from an input Lagrangian or Eulerian displacement gradient tensor. Besides this, the function returns practical measures of strain such as principal elongations, dilatation, and magnitude and orientation of maximum shear strain.

```
function [eps,pstrains,dilat,maxsh] = FinStrain(e,frame)
%FinStrain computes finite strain from an input displacement
%gradient tensor
%
% [eps,pstrains,dilat,maxsh] = FinStrain(e,frame)
%
% e = 3 x 3 Lagrangian or Eulerian displacement gradient tensor
% frame = Reference frame. Enter 0 for undeformed (Lagrangian) state, or
%         1 for deformed (Eulerian) state
% eps = 3 x 3 Lagrangian or Eulerian strain tensor
% pstrains = 3 x 3 matrix with magnitude (column 1), trend (column 2) and
%            plunge (column 3) of maximum (row 1), intermediate (row 2),
%            and minimum (row 3) elongations
% dilat = dilatation
% maxsh = 1 x 2 vector with max. shear strain and orientation with
%         respect to maximum principal strain direction. Only valid in 2D
%
% NOTE: Output angles are in radians
%
% FinStrain uses function CartToSph

%Initialize variables
eps = zeros(3,3);
pstrains = zeros(3,3);
maxsh = zeros(1,2);
%Compute strain tensor (Eqs. 9.4 and 9.5)
for i=1:3
    for j=1:3
        eps(i,j)=0.5*(e(i,j)+e(j,i));
        for k=1:3
            %If undeformed reference frame: Lagrangian strain tensor
            if frame == 0
                eps(i,j) = eps(i,j) + 0.5*(e(k,i)*e (k,j));
            %If deformed reference frame: Eulerian strain tensor
            elseif frame == 1
                eps(i,j) = eps(i,j) - 0.5*(e(k,i)*e (k,j));
            end
        end
    end
end
```

```
%Compute principal elongations and orientations. Here we use the MATLAB
%function eig
[V,D] = eig(eps);

%Principal elongations
for i=1:3
    ind = 4-i;
    %Magnitude
    %If undeformed reference frame: Lagrangian strain tensor (Eq. 9.14)
    if frame == 0
        pstrains(i,1) = sqrt(1.0+2.0*D(ind,ind))-1.0;
    %If deformed reference frame: Eulerian strain tensor (Eq. 9.16)
    elseif frame == 1
        pstrains(i,1) = sqrt(1.0/(1.0-2.0*D(ind, ind)))-1.0;
    end
    %Orientations
    [pstrains(i,2),pstrains(i,3)] = CartToSph(V(1, ind),V(2,ind),V(3,ind));
end

%dilatation (Eq. 9.18)
dilat = (1.0+pstrains(1,1))*(1.0+pstrains(2,1))*(1.0+pstrains(3,1)) - 1.0;

%Maximum shear strain: This only works if plane strain
lmax = (1.0+pstrains(1,1))^2; %Maximum quadratic elongation
lmin = (1.0+pstrains(3,1))^2; %Minimum quadratic elongation
%Maximum shear strain: Ragan (1967) Eq. 3.46
maxsh(1,1) = (lmax-lmin)/(2.0*sqrt(lmax*lmin));
%Angle of maximum shear strain with respect to maximum principal strain
%Ragan (1967) Eq. 3.45
%If undeformed reference frame
if frame == 0
    maxsh(1,2) = pi/4.0;
%If deformed reference frame
elseif frame == 1
    maxsh(1,2) = atan(sqrt(lmin/lmax));
end
end
```

9.8 THE ROTATION AND STRETCH TENSORS

With finite strain, we can no longer decompose the displacement gradient into the sum of a symmetric strain tensor and an antisymmetric rotation tensor as we did for infinitesimal strain. But, you can have a multiplicative (or polar) decomposition of the deformation gradient, $\mathbf{F}$, into the product of two tensors:

$$\mathbf{F} = \mathbf{R} \cdot \mathbf{U} = \mathbf{V} \cdot \mathbf{R} \tag{9.20}$$

$$d\mathbf{x} = (\mathbf{R} \cdot \mathbf{U}) \cdot d\mathbf{X} = (\mathbf{V} \cdot \mathbf{R}) \cdot d\mathbf{X} \tag{9.21a}$$

Figure 9.2 The rotation tensor, R_{ij}, that relates the eigenvectors of the principal axes of the finite deformation tensor (or the strain tensor) in the deformed state, $\hat{n}_1$, $\hat{n}_2$, and $\hat{n}_3$, to those of the equivalent deformation (or strain) tensor in the undeformed state, $\hat{N}_1$, $\hat{N}_2$, and $\hat{N}_3$. The components of R_{ij} are the direction cosines of the nine angles between new and old eigenvectors. Note that neither **n** nor **N** represent a coordinate system.

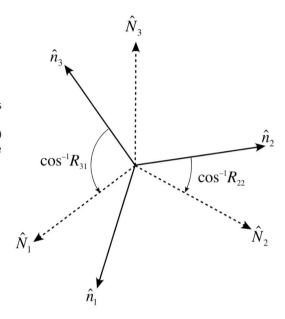

In summation notation:

$$dx_i = R_{ik} U_{kj}\, dX_j = V_{ik} R_{kj}\, dX_j \tag{9.21b}$$

R is the orthogonal rotation tensor that defines the rotation of the principal axes. Basically, **R** rotates the principal axes of **C** in the initial state (**X**) into the principal axes of $\bar{\mathbf{C}}$ or $\mathbf{B}^{-1}$ in the final state (**x**). **U** is known as the *right stretch tensor* and **V** is called the *left stretch tensor*, both are symmetric tensors. As it turns out, they are rather simply related to the Green deformation tensor:

$$\mathbf{U} = \mathbf{C}^{\frac{1}{2}} = \left(\mathbf{F}^T \cdot \mathbf{F}\right)^{\frac{1}{2}} \quad \text{and} \quad \mathbf{V} = \left(\mathbf{C}^{-1}\right)^{\frac{1}{2}} = \left(\mathbf{F} \cdot \mathbf{F}^T\right)^{\frac{1}{2}} \tag{9.22}$$

The rotation tensor, **R**, gives the difference between the initial and final orientations of the principal axes, as shown in Figure 9.2. $\hat{N}_1 \geq \hat{N}_2 \geq \hat{N}_3$ are the eigenvectors of the principal axes in the initial state and $n_1 \geq n_2 \geq n_3$ the eigenvectors of the principal axes in the deformed or the final state. Thus, we can write

$$\hat{\mathbf{n}}_\alpha = \mathbf{R} \cdot \hat{\mathbf{N}}_\alpha \quad \text{or} \quad n_{\alpha i} = R_{ij} N_{\alpha j} \tag{9.23a}$$

where α is the index of the principal axis, not a summation counter like i and j. Or if we know **n** and **N**, we can calculate **R**:

$$R_{ij} = n_{\alpha i} N_{\alpha j} \tag{9.23b}$$

This may look, superficially at least, like a rotation of axes or a tensor transformation but it is not. *Our reference axes, which are not shown in the above diagram, do not change.* In general, neither the initial nor the final orientations of the principal axes will be parallel to the axes of the coordinate system. Nonetheless, the rotation tensor, **R**, is an orthogonal matrix like the transformation matrix, **a**, that we saw earlier and it works in much the same way. The nine components of **R** are the direction cosines of the angles between the axes $\hat{N}_\alpha$ and $\hat{n}_\alpha$.

In infinitesimal strain, the order in which the rotation and the strain occur does not matter, so we can write

$$e_{ij} = \varepsilon_{ij} + \omega_{ij} = \omega_{ij} + \varepsilon_{ij}$$

9.8 The rotation and stretch tensors

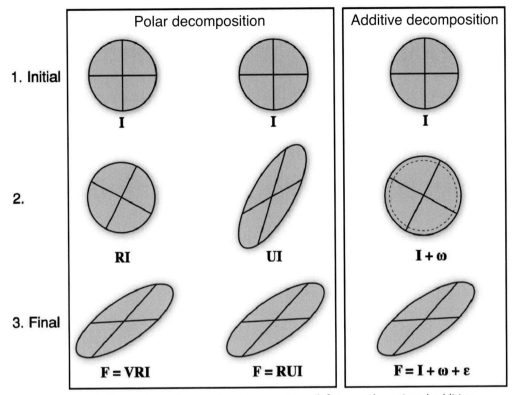

Figure 9.3 Comparison of the polar decomposition (left two columns) and additive decomposition (right column) of the deformation gradient tensor, **F**. Because **F** is the same for all three cases, both the initial and final for those cases must also be the same; only the middle state, 2, differs. **I** is the identity matrix, which represents the initial state; **V** is the left stretch tensor; **U** is the right stretch tensor. In the additive decomposition case, note 10% dilation of circle in intermediate state. Figure is modified from Cladouhos and Allmendinger (1993).

In finite strain, the order is important. As described by Malvern (1969), the same final deformation can be represented by (Fig. 9.3):

1. a stretch defined by **U**,
2. a rigid body rotation, **R**, and
3. a translation.

or,

1. a translation,
2. a rigid body rotation, **R**, and
3. a stretch defined by **V**.

The left stretch tensor, **V**, defines the strain of the deformed region in the deformed state and is most commonly preferred for geological analysis. Because U describes the strain in a state that the geologist never sees (i.e., the initial state), it is not particularly useful (as we will see below for the Mohr circle for finite strain in the deformed state).

9.9 MULTIPLE DEFORMATIONS

In infinitesimal strain, we saw that the displacement gradient tensor, **e**, of each increment of deformation – due to superposed deformations such as movement of several faults – could be added together to provide a picture of the total strain:

$$^{total}\mathbf{e} = \sum {}^{n}\mathbf{e} = {}^{1}\mathbf{e} + {}^{2}\mathbf{e} + \cdots + {}^{n}\mathbf{e} \tag{9.24}$$

This is no longer true for the case of finite strain. As before, we start with the deformation gradient tensor, **F**. We start with the first deformation, indicated by the leading superscript 1:

$$^{1}d\mathbf{x} = {}^{1}\mathbf{F} \cdot d\mathbf{X}$$

If we now superimpose a second deformation ($^{2}\mathbf{F}$), the deformed state for the first deformation ($^{1}\mathbf{F}$) becomes the starting state for the second deformation. That is,

$$^{2}d\mathbf{X} = {}^{1}d\mathbf{x} = {}^{1}\mathbf{F} \cdot d\mathbf{X}$$

and now

$$^{2}d\mathbf{x} = {}^{2}\mathbf{F} \cdot {}^{2}d\mathbf{X} = {}^{2}\mathbf{F} \cdot ({}^{1}\mathbf{F} \cdot d\mathbf{X}) = {}^{2}\mathbf{F} \cdot {}^{1}\mathbf{F} \cdot d\mathbf{X} \tag{9.25}$$

We can compare this to the infinitesimal strain formulation in Equation 9.24 by recalling the relation between the displacement and deformation gradient tensors:

$$\mathbf{e} \equiv \mathbf{F} - \mathbf{I} = F_{ij} - \delta_{ij}$$

Writing the above equation in terms of **e**:

$$^{2}\mathbf{F} \cdot {}^{1}\mathbf{F} = ({}^{2}\mathbf{e} + \mathbf{I})({}^{1}\mathbf{e} + \mathbf{I}) = {}^{1}\mathbf{e} + {}^{2}\mathbf{e} + {}^{2}\mathbf{e} \cdot {}^{1}\mathbf{e} + \mathbf{I} \tag{9.26}$$

You can see that this is equivalent to the summation of the displacement gradient tensors *except* for the higher order term, $^{2}\mathbf{e} \cdot {}^{1}\mathbf{e}$. Because matrix multiplication is non-commutative, in general $^{2}\mathbf{e} \cdot {}^{1}\mathbf{e} \neq {}^{1}\mathbf{e} \cdot {}^{2}\mathbf{e}$. *Therefore, for finite strains you must know the order in which the deformations occur.* We will make extensive use of equations like 9.26 in the next chapter. If you are applying finite strain to the analysis of fault data (either a measured fault population or faults in a thrust belt), you must know the order in which every single fault formed (Cladouhos and Allmendinger, 1993). Although this is feasible for larger faults in a thrust belt, it is virtually impossible for fault slip data.

9.10 MOHR CIRCLE FOR FINITE STRAIN

Because we most often want to use the Mohr circle to learn about the deformed state (that is, we usually want to determine the orientation of the principal axes when we know how three randomly orientated lines have been deformed, etc.), we'll use one of the tensors referred to the spatial coordinates (i.e., the present-day coordinates). The one most commonly used is the Cauchy deformation tensor:

$$\bar{C}_{ij} = \begin{bmatrix} \bar{C}_1 & 0 & 0 \\ 0 & \bar{C}_2 & 0 \\ 0 & 0 & \bar{C}_3 \end{bmatrix} \quad \text{and} \quad \text{the transformation matrix } a_{ij} = \begin{pmatrix} \cos\theta & 0 & \sin\theta \\ 0 & 1 & 0 \\ -\sin\theta & 0 & \cos\theta \end{pmatrix}.$$

The tensor transformation equation is

$$\bar{C}'_{ij} = a_{ik} a_{jl} \bar{C}_{kl} \tag{9.27}$$

9.10 Mohr circle for finite strain

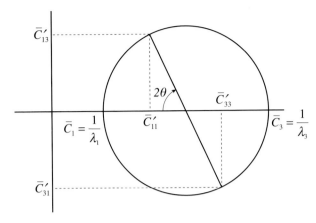

Figure 9.4 Mohr circle for finite strain in the deformed state

so

$$\bar{C}'_{ij} = \begin{bmatrix} \bar{C}'_{11} & 0 & \bar{C}'_{13} \\ 0 & \bar{C}_2 & 0 \\ \bar{C}'_{31} & 0 & \bar{C}'_{33} \end{bmatrix} = \begin{bmatrix} \left(\bar{C}_1 \cos^2\theta + \bar{C}_3 \sin^2\theta\right) & 0 & ((\bar{C}_3 - \bar{C}_1)\cos\theta\sin\theta) \\ 0 & \bar{C}_2 & 0 \\ ((\bar{C}_1 - \bar{C}_3)\cos\theta\sin\theta) & 0 & \left(\bar{C}_1 \sin^2\theta + \bar{C}_3 \cos^2\theta\right) \end{bmatrix}$$

From this we get the familiar equations for the Mohr circle (Fig. 9.4):

$$\bar{C}'_{11} = \frac{(\bar{C}_1 + \bar{C}_3)}{2} + \frac{(\bar{C}_1 - \bar{C}_3)}{2}\cos 2\theta \quad \text{and} \quad \bar{C}'_{13} = \frac{(\bar{C}_1 - \bar{C}_3)}{2}\sin 2\theta \tag{9.28}$$

Remember that we derived an equation for the stretch of lines in the final reference state:

$$\frac{1}{S^2_{(1)}} = \frac{1}{\lambda_{(1)}} = \frac{dx_i}{ds}\bar{C}_{ij}\frac{dx_j}{ds}$$

and, when a line is parallel to the x_1 axis of the coordinate system:

$$\frac{1}{S^2_{(1)}} = \frac{1}{\lambda_{(1)}} = \bar{C}_{11}$$

So, substituting in the above equation and using the reciprocal quadratic elongation ($\lambda' = 1/\lambda$), we get

$$\lambda' = \frac{(\lambda'_1 + \lambda'_3)}{2} + \frac{(\lambda'_1 - \lambda'_3)}{2}\cos 2\theta \tag{9.29a}$$

Many of you will recognize Equation 9.29a as one of the two equations for the Mohr circle for finite strain (referred to the deformed state). The other equation is

$$\gamma' = \frac{\gamma}{\lambda} = \frac{(\lambda'_1 - \lambda'_3)}{2}\sin 2\theta. \tag{9.29b}$$

Thus, you can see that the component $\bar{C}_{ij}$ (where $i \neq j$) of the Cauchy deformation tensor is equal to γ/λ.

There is one particularly useful property of the Mohr circle for strain that is well worth a mention here: the concept of the *pole to the Mohr circle*. Mohr circle constructions can be confusing to first time users because it is difficult to relate physical orientation to the points on the circle. The Mohr diagram for finite strain (Fig. 9.5) contains a unique point on the Mohr circle (P) such that all lines that connect P to points on the circle are parallel to the orientation of longitudinal strain represented by those lines in the physical plane (Allison, 1984; Cutler and

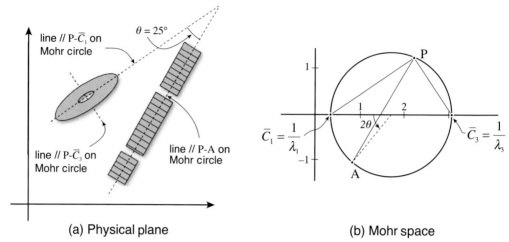

Figure 9.5 Illustration of the pole to the Mohr circle for the case of finite strain in the deformed state. (a) Cross section (left) and longitudinal section (right) of deformed crinoid stems that lie in a plane perpendicular to $\bar{C}_2$. (b) Mohr circle for finite strain. P is the pole to the Mohr circle; lines from the pole to the principal axes are parallel to the principal axes' orientation in physical space.

Elliott, 1983; Ragan, 2009). This is more easily visualized with an illustration. Figure 9.5a shows two crinoid stems – one perpendicular to the section (the ellipse) that gives the orientation and ratios of the principal stretches and a second parallel to the cross section that gives a longitudinal stretch of 1.12 in an orientation that is 25° from the orientation of maximum stretch (point **A**). A Mohr circle (Fig. 9.5b) is constructed based on the ratios of the principal stretches. Lines from the points representing the principal strains and parallel to the principal strain orientations (major and minor axes of the elliptical crinoid section) are traced. These lines intersect at point **P**, which is the pole of the Mohr circle (Fig. 9.5b). From point **P**, a line parallel to the crinoid stem in the plane of the section is drawn so that it intersects the circle at point **A**. This point represents the state of strain of the crinoid stem. All lines that contain **P** intersect the circle at the longitudinal (λ') and shear over longitudinal (γ/λ) strain values of those lines in the physical plane.

As we discussed in Chapter 6, the pole to the Mohr circle can also be used to describe the relationship between the Mohr diagram for stress and the planes on which tractions act in the physical plane (Mandl and Shippam, 1981). In fact, there is a pole for a Mohr circle construction of any second rank tensor, and the properties of the pole are invaluable for quickly relating points on the Mohr circle to the orientations of those attributes in the physical plane.

9.11 COMPATIBILITY EQUATIONS

The equations, such as the infinitesimal strain tensor,

$$\varepsilon_{ij} = \frac{1}{2}\left(\frac{\partial u_i}{\partial X_j} + \frac{\partial u_j}{\partial X_i}\right) \tag{9.30}$$

that we have developed in our understanding of strain work very well when we know the displacements and we want to calculate strain. Any set of displacements, u_i, that you choose will result in strain. However, the reverse is not so easy: How do we know if any particular strain

9.11 Compatibility equations

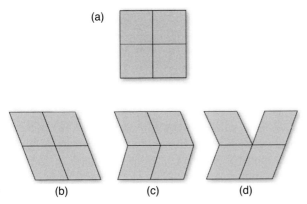

Figure 9.6 Illustration of different types of finite strain. (a) Initial geometry; (b) homogeneous, continuous deformation; (c) heterogeneous, continuous deformation; and (d) heterogeneous, discontinuous deformation caused by lack of strain compatibility which has produced the gap between the upper two blocks.

we specify is the plausible result of a real set of displacements? The expression for the infinitesimal strain tensor, above, represents six partial differential equations, but there are only three u_i. So, you could easily specify ε_{ij} such that there are no real values of u_i that would satisfy all six equations.

In structural geology, we are introduced qualitatively to the idea of strain compatibility: All the pieces must fit together without any gaps or overlaps (Fig. 9.6). The deformation in Figure 9.6d is not compatible because a gap has opened up between the two top blocks. Mathematically, we need to find some condition such that Equation 9.30 can be integrated and there exist a continuous, single valued set of displacements across the volume (in three dimensions). Such conditions are known as Saint-Venant's compatibility equations.

Note that we are only dealing with the symmetric part of the displacement gradient tensor here because rigid body translations and rotations do not affect compatibility. Also, although we show the equations for infinitesimal strain, compatibility also applies to finite strain. To see how the compatibility equations are derived, see Malvern (1969, p. 185) or any other good continuum mechanics textbook. All six equations are repeated here:

$$-S_{33} \equiv \frac{\partial^2 \varepsilon_{11}}{\partial X_2^2} + \frac{\partial^2 \varepsilon_{22}}{\partial X_1^2} - 2\frac{\partial^2 \varepsilon_{12}}{\partial X_1 \partial X_2} = 0 \tag{9.31a}$$

$$-S_{11} \equiv \frac{\partial^2 \varepsilon_{22}}{\partial X_3^2} + \frac{\partial^2 \varepsilon_{33}}{\partial X_2^2} - 2\frac{\partial^2 \varepsilon_{23}}{\partial X_2 \partial X_3} = 0 \tag{9.31b}$$

$$-S_{22} \equiv \frac{\partial^2 \varepsilon_{33}}{\partial X_1^2} + \frac{\partial^2 \varepsilon_{11}}{\partial X_3^2} - 2\frac{\partial^2 \varepsilon_{31}}{\partial X_3 \partial X_1} = 0 \tag{9.31c}$$

$$-S_{23} \equiv -\frac{\partial^2 \varepsilon_{11}}{\partial X_2 \partial X_3} + \frac{\partial}{\partial X_1}\left(-\frac{\partial \varepsilon_{23}}{\partial X_1} + \frac{\partial \varepsilon_{31}}{\partial X_2} + \frac{\partial \varepsilon_{12}}{\partial X_3}\right) = 0 \tag{9.31d}$$

$$-S_{31} \equiv -\frac{\partial^2 \varepsilon_{22}}{\partial X_3 \partial X_1} + \frac{\partial}{\partial X_2}\left(\frac{\partial \varepsilon_{23}}{\partial X_1} - \frac{\partial \varepsilon_{31}}{\partial X_2} + \frac{\partial \varepsilon_{12}}{\partial X_3}\right) = 0 \tag{9.31e}$$

$$-S_{12} \equiv -\frac{\partial^2 \varepsilon_{33}}{\partial X_1 \partial X_2} + \frac{\partial}{\partial X_3}\left(\frac{\partial \varepsilon_{23}}{\partial X_1} + \frac{\partial \varepsilon_{31}}{\partial X_2} - \frac{\partial \varepsilon_{12}}{\partial X_3}\right) = 0 \tag{9.31f}$$

If you are only dealing with horizontal, two-dimensional or plane strain note that only Equation 9.31a does not have a term related to the X_3 axis on the right-hand side. This is the equation that you would use, then, to make the horizontal strains compatible.

When might you need to use these equations? The answer is pretty much whenever you are combining strain data from a variety of different sources. A prominent example would be the world strain map (Kreemer *et al.*, 2003) which integrates deformation from GPS, earthquake focal mechanisms, Quaternary fault slip rates, and other sources to produce smoothed strain maps throughout the globe. Holt *et al.* (Holt *et al.*, 2000) describe how they incorporate Saint-Venant's equations in their calculations. We will return to the concept of strain compatibility in Chapter 11, where we use kinematic models to approach fault related folding.

9.12 EXERCISES

1. Two brachiopods on a bedding plane (Fig. 9.7) have experienced an angular shear (ψ) that has distorted the bilateral symmetry typical of undeformed specimens so that the hinge line is no longer parallel to the median line. Two others are perpendicular. (a) What is the ratio of maximum to minimum stretch in the bedding plane? (b) What is the area change in the bedding plane if the long axis of the lower brachiopod is a line of no finite elongation?
2. Figure 9.8 shows a bed of sandstone with a clastic dike that has propagated into the adjacent shale. Assume that the clastic dike was originally perpendicular to bedding and the cleavage, shown by the dashed lines, is the direction of maximum finite extension. A quartz vein has extended during deformation and broken into four sections in the shale. What are the magnitudes of maximum/minimum stretch and the area change in the shale layer.

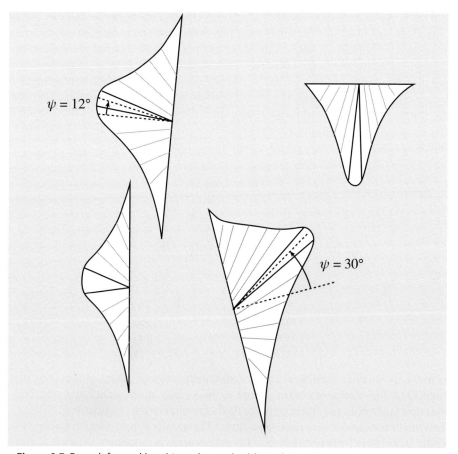

Figure 9.7 Four deformed brachiopods on a bedding plane.

9.12 Exercises   181

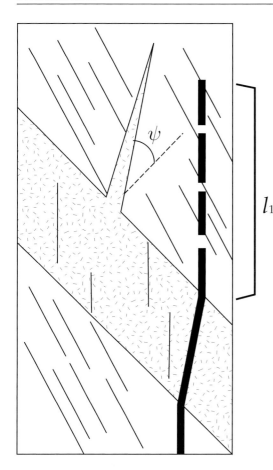

Figure 9.8 Sketch of a sand bed (stippled) with a weak vertical cleavage interbedded with shale that has a stronger, inclined cleavage. ψ is the angular shear of a sandstone dike, and l_1 is the final length of a boudinaged quartz vein.

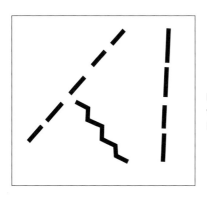

Figure 9.9 Three rutile needles that are embedded within a quartz grain (e.g., the Cambrian Weaverton Formation of the Appalachians; Mitra, 1978).

3. Figure 9.9 shows three rutile needles embedded within a quartz grain that has experienced intragranular strain (see Mitra, 1978). Measure the longitudinal strain of the three grains to determine the principal stretches and the orientation of the maximum stretching direction. Hint: Assume that the pole to the Mohr circle has an orientation and elongation defined by the needle on the right, with a stretch that lies between the other two needles (e.g., Lisle and Ragan, 1988).

4. In forward structural modeling, it is common to place markers such as regular grids, circles, etc. to track the evolution of deformation in a model. As we saw in Section 8.8.2, a minimum of four stations or displacement points are necessary to compute the strain in three dimensions. Write a computer program or MATLAB function to compute three-dimensional, finite strain from the initial and final coordinates of the nodes of a regular tetrahedron.
5. Modify function **GridStrain** such that it computes the two-dimensional finite strain of a network of displacement points or stations. Notice that in this case the reference frame is important. Do all your calculations in the deformed reference frame. Hint: Use function **FinStrain** instead of **InfStrain** in **GridStrain**, and modify **GridStrain** accordingly.

CHAPTER

TEN

Progressive strain histories and kinematics

10.1 FINITE VERSUS INCREMENTAL STRAIN

The limitation of finite strain methods is that they do not consider the kinematics, or the displacement paths of particles during deformation. As shown in Chapter 9, superposition of large deformations is not commutative; the sequence of events matters, and the finite strain can arise by an infinite number of paths. Moreover, when approaching the structure and tectonic history of a region, it is routine to ask questions that require knowledge of the kinematics such as: What is the sense and direction of shear in a fault zone? What is the strain history near plate boundaries? Or, what is the appropriate kinematic fold model for a specific structure? Kinematic analysis can in some cases provide critical tests against the predictions of geodynamic models for processes such as folding and mountain-building. In this chapter, we evaluate different types of progressive deformation. We provide examples of how a strain history can be quantified using geologic observations.

10.1.1 Progressive strain histories in two dimensions: Pure shear

Let's start with a simple example of a square that is deformed into a rectangle (Fig. 10.1). The intermediate principal stretch $S_2 = 1$, and the deformation can be completely described in two dimensions. If we ignore the translation component of deformation, the deformed position vector (**x**) can be written in terms of the initial position vector (**X**) by the equations

$$x_i = \frac{\partial x_i}{\partial X_j} X_j$$

or

$$\begin{aligned} x_1 &= S_1 X_1 \\ x_2 &= X_2 \\ x_3 &= S_3 X_3 \end{aligned} \tag{10.1}$$

Figure 10.1 Pure shear deformation of a square after a maximum stretch of 2.5, with the maximum incremental and finite stretch orientation parallel to the X_1 axis. The displacement path of points around the outside of the box is subdivided into 10 increments of strain. Large white and gray circles are initial and final positions, respectively. The eigenvectors of the displacement field are parallel to the principal axes of finite strain.

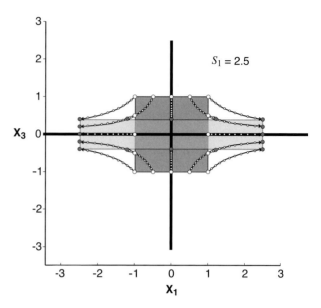

where S_1 and S_3 are the maximum and minimum principal stretches. The matrix, $\frac{\partial x_i}{\partial X_j}$, was introduced in Equation 7.9 as the deformation gradient tensor, $\mathbf{F}$. In this case, it forms a 3×3 matrix, $^{ps}\mathbf{F}$, such that

$$\begin{bmatrix} x_1 \\ x_2 \\ x_3 \end{bmatrix} = \begin{bmatrix} S_1 & 0 & 0 \\ 0 & 1 & 0 \\ 0 & 0 & S_3 \end{bmatrix} \begin{bmatrix} X_1 \\ X_2 \\ X_3 \end{bmatrix} \tag{10.2}$$

or

$$\mathbf{x} = {}^{ps}\mathbf{F} \cdot \mathbf{X} \tag{10.3}$$

In Figure 10.1, the finite strain has been decomposed into 10 infinitesimal strain increments, defined by incremental deformation gradient tensors that can be indexed so that the final position vector of an increment is the initial position vector for the next increment:

$$\begin{aligned} {}^1x_i &= {}^1F_{ij}{}^1X_j \\ {}^2X_i &= {}^1x_i \\ {}^2x_i &= {}^2F_{ij}{}^2X_j \\ {}^3X_i &= {}^2x_i \\ {}^3x_i &= {}^3F_{ij}{}^3X_j \\ &\vdots \\ {}^nx_i &= {}^nF_{ij}{}^nX_j \end{aligned} \tag{10.4}$$

The incremental deformation gradient tensors define the maximum stretch direction and magnitude for each increment of strain, and in this case, the maximum and minimum stretching directions are parallel to the X_1 and X_3 axes of the coordinate reference frame. The principal stretches used for all the strain increments in Figure 10.1 are $S_1^{1/10}$ and $S_3^{1/10}$, respectively. Variation in the orientation and magnitude of incremental stretches through time defines a *cumulative incremental strain history*.

10.1 Finite versus incremental strain

The deformation gradient tensor at the end of the strain history, ^{ps}F, results from the superposition of each of the incremental deformations so that

$$x_i = {}^nF_{ij} \ldots {}^3F_{ij}{}^2F_{ij}{}^1F_{ij}{}^1X_j \tag{10.5}$$

After one increment of strain, the incremental and finite deformation gradient tensors are the same. But for any later point in the strain history, the deformation gradient tensor for the finite strain is equal to the product of all the preceding incremental deformation gradient tensors.

The displacement field in Figure 10.1 is symmetric and consists of hyperbolic paths. As is clear from Equation 5.22 and inspection of Figure 10.1, the X_1 and X_3 axes are flow apophyses (Ramberg, 1975) defined by the *eigenvectors* of the deformation gradient tensor, or the vectors that change length but not orientation in response to the linear transformation ^{ps}F. The amount of length change is indicated by the principal stretches, or the *eigenvalues*. The eigenvectors for the maximum and minimum stretches define stable and unstable equilibriums with respect to the orientation of passive line markers. Much like a ball at rest is stable at the bottom of a trough but unstable at the top of a hill, stability is defined in terms of the response of an equilibrium to perturbations. In the example here, the eigenvector parallel to the maximum stretch (or the X_1 axis) defines a stable orientation, and a slight perturbation will result in rotation towards the equilibrium orientation. The other eigenvector defines an unstable equilibrium orientation; any perturbation leads to rotation away from this orientation and towards the stable equilibrium.

For the determination of finite strain, we can evaluate the orientation of lines in the deformed reference frame using the Cauchy deformation tensor for pure shear, $\bar{C}_{ij}$. As shown in Chapter 9:

$$\frac{|\mathbf{X}|^2}{|\mathbf{x}|^2} = \frac{1}{\lambda} = \lambda' = \frac{\bar{C}_{kl}X_kX_l}{|\mathbf{x}|^2} = \bar{C}_{kl}\alpha'_k\alpha'_l$$

where α_k' and α_l' are the direction cosines that define the orientation of a vector after deformation. This equation can be rewritten in terms of θ', or the orientation of the line in the deformed reference frame relative to the direction of maximum stretch:

$$\lambda' = \lambda'_1\cos^2\theta' + \lambda'_3\sin^2\theta'$$

or

$$\lambda' = (\lambda'_1 - \lambda'_3)\cos^2\theta' + \lambda'_3$$

Solving for $\lambda' = 1$, we derive the *orientation of the lines of no finite elongation*:

$$\cos^2\theta'_n = \frac{(1 - \lambda'_3)}{(\lambda'_1 - \lambda'_3)} \tag{10.6}$$

There are two roots for θ'_n that satisfy this equation and correspond to two orientations symmetrical about the X_1 axis. The limit of this function is 1 as λ'_3 approaches infinity, which corresponds to a θ'_n value of $0°$. If there is no area change, the limit is ½ as λ'_1 and λ'_3 approach 1, which corresponds to a θ'_n value of $\pm 45°$, or the orientation of the lines of no infinitesimal elongation. Any line oriented so that θ' is less than θ'_n has extended. Lines oriented so that θ' is greater than θ'_n have shortened. Combining this result with the orientation of the lines of no infinitesimal elongation leaves us with the three zones of Ramsay (1967; Figure 10.2a) that indicate whether a line of a specific orientation (θ') is shortening (infinitesimal) and has shortened (finite) (i.e., zone 3 at $\theta' > 45°$), is lengthening (infinitesimal) but has shortened (finite) (zone 2, $\theta'_n < \theta' < 45°$), and is lengthening (infinitesimal) and has lengthened (finite) (zone 1, $\theta' < \theta'_n$).

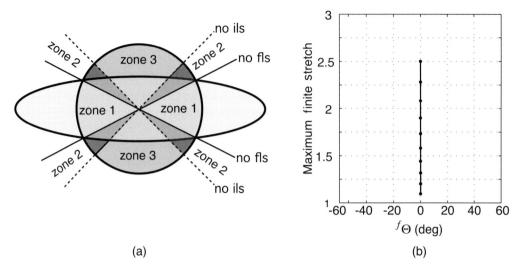

Figure 10.2 (a) Zones that describe the infinitesimal and finite longitudinal strain of passive line markers during pure shear (after Ramsay, 1967). (b) Progressive finite strain history for pure shear, which shows variations in the orientation relative to X_1 ($^f\Theta$) and magnitude (S_1) of maximum finite stretch as strain accumulates.

The principal axes for the Cauchy deformation tensor are $[1/S_1^2 \ 0 \ 0]$, $[0 \ 1 \ 0]$, and $[0 \ 0 \ 1/S_3^2]$ and the eigenvectors parallel to these axes do not change as the strain accumulates. In these circumstances, the principal axes of incremental and finite strain are parallel (finite stretch orientation with respect to X_1, $^f\Theta$, is constant throughout the deformation, Fig. 10.2b). When the finite strain axes do not rotate relative to the incremental strain axes, strain is *irrotational*, and deformation can be defined as *pure shear*. Variations in the orientation and magnitude of finite strain characterize the *progressive finite strain history*, which in the case of pure shear is not very interesting (Fig. 10.2b). The MATLAB® function **PureShear**, below, computes and plots the displacement paths (Fig. 10.1) and progressive finite strain history (Fig. 10.2b) for pure shear. To reproduce Figures 10.1 and 10.2(b), type in MATLAB:

```
pts = [-1 -1;-1 -0.5;-1 0;-1 0.5;-1 1;-0.5 1;0 1;0.5 1;1 1;1 0.5;1 0;...
    1 -0.5;1 -1;0.5 -1;0 -1;-0.5 -1]; %Initial points coordinates
[paths,psf] = PureShear(pts,2.5,10);

function [paths,pfs] = PureShear(pts,st1,ninc)
%PureShear computes and plots displacement paths and progressive finite
%strain history for pure shear with maximum stretching parallel to the
%X1 axis
%
%   USE: [paths,pfs] = PureShear(pts,st1,ninc)
%
%   pts: npoints x 2 matrix with X1 and X3 locations of points
%   st1 = Maximum principal stretch
%   ninc = number of strain increments
%   paths = displacement paths of points
%   pfs = progressive finite strain history. column 1 = orientation of
%         maximum stretch with respect to X1 in degrees, column 2 = maximum
```

10.1 Finite versus incremental strain

```
%           stretch magnitude
%
%   NOTE: Intermediate principal stretch is 1.0 (Plane strain)
%         Output orientations are in radians

%Compute minimum principal stretch and incremental stretches
st1inc=st1^(1.0/ninc);
st3=1.0/st1;
st3inc=st3^(1.0/ninc);

%Initialize displacement paths
npts = size(pts,1); %Number of points
paths = zeros(npts,2,ninc+1);
paths(:,:,1) = pts; %Initial points of paths are input points

%Calculate incremental deformation gradient tensor
F = [(st1inc) 0.0; 0.0 (st3inc)];

%Create a figure and hold
figure;
hold on;

%Compute displacement paths
for i=1:npts %for all points
    for j=2:ninc+1 %for all strain increments
        %Equation 10.2-10.5
        for k=1:2
            for L=1:2
                paths(i,k,j) = F(k,L)*paths(i,L,j-1) + paths(i,k,j);
            end
        end
    end
    %Plot displacement path of point. Use MATLAB function squeeze to reduce
    %the 3D matrix to one vector in X1 and another in X3
    xx = squeeze(paths(i,1,:));
    yy = squeeze(paths(i,2,:));
    plot(xx,yy,'k.-');
end

%Release plot and set axes
hold off;
axis equal;
xlabel('X1'); ylabel('X3');
grid on;

%Initalize progressive finite strain history
pfs = zeros(ninc+1,2);
pfs(1,:) = [0 1.0]; %Initial state

%Calculate progressive finite strain history
for i=1:ninc
    %First determine the finite deformation gradient tensor
```

```
    finF = F^i;
    %Determine Green's deformation tensor
    G = finF*finF';
    %Stretch magnitude and orientation: Maximum eigenvalue and their
    %corresponding eigenvectors of Green's tensor. Use MATLAB function eig
    [V,D] = eig(G);
    pfs(i+1,1) = atan(V(2,2)/V(1,2));
    pfs(i+1,2) = sqrt(D(2,2));
end

%Plot progressive finite strain history
figure;
plot(pfs(:,1)*180/pi,pfs(:,2),'k.-');
xlabel('Theta finite deg');
ylabel('Maximum finite stretch');
axis([-90 90 1 max(pfs(:,2))+0.5]);
grid on;
end
```

10.1.2 Progressive strain histories in two dimensions: Simple shear

Pure shear may be an appropriate description of two-dimensional deformation in some cases, but deformation is commonly localized in shear zones where the displacement field is best described by straight line displacements parallel to the shear zone boundaries, which we assume to be the X_1 axis in Figure 10.3. The new coordinates in terms of old coordinates are given by:

$$\begin{aligned} x_1 &= X_1 + \gamma X_3 \\ x_2 &= X_2 \\ x_3 &= X_3 \end{aligned} \quad (10.7)$$

The deformation gradient tensor is given by $^{ss}\mathbf{F}$:

$$^{ss}F_{ij} = \begin{bmatrix} 1 & 0 & \gamma \\ 0 & 1 & 0 \\ 0 & 0 & 1 \end{bmatrix}, \quad (10.8)$$

where γ is the engineering shear strain introduced earlier. It should be no surprise that the matrix is asymmetric given the displacement field depicted in Figure 10.3. The figure shows a progressive strain history consisting of 10 increments of infinitesimal strain, with an engineering shear strain of $\gamma/10$ for each incremental deformation gradient tensor. The deformation gradient tensor for the finite strain is obtained by superposition of incremental deformation gradient tensors as in Equation 10.5.

There are two eigenvectors, one parallel to the intermediate stretch and one parallel to the X_1 axis. As expected, the X_1 axis is an orientation that experiences no rotation or longitudinal strain due to the deformation gradient tensor, $^{ss}\mathbf{F}$. In this example, the equilibrium defined by the X_1 axis orientation is stable for passive line markers rotated or perturbed in a counterclockwise sense and unstable for markers rotated clockwise, much like a ball at rest on a ledge. As we will see in the next section, when we combine this deformation field with pure shear, the dual behavior

10.1 Finite versus incremental strain

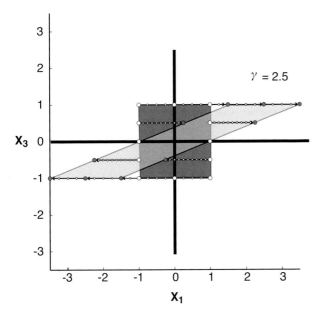

Figure 10.3 Simple shear deformation of a square after a γ of 2.5, with the shear direction parallel to the X_1 axis. The displacement path of points around the outside of the box is subdivided into 10 increments of strain. Large white and gray circles are initial and final positions, respectively.

of this orientation occurs when the acute angle between the stable and unstable eigenvectors of the deformation gradient tensor for a general deformation approaches 0. Because the eigenvectors are parallel to lines of no finite elongation, the eigenvalues are equal to 1.

For the finite strain, we forward model the deformation using the undeformed coordinates, so we use the Green deformation tensor, C_{ij}.

For simple shear:

$$C_{ij} = \begin{bmatrix} 1+\gamma^2 & 0 & \gamma \\ 0 & 1 & 0 \\ \gamma & 0 & 1 \end{bmatrix} \tag{10.9}$$

Solving for the eigenvalues (λ), or $\dfrac{|\mathbf{x}|^2}{|\mathbf{X}|^2}$ parallel to the principal stretches:

$$\begin{aligned} \lambda_1 &= \frac{\gamma^2 + 2 + \gamma\sqrt{\gamma^2+4}}{2} \\ \lambda_2 &= 1 \\ \lambda_3 &= \frac{\gamma^2 + 2 - \gamma\sqrt{\gamma^2+4}}{2} \end{aligned} \tag{10.10}$$

The orientations of the eigenvectors, or the orientations of the maximum and minimum principal finite strain, are

$$\begin{bmatrix} \dfrac{\gamma+\sqrt{\gamma^2+4}}{2} & 0 & 1 \end{bmatrix} \quad \begin{bmatrix} 0 & 1 & 0 \end{bmatrix} \quad \begin{bmatrix} \dfrac{\gamma-\sqrt{\gamma^2+4}}{2} & 0 & 1 \end{bmatrix}$$

Note that these are not unit vectors, and we have simplified the expression for the eigenvectors by setting the component in the X_3 direction equal to 1. The orientation of the maximum principal stretch relative to the X_1 axis is

$$\Theta_1 = \tan^{-1}\left(\frac{2}{\gamma+\sqrt{\gamma^2+4}}\right) \tag{10.11}$$

For very small magnitudes of shear strain (as γ approaches 0), the eigenvectors are parallel to the principal axes of the incremental strain ellipse with orientations $[1\ 0\ 1]$ and $[-1\ 0\ 1]$ (i.e., Θ_1 and Θ_3 approach $45°$ and $-45°$). For large strain magnitudes, the eigenvector for the maximum stretch approaches $[1\ 0\ 0]$. There is no infinitesimal or finite longitudinal strain parallel to the shear plane and, since there are two lines of no finite elongation symmetric about the maximum stretch direction, the other orientation of these lines (θ'_n) is $2\Theta_1$ (Fig. 10.4a). We define this progressive finite strain history as *rotational* because the principal axes of finite strain rotate relative to the principal axes of infinitesimal or incremental strain ($^f\Theta$ is not constant throughout the deformation, Fig. 10.4b). This displacement field, with displacement paths parallel to the shear direction and rotational strain histories, is characteristic of *simple shear*.

The asymmetric deformation gradient tensor for simple shear can be decomposed into a symmetric second rank tensor that reflects the pure strain component of the deformation and a rigid body rotation (Malvern, 1969):

$$^{ss}\mathbf{F} = \mathbf{R} \cdot \mathbf{U} \tag{10.12}$$

where $\mathbf{U}$ is the right stretch tensor (e.g., Chapter 9.8) that defines the pure strain component and $\mathbf{R}$ is the rotation tensor. We first calculate $\mathbf{U}$:

$$\mathbf{U} = \sqrt{^{ss}\mathbf{F} \cdot \left(^{ss}\mathbf{F}^T\right)} \tag{10.13}$$

The eigenvalues and eigenvectors of $\mathbf{U}$ depict the magnitude and orientation of the principal finite stretches, and the rotation tensor can be calculated by removing the pure strain from the finite deformation gradient tensor:

$$\mathbf{R} = {^{ss}\mathbf{F}} \cdot \mathbf{U}^{-1}$$

The *internal rotation* (ω_i) within the rotation matrix $\mathbf{R}$ is equal to $\theta/2$ for small strains.

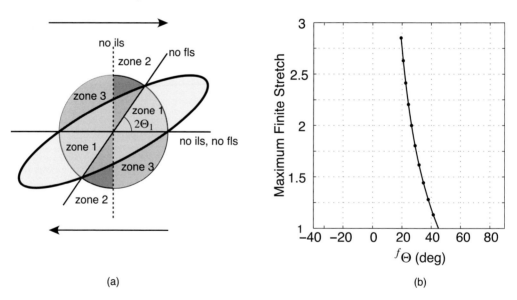

Figure 10.4 (a) Zones that describe the infinitesimal and finite longitudinal strain of passive line markers during simple shear (after Ramsay, 1967). (b) Progressive finite strain history for simple shear, which shows variations in the orientation ($^f\Theta$) and magnitude (S_1) of maximum finite stretch as strain accumulates.

10.1 Finite versus incremental strain

The function **SimpleShear**, below, computes and plots the displacement paths (Fig. 10.3) and progressive finite strain history (Fig. 10.4b) for simple shear. To reproduce Figures 10.3 and 10.4b, type in MATLAB:

```
pts=[-1 -1;-1 -0.5;-1 0;-1 0.5;-1 1;-0.5 1;0 1;0.5 1;1 1;1 0.5;1 0;...
    1 -0.5;1 -1;0.5 -1;0 -1;-0.5 -1]; %Initial points coordinates
[paths,psf] = SimpleShear(pts,2.5,10);

function [paths,pfs] = SimpleShear(pts,gamma,ninc)
%SimpleShear computes and plots 2D displacement paths and progressive finite
%strain history for simple shear parallel to the X1 axis
%
%   USE: [paths,pfs] = SimpleShear(pts,gamma,ninc)
%
%   pts: npoints x 2 matrix with X1 and X3 locations of points
%   gamma = Engineering shear strain
%   ninc = number of strain increments
%   paths = displacement paths of points
%   pfs = progressive finite strain history. column 1 = orientation of
%         maximum stretch with respect to X1 in degrees, column 2 = maximum
%         stretch magnitude
%
%   NOTE: Intermediate principal stretch is 1.0 (Plane strain)
%         Output orientations are in radians

%Incremental engineering shear strain
gammainc = gamma/ninc;

%Initialize displacement paths
npts = size(pts,1); %Number of points
paths = zeros(npts,2,ninc+1);
paths(:,:,1) = pts; %Initial points of paths are input points

%Calculate incremental deformation gradient tensor
F = [1.0 gammainc; 0.0 1.0];

%Create a figure and hold
figure;
hold on;

%Compute displacement paths
for i=1:npts %for all points
    for j=2:ninc+1 %for all strain increments
        %Equation 10.2-10.5
        for k=1:2
            for L=1:2
                paths(i,k,j) = F(k,L)*paths(i,L,j-1) + paths(i,k,j);
            end
        end
    end
end
```

```
    %Plot displacement path of point. Use MATLAB function squeeze to reduce
    %the 3D matrix to one vector in X1 and another in X3
    xx = squeeze(paths(i,1,:));
    yy = squeeze(paths(i,2,:));
    plot(xx,yy,'k.-');
end

%Release plot and set axes
hold off;
axis equal;
xlabel('X1'); ylabel('X3');
grid on;

%Initalize progressive finite strain history
pfs = zeros(ninc+1,2);
%Initial state: Maximum extension is at 45 deg from shear zone
pfs(1,:) = [pi/4.0 1.0];

%Calculate progressive finite strain history
for i=1 :ninc
    %First determine the finite deformation gradient tensor
    finF = F^i;
    %Determine Green's deformation tensor
    G = finF*finF';
    %Stretch magnitude and orientation: Maximum eigenvalue and their
    %corresponding eigenvectors of Green's tensor. Use MATLAB function eig
    [V,D] = eig(G);
    pfs(i+1,1) = atan(V(2,2)/V(1,2));
    pfs(i+1,2) = sqrt(D(2,2));
end

%Plot progressive finite strain history
figure;
plot(pfs(:,1)*180/pi,pfs(:,2),'k.-');
xlabel('Theta finite deg');
ylabel('Maximum finite stretch');
axis([-90 90 1 max(pfs(:,2))+0.5]);
grid on;
end
```

10.1.3 General shear: Combinations of pure shear and simple shear

Pure shear and simple shear are useful end members for evaluating the kinematics in two dimensions, but a more complete treatment of strain requires consideration of *sub-simple shear* (DePaor, 1983), which is a *general shear* where deformation lies within the spectrum of behavior between simple shear and pure shear. To characterize the displacement paths during a sub-simple shear, we first need to define the pure shear reference frame (the orientation of the principal stretches) in terms of the kinematic frame for simple shear (the shear plane and shear direction). At this point, we restrict the discussion to the two dimensions that contain the

10.1 Finite versus incremental strain

maximum and minimum stretches, and we consider two cases: (1) sub-simple shear where the principal shortening direction for coaxial deformation is oriented perpendicular to the shear plane and the principal extension direction is parallel to the shear direction; and (2) sub-simple shear where the principal axis of extension lies perpendicular to the shear plane and the principal shortening direction lies parallel to the shear direction.

We showed in Chapter 9 that the sequence in which we apply strain increments is important for the kinematics. For sub-simple shear, the final position vector **x** is not the same for simple shear followed by pure shear and pure shear followed by simple shear (Fossen and Tikoff, 1993). We are interested in the progressive strain history for *simultaneous* simple shearing and pure shearing. The deformation gradient rate tensor for this state is given by Ramberg (1975):

$$^{gs}\dot{\mathbf{F}} = \begin{bmatrix} \exp(\dot{\varepsilon}_1 t) & \frac{\dot{\gamma}}{2\dot{\varepsilon}_1}(\exp(\dot{\varepsilon}_1 t) - \exp(\dot{\varepsilon}_3 t)) \\ 0 & \exp(\dot{\varepsilon}_3 t) \end{bmatrix} \quad (10.14)$$

where $\dot{\varepsilon}_1$ and $\dot{\varepsilon}_3$ are the rate of elongation parallel to the $\mathbf{X}_1$ and $\mathbf{X}_3$ directions, respectively, and $\dot{\gamma}$ is the engineering shear strain rate. Note that the components of this matrix and the contributions of pure and simple shear are given in terms of strain rates and not strains. We can express this matrix in a form that is independent of time for a deformation that occurs at a steady rate. If the maximum extension direction is parallel to the $\mathbf{X}_1$ axis and area is constant, then $S_1 = \exp(\dot{\varepsilon}_1 t)$, $S_3 = \exp(\dot{\varepsilon}_3 t)$ and $\gamma = \dot{\gamma}t$. Consequently, $\frac{\dot{\gamma}}{\dot{\varepsilon}_1} = \frac{\gamma}{\ln S_1}$, $\ln(S_1) = \dot{\varepsilon}t$, and the expression for $^{gs}\mathbf{F}$ can be simplified (Merle, 1986):

$$^{gs}\mathbf{F} = \begin{bmatrix} S_1 & \frac{\gamma(S_1 - S_3)}{2\ln S_1} \\ 0 & S_3 \end{bmatrix} \quad (10.15)$$

With superposition of infinitesimal strain increments, this deformation gradient tensor is used in Figure 10.5 to construct displacement paths associated with progressive deformations for the case of sub-simple shear where the maximum shortening direction for pure shear is perpendicular to the shear plane (Fig. 10.5).

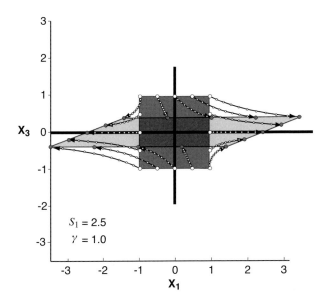

Figure 10.5 Sub-simple shear, or simultaneous pure and simple shear, with pure shear shortening perpendicular to the shear zone boundaries and pure shear extension parallel to the shear zone boundaries. No area change, with a maximum stretch of 2.5 for the pure shearing component, and a γ of 1.0 for the simple shear component.

Now we are in a position where we can evaluate the displacement field in response to variations in the ratio of pure shear to simple shear (Fig. 10.6). As in the case of pure shear, there are two eigenvectors for the deformation gradient tensor. These are parallel to apophyses that define stable and unstable equilibrium marker orientations. The stable orientation is parallel to the X_1 axis $[1 \quad 0]$, whereas the unstable orientation $\left[\dfrac{-\gamma}{2\ln S_1} \quad 1\right]$ varies as a function

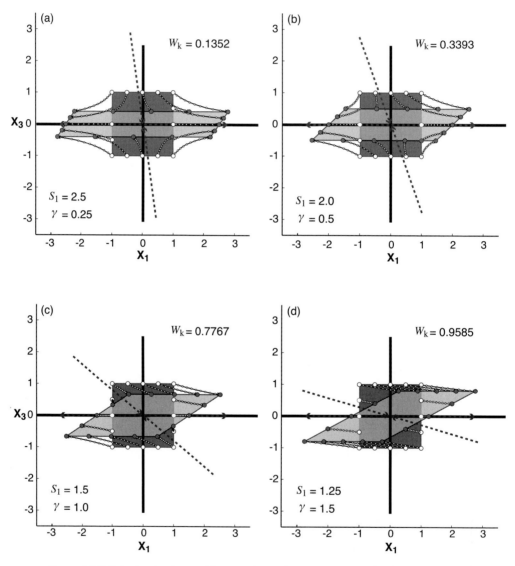

Figure 10.6 Sub-simple shear with pure shear shortening perpendicular to the shear zone boundaries, no area change, and variations in the relative proportion of simple shear (γ) and pure shear (S_1). (a) $S_1 = 2.5$, $\gamma = 0.25$; (b) $S_1 = 2.0$, $\gamma = 0.5$; (c) $S_1 = 1.5$, $\gamma = 1.0$; and (d) $S_1 = 1.25$, $\gamma = 1.5$. The displacement path of points around the outside of the box is subdivided into 10 increments of strain. Large white and gray circles are initial and final positions, respectively. Eigenvectors are shown by the dashed arrows. W_k is the kinematic vorticity number.

10.1 Finite versus incremental strain

of the ratio of simple shear to pure shear. In two dimensions, a simple dimensionless measure of the ratio of pure to simple shearing is the cosine of the acute angle between the eigenvectors of the deformation gradient tensor (Bobyarchick, 1986), or the *kinematic vorticity number*, W_k (Truesdell, 1953) (Fig. 10.6). Sub-simple shear is characterized by a kinematic vorticity number that lies between 0 and 1. For sub-simple shear with shortening perpendicular to the shear zone boundaries, the eigenvectors bracket the orientations where passive line markers rotate with the sense of shear and the orientations where line markers rotate opposite to the sense of shear.

The Green deformation tensor for a general shear with shortening across the shear zone is

$$\mathbf{C} = \begin{bmatrix} S_1^2 + \dfrac{\gamma^2(S_1 - S_3)^2}{4 \ln S_3} & \dfrac{\gamma(S_1 - S_3)}{2 \ln S_1} \\ \dfrac{\gamma(S_1 - S_3)}{2 \ln S_1} & S_3^2 \end{bmatrix} \quad (10.16)$$

The eigenvectors for the Green deformation tensor are parallel to the maximum finite stretch, and approach the shear plane as strain accumulates, but the initial orientation of the maximum stretch varies from 0° to 45° as W_k varies from 0 to 1 (Fig. 10.7). In practice, a lineation that reflects the direction of maximum stretch will rotate into parallelism with the shear direction with increasing finite strain magnitude.

In the case where the direction of maximum stretching for pure shear is perpendicular to the shear plane, the maximum stretch associated with the pure shear component of deformation, S_1, is parallel to the X_3 axis, and S_3 is parallel to the X_1 axis, so the gradient tensor for simultaneous pure and simple shearing is

$$^{gs}\mathbf{F} = \begin{bmatrix} S_3 & \dfrac{\gamma(S_3 - S_1)}{2 \ln S_3} \\ 0 & S_1 \end{bmatrix} \quad (10.17)$$

Figure 10.8 shows deformation of boxes in response to simultaneous pure shear and simple shear. In this case, the unstable orientation for passive line markers lies parallel to the shear plane and the stable orientation is oblique to the shear plane. Passive line markers reorient during deformation into parallelism with the stable eigenvector. The finite strain ellipse initially has an orientation between 45° and 90° as the kinematic vorticity number varies from 0 to 1; but in contrast to the case with pure shear stretching parallel to the shear zone boundaries,

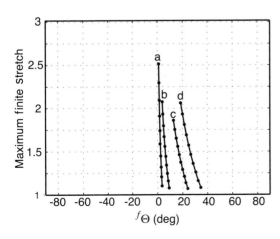

Figure 10.7 Progressive finite strain histories for sub-simple shear and the examples in Figure 10.6.

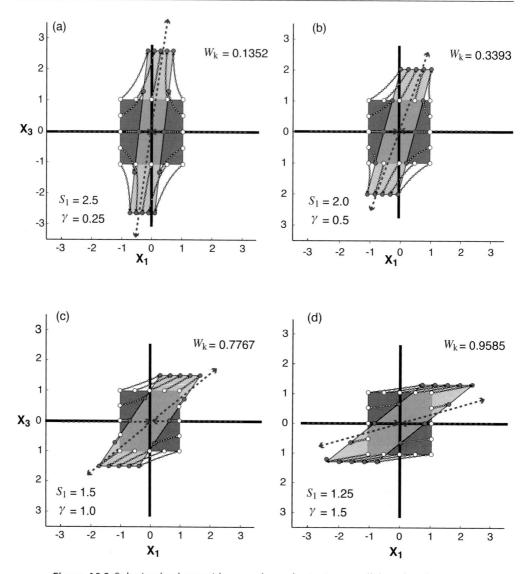

Figure 10.8 Sub-simple shear with pure shear shortening parallel to the shear zone boundaries, no area change, and variations in the relative proportion of simple shear (γ) and pure shear (S_1). (a) $S_1 = 2.5$, $\gamma = 0.25$; (b) $S_1 = 2.0$, $\gamma = 0.5$; (c) $S_1 = 1.5$, $\gamma = 1.0$; and (d) $S_1 = 1.25$, $\gamma = 1.5$. The displacement path of points around the outside of the box is subdivided into 10 increments of strain. Large white and gray circles are initial and final positions, respectively. Eigenvectors are shown by the dashed arrows. W_k is the kinematic vorticity number.

the maximum finite stretch direction at large strains varies significantly for different kinematic vorticity numbers (Fig. 10.9).

The function **GeneralShear**, below, computes displacement paths, kinematic vorticity number, and progressive finite strain history for general shear with a pure shear stretch, no area change, and a simple shear strain. Maximum finite stretch can be parallel (kk = 0) or perpendicular (kk = 1) to the shear zone. For example, to reproduce Figures 10.6b and 10.7b, type in MATLAB:

10.1 Finite versus incremental strain

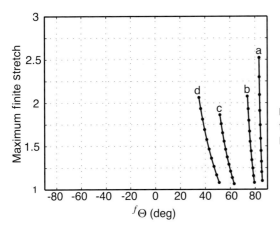

Figure 10.9 Progressive finite strain histories for sub-simple shear and the examples in Figure 10.8.

```
pts=[-1 -1;-1 -0.5;-1 0;-1 0.5;-1 1;-0.5 1;0 1;0.5 1;1 1;1 0.5;1 0;...
    1 -0.5;1 -1;0.5 -1;0 -1;-0.5 -1]; %Initial points coordinates
[paths,wk,psf] = GeneralShear(pts,2.0,0.5,0,10);
```

And to recreate Figures 10.8b and 10.9b:

```
[paths,wk,psf] = GeneralShear(pts,2.0,0.5,1,10);

function [paths,wk,pfs] = GeneralShear(pts,st1,gamma,kk,ninc)
%GeneralShear computes displacement paths, kinematic vorticity numbers
%and progressive finite strain history, for a general shear with a pure
%shear stretch, no area change, and a single shear strain
%
%   USE: [paths,wk,pfs] = GeneralShear(pts,st1,gamma,kk,ninc)
%
%   pts: npoints x 2 matrix with X1 and X3 locations of points
%   st1: Pure shear stretch parallel to shear zone
%   gamma = Engineering shear strain
%   kk = An integer that indicates whether the maximum finite stretch is
%        parallel (kk = 0), or perpendicular (kk = 1) to the shear direction
%   ninc = number of strain increments
%   paths = displacement paths of points
%   wk = Kinematic vorticity number
%   pfs = progressive finite strain history. column 1 = orientation of
%         maximum stretch with respect to X1 in degrees, column 2 = maximum
%         stretch magnitude
%
%   NOTE: Intermediate principal stretch is 1.0 (Plane strain)
%         Output orientations are in radians

%Compute minimum principal stretch and incremental stretches
st1inc=st1^(1.0/ninc);
st3=1.0/st1;
st3inc=st3^(1.0/ninc);
```

```matlab
%Incremental engineering shear strain
gammainc = gamma/ninc;

%Initialize displacement paths
npts = size(pts,1); %Number of points
paths = zeros(npts,2,ninc+1);
paths(:,:,1) = pts; %Initial points of paths are input points

%Calculate incremental deformation gradient tensor
%If max. finite stretch parallel to shear direction (Eq. 10.15)
if kk == 0
    F = [st1inc (gammainc*(st1inc-st3inc))/(2.0*log (st1inc));0.0 st3inc];
%If max. finite stretch perpendicular to shear direction (Eq. 10.17)
elseif kk == 1
    F = [st3inc (gammainc*(st3inc-st1inc))/(2.0*log (st3inc));0.0 st1inc];
end

%Create a figure and hold
figure;
hold on;

%Compute displacement paths
for i=1:npts %for all points
    for j=2:ninc+1 %for all strain increments
        %Equations 10.2-10.5
        for k=1:2
            for L=1:2
                paths(i,k,j) = F(k,L)*paths(i,L,j-1) + paths(i,k,j);
            end
        end
    end
    %Plot displacement path of point. Use MATLAB function squeeze to reduce
    %the 3D matrix to one vector in X1 and another in X3
    xx = squeeze(paths(i,1,:));
    yy = squeeze(paths(i,2,:));
    plot(xx,yy,'k.-');
end

%Release plot and set axes
hold off;
axis equal;
xlabel('X1'); ylabel('X3');
grid on;

%Determine the eigenvectors of the flow (apophyses)
[V,D] = eigs(F);
%If max. finite stretch parallel to shear direction
if kk == 0
    theta2=atan(V(2,2)/V(1,2));
```

```
%If max. finite stretch perpendicular to shear direction
elseif kk == 1
    theta2=atan(V(2,1)/V(1,1));
end
wk = cos(theta2);

%Initalize progressive finite strain history. We are not including the
%initial state
pfs = zeros(ninc);

%Calculate progressive finite strain history
for i=1:ninc
    %First determine the finite deformation gradient tensor
    finF = F^i;
    %Determine Green's deformation tensor
    G = finF*finF';
    %Stretch magnitude and orientation: Maximum eigenvalue and their
    %corresponding eigenvectors of Green's tensor. Use MATLAB function eig
    [V,D] = eig(G);
    pfs(i,1) = atan(V(2,2)/V(1,2));
    pfs(i,2) = sqrt(D(2,2));
end

%Plot progressive finite strain history
figure;
plot(pfs(:,1)*180/pi,pfs(:,2),'k.-');
xlabel('Theta finite deg');
ylabel('Maximum finite stretch');
axis([-90 90 1 max(pfs(:,2))+0.5]);
grid on;
end
```

10.2 DETERMINATION OF A STRAIN HISTORY

Finite strain can be quantified from a range of deformed objects using measurements of angular shear, longitudinal strain, or some combination of the two, but quantitative reconstructions of kinematics require some assessment of the incremental and progressive finite strain histories, a more difficult objective. One approach is to substitute space for time; this is our aim whenever we assume that a larger finite strain magnitude, a tighter fold, or a fault with more slip represents a later stage in some characteristic structural evolution.

Some deformation fabrics such as fibrous pressure shadows and porphyroblasts with inclusion trails allow for the quantitative depiction of the strain history (Elliott, 1972). Measurement of the distribution of strain histories places a constraint on the structural evolution of regions or individual structures and can provide a test of the space-for-time assumption. In the following section, we describe a method whereby syntectonic fiber growths can be used to quantify the displacement path, which can then be used to calculate the cumulative incremental and progressive finite strain histories.

10.2.1 Syntectonic fibers: Indicators of external and internal rotation

For an increment of strain, the external rotation rate, or external vorticity ($\dot{\omega}_e$), is determined by the spin, or angular velocity of the stretching axes relative to the material ($\dot{\omega}_{sp}$), and the internal velocity ($\dot{\omega}_i$).

$$\dot{\omega}_e = \dot{\omega}_{sp} + \dot{\omega}_i \qquad (10.18)$$

To illustrate the relative contributions of spin and internal rotation to the external rotation, consider Figure 10.10. Structural geologists trained in the northeastern United States will recognize this photo as a fossilized Lycopsid trunk from the Llewellyn Formation of Bear Valley, Pennsylvania (Nickelsen, 1979). The trunk is oriented vertically on a 36°-dipping fold limb within the Appalachian Valley and Ridge fold belt.

A simple explanation for the vertical orientation of the trunk before and after deformation is that the bedding was a passive line marker during irrotational strain, and the stump happened to be originally oriented parallel to the stable eigenvector for pure shear; both the ω_{sp} and ω_i were zero. This seems unlikely given that the cleavage, although steeper than bedding, is not close to vertical (Fig. 10.10). Moreover, there are slickenlines on bedding planes and numerous shear zones in finer-grained layers at this locality, probably due to flexural shear and flexural slip folding. So, the more likely explanation is that the external rotation (ω_e) of the long axis of the tree trunk is negligible because the rigid rotation of beds through the stretching axes during tilting was equal and opposite in sign to the angular shear associated with flexure. In either case, a tree that was vertical when sediments were deposited 300 million years ago is still vertical after ~50% shortening of the Appalachian fold and thrust belt!

The separation of an external rotation into the components of rigid and internal rotation requires an assessment of the incremental strain history. That is, we need a record of how the magnitude and orientation of stretch varies through time relative to both an external reference frame, such as geographic coordinates, and an internal reference frame, such as shear zone boundaries or bedding layers with contrasting strength. We do this here using syntectonic fibers that grow in pressure shadows around rigid objects and record the displacement of the matrix relative to the rigid host.

Syntectonic fiber growth can be syntaxial or antitaxial, depending on the composition of the fibers and the rigid object (Durney and Ramsay, 1973). In the case of syntaxial fibers, the fibers nucleate on the margins of a rigid host of the same composition and grow outward towards the matrix (e.g., the crinoid method of Durney and Ramsay, 1973) so the most recent increment of strain is at the tip of the pressure shadow. For antitaxial fibers, the fibers nucleate on grains in the deforming matrix and grow towards a host of different composition (e.g., the pyrite method of Durney and Ramsay, 1973) so the most recent strain increment is at the interface between the pyrite and the pressure shadow. Since the curvature of fibers is a measure of the external rotation, coaxial strain histories are represented by straight fibers and non-coaxial strain histories are depicted by curved fibers. An analysis of the incremental strain history relative to the external and internal reference frames allows us to determine if fiber curvature reflects rotation of the rock body through a fixed irrotational stretching direction or simple shear (i.e., internal rotation).

In the following sections, we outline two techniques for quantifying a non-coaxial strain history by assuming that a rotational strain history recorded by curved fibers can be approximated by a series of small irrotational strain increments separated by rigid body rotations.

10.2 Determination of a strain history

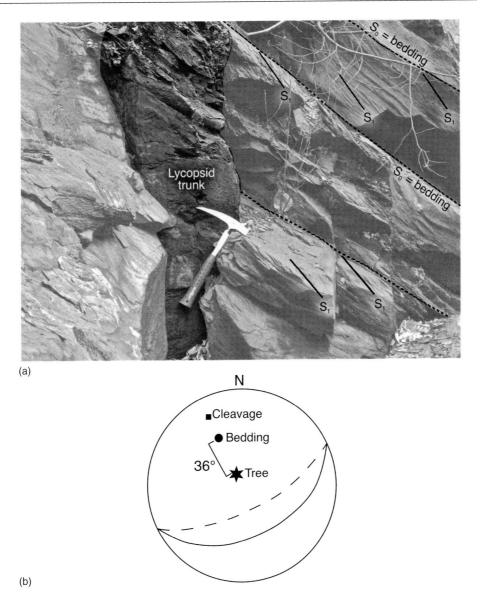

Figure 10.10 (a) Photograph of a Lycopsid trunk described by Nickelsen (1979) in the Pennsylvanian Llewellyn Formation in the Valley and Ridge Province of northeastern Pennsylvania. (b) Stereonet showing the poles and great circles for bedding (circle), cleavage (square), and the long axis of the tree stump (modified after Nickelsen, 1979).

10.2.2 Cleavage-parallel displacements and passively deformed fibers

We start with a pyrite pressure shadow with antitaxial fibers as viewed in a thin section cut parallel to cleavage from the metamorphic hinterland of the Taiwan arc–continent collision (Fig. 10.11a). There is no evidence for shortening in the direction perpendicular to the long axis of fibers, because all fibers in the pressure shadow are parallel to each other and the pressure shadow brackets the rigid pyrite sphere such that adjacent fibers do not converge or diverge as they are displaced with the matrix away from the pyrite. There is, however, evidence for

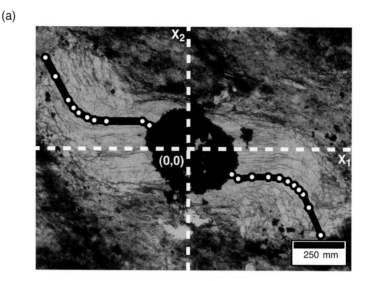

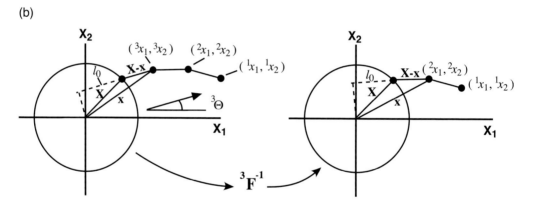

Figure 10.11 (a) Photomicrograph of a pyrite pressure shadow viewed downward on the cleavage plane from a Cretaceous phyllite near the eastward-facing mountain front of the Central Range of Taiwan. (b) Sketch showing parameters used in strain history calculations. l_0 is projection of initial position vector **X** on line parallel to fiber segment (**x**–**X**). Θ is angle between maximum stretch and X_1 axis. F^{-1} from increment 3 is used to unstrain $^3\mathbf{x}$ to $^3\mathbf{X}$, leaving two increments remaining (right).

deformation of early-formed fibers at the outer edge of the pressure shadow, with recrystallization and thickening of fibers. The thin section contains the maximum stretch (the long axis of fibers) and the intermediate stretch (perpendicular to the long axis of fiber segments), which is approximately equal to 1. We subdivide the fiber into a number of increments so that the curved displacement path is characterized by a series of short, straight, line segments (Fig. 10.11b). Since the fiber growth is antitaxial (Ramsay and Huber, 1983), the last increment of fiber growth is recorded by the portion of the fiber closest to the pyrite. The center of the pyrite serves as the origin of a Cartesian coordinate system, and an arbitrary external reference frame (e.g., horizontal or a structural lineation) is chosen as the X_1 axis.

In this example, we assume that fibers deform passively based on the evidence for deformation of early-formed fibers. Such a problem is well suited for an inverse approach where we

10.2 Determination of a strain history

quantify the last increment of strain and use that deformation gradient tensor to undeform all the remaining fiber segments. The deformation gradient tensor for the last increment, $^n\mathbf{F}$, or $^n\left(\frac{\partial x_i}{\partial X_j}\right)$ can be written in a coordinate system that parallels the principal maximum and intermediate stretches:

$$\begin{aligned} ^{pn}x_1 &= {^nS_1}\,^{pn}X_1 \\ ^{pn}x_2 &= {^{pn}X_2} \end{aligned} \tag{10.19}$$

or

$$^{pn}\mathbf{F} = \begin{bmatrix} ^nS_1 & 0 \\ 0 & 1 \end{bmatrix} \tag{10.20}$$

where the superscript n is indexed to the increment number and the superscript p refers to a reference frame parallel to the principal axes. All displacements occur parallel to the maximum stretch direction, so the orientation of extension for the last increment, $^n\Theta$, equals (Fig. 10.11b)

$$^n\Theta = \tan^{-1}\left(\frac{^nx_2 - {^nX_2}}{^nx_1 - {^nX_1}}\right) \tag{10.21}$$

To determine the stretch for the nth increment, we first calculate the projection of the initial position vector onto a line parallel to the stretch direction so that (Fig. 10.11b)

$$l_0 = |\mathbf{X}| \cos\theta \tag{10.22}$$

The angle θ is the arccosine of the dot product of the unit vectors parallel to $\mathbf{X}$ and $\mathbf{x} - \mathbf{X}$:

$$\theta = \cos^{-1}\left(\left(\frac{\mathbf{X}}{|\mathbf{X}|}\right) \cdot \left(\frac{\mathbf{x} - \mathbf{X}}{|\mathbf{x} - \mathbf{X}|}\right)\right) \tag{10.23}$$

The maximum incremental stretch for increment n is

$$^nS_1 = \frac{l_0 + |\mathbf{x} - \mathbf{X}|}{l_0} \tag{10.24}$$

We can now evaluate $\mathbf{F}$ in the arbitrary reference frame that we have established parallel to the X_1 axis using a tensor transformation (Eq. 5.12):

$$^nF_{ij} = R_{ik}R_{jl}\,^{pn}F_{kl}$$

or

$$\mathbf{F} = \mathbf{R} \cdot {^{pn}\mathbf{F}} \cdot \mathbf{R}^T$$

Since the fiber deforms in response to each increment of deformation, we can use the deformation gradient tensor for the last increment to restore all the points along the fiber to positions prior to that strain.

The inverse of the tensor $^n\mathbf{F}$ has the property of displacing the deformed position of the last increment back to its initial position:

$$^n\mathbf{X} = {^n\mathbf{F}^{-1}} \cdot {^n\mathbf{x}} \tag{10.25}$$

All other points along the fiber can be similarly restored.

Now we are left with $n-1$ increments, and the initial position vector for increment $n-1$, $^{n-1}(X_1, X_2)$, is the same as $^n(X_1, X_2)$ (Fig. 10.11b). In other words, all fiber segments on a single fiber originate from the same point on the pyrite surface. The other points along the displacement path are restored to new positions prior to the strain associated with $^{n-1}\mathbf{F}$. Using Equations 10.19 through 10.25, $^{n-1}\mathbf{F}^{-1}$ is determined in the same manner as for the increment n, and then is used to restore all the points to positions prior to the increment $n-1$. This process is repeated until the first increment of strain is restored and there is only one point remaining at the surface of the pyrite grain (Fig. 10.11b). At this stage, we have $n\ ^i\Theta$-values that represent the maximum stretch directions for each increment and we have $n\ ^iS$-values for the incremental stretches, so we can plot a cumulative incremental strain history (Fig. 10.12a). This diagram shows variations in the orientation of incremental stretching as strain accumulates (Fig. 10.12a). If the strain rate is constant and the strain increments are very small, we can treat the vertical axis as equivalent to time. A vertical path on this diagram is an irrotational deformation, whereas a horizontal path is a rigid body rotation.

The progressive finite strain history is obtained by determining the orientation and magnitude of stretch after each strain increment. First, we multiply each deformation gradient tensor in sequence so that

$$\text{finite } ^1\mathbf{F} = {}^1\mathbf{F}$$
$$\text{finite } ^2\mathbf{F} = {}^2\mathbf{F} \cdot {}^1\mathbf{F}$$
$$\text{finite } ^3\mathbf{F} = {}^3\mathbf{F} \cdot {}^2\mathbf{F} \cdot {}^1\mathbf{F} \tag{10.26}$$
$$\vdots$$
$$\text{finite } ^n\mathbf{F} = {}^n\mathbf{F} \ldots \cdot {}^3\mathbf{F} \cdot {}^2\mathbf{F} \cdot {}^1\mathbf{F}$$

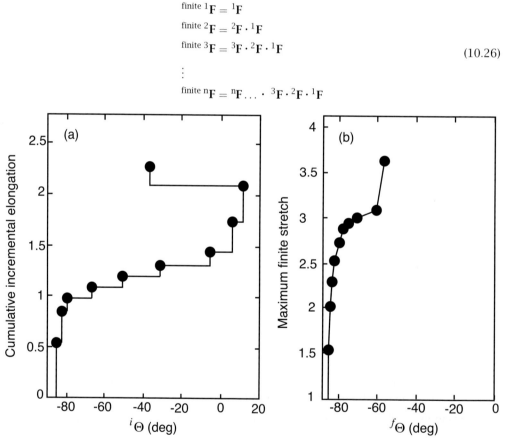

Figure 10.12 (a) Cumulative incremental strain history showing cumulative elongations vs. orientation of incremental extension and (b) progressive finite strain history showing magnitude vs. orientation of maximum finite stretch for the example in Figure 10.11. The cumulative incremental strain history consists of pure shear (coaxial) strain increments separated by rigid body rotations. The progressive finite strain history is the same as the cumulative incremental strain history for the first increment but, subsequently, it shows the response of the finite strain ellipse to variations in the orientation and magnitude of the incremental strain.

10.2 Determination of a strain history

The finite deformation gradient tensors from the second increment to the nth increment are not symmetric because the curved fiber is approximated by a series of irrotational strain increments separated by rigid body rotations. We evaluate the finite strain by determining the eigenvectors and eigenvalues associated with the Green deformation tensor:

$$\mathbf{C} = \mathbf{F} \cdot \mathbf{F}^T$$

By calculating a new Green deformation tensor after each increment of strain, we can depict progressive variations in the orientation of maximum finite stretch (i.e., the orientation of the eigenvector of $\mathbf{C}$ that defines the maximum stretch) and the magnitude of finite stretch (i.e., square root of the eigenvalues of $\mathbf{C}$). These values are used to construct a progressive finite strain history (Fig. 10.12b).

10.2.3 Geological applications of strain histories in cleavage planes: An example from Taiwan

So what do we do with these histories? The X_1 axis for both plots was chosen parallel to horizontal. The cumulative incremental strain history begins with three increments of elongation at 80–85° to horizontal, or downdip stretch on cleavage planes (Fig. 10.12a). The progressive finite strain history depicts a maximum stretch of about 2.3 after this early history with little external rotation (Fig. 10.12b). Then, there was a series of four increments with 85–90° of counterclockwise external rotation with little elongation, followed by two large elongations parallel to strike, and finally a small, late oblique elongation. The magnitude of the maximum finite stretch was 3.7.

There is a lot of information here, but what does it mean? One would like to see more examples to evaluate the spatial heterogeneity of strain histories, but, nevertheless, there are some features of this sample that are interesting in light of the regional distribution of structural fabrics in Taiwan (Fig. 10.13). In the collisional mountain belt of Taiwan, there is a downdip stretching lineation in the foreland, and along-strike stretching lineation in the hinterland. This spatial variation in finite strain could be explained by partitioning of oblique convergence into downdip stretching in the foreland (the pro-wedge that faces the incoming flux of material from the Asian passive margin), and strike slip shearing in the hinterland (the retro-wedge that faces the colliding volcanic arc (Fig. 10.13)). One intriguing possibility is that the strain history of the sample, early downdip extension followed by late along-strike extension, reflects the change from downdip shearing to along-strike shearing when the rock is advected through this fixed displacement field within the mountain belt.

10.2.4 Cleavage-perpendicular sections and passive fibers

For an example of a non-coaxial strain history in a cleavage-perpendicular section, we turn to a pyrite pressure shadow from a 25°-dipping fold limb within the Marcellus Shale of central Pennsylvania (Fig. 10.14a). We select an external reference frame for the X_1 axis (cleavage or bedding) and then center the origin on the pyrite host. The assumption is that this section contains the maximum shortening and maximum extension direction (the X_1–X_3 plane). In this example, we assume that the antiaxial syntectonic fibers have deformed passively, and each increment of strain affects the length and orientation of the fiber segments that grew during prior increments. As in the cleavage-parallel example, we work to progressively "undeform" the pressure shadow, beginning with the last increment. For the last increment (the fiber segment

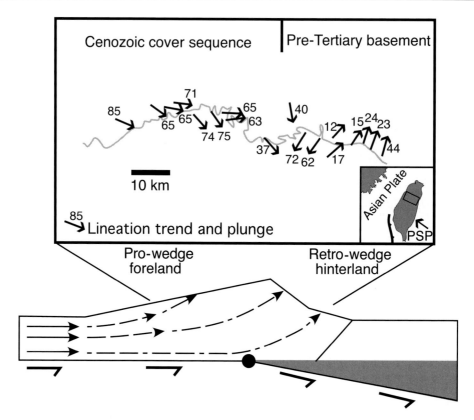

Figure 10.13 Map depicting orientation of tectonic lineations along the Central Cross-Island Highway of Taiwan. Inset shows transect location. PSP is Philippine Sea plate. Lower diagram shows an interpretation for the advective flow paths through a double-sided mountain belt, given accretion in the foreland and erosion off the surface (after Willett et al., 1993).

adjacent to the pyrite surface for antiaxial fibers), the initial points $^n(X_1, X_3)$ relate to the position after deformation $^n(x_1, x_3)$ by the equation

$$^n x_i = \left(\frac{\partial x_i}{\partial X_j}\right) {}^n X_j$$

or in matrix form:

$$^n \mathbf{x} = {}^n \mathbf{F} \cdot {}^n \mathbf{X}$$

The undeformed ($^n\mathbf{X}$) and deformed ($^n\mathbf{x}$) coordinates in our arbitrary reference frame X_1–X_3 can be rewritten in terms of the coordinates in a reference frame that is parallel to the maximum stretch ($^{pn}\mathbf{X}$) and ($^{pn}\mathbf{x}$) through a vector coordinate transformation (Chapter 3):

$$\begin{aligned} ^{pn}X_i &= {}^n a_{ij} {}^n X_j \\ ^{pn}x_i &= {}^n a_{ij} {}^n x_j \end{aligned} \qquad (10.27)$$

or

$$\begin{aligned} ^{pn}\mathbf{X} &= {}^n \mathbf{a} \cdot {}^n \mathbf{X} \\ ^{pn}\mathbf{x} &= {}^n \mathbf{a} \cdot {}^n \mathbf{x} \end{aligned}$$

10.2 Determination of a strain history

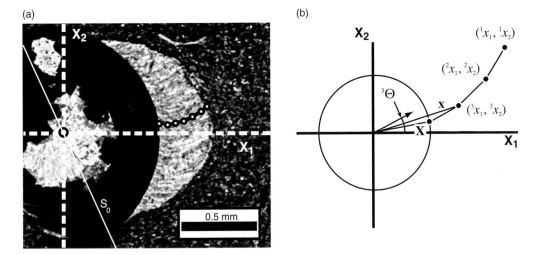

Figure 10.14 (a) Photomicrograph of pressure shadow from a cleavage-perpendicular section of Marcellus Formation in central Pennsylvania. Curved antitaxial calcite fibers developed on near-spherical pyrite framboids. Dashed line shows the orientation of core (vertical). (b) Points used in Equation 10.30. Note that S_1 for third increment in diagram is not parallel to x–X as it was in the cleavage-parallel case. All displacements are parallel to pure shear displacement paths from Figure 10.1, but S_1 direction is not parallel to the X_1 axis.

where the superscript p refers to a reference frame parallel to the maximum and minimum principal stretches $^p(X_1 - X_3)$. a allows a change of coordinates from our arbitrary reference frame to a reference frame parallel to the principal stretches, and the $^n\Theta$ contained within a is the difference in orientation between the two reference frames. Now we have four equations and five unknowns, $^nX_1$, $^nX_3$, $^nx_1$, $^nx_3$, and $^n\Theta$. In the reference frame parallel to the maximum principal stretch:

$$^{pn}\mathbf{x} = {^{pn}\mathbf{F}} \cdot {^{pn}\mathbf{X}}$$

or

$$^{pn}\begin{bmatrix} x_1 \\ x_3 \end{bmatrix} = {^n\begin{bmatrix} S_1 & 0 \\ 0 & S_3 \end{bmatrix}} {^{pn}\begin{bmatrix} X_1 \\ X_3 \end{bmatrix}}$$

$$^nS_1 = \frac{^{pn}x_1}{^{pn}X_1}$$
$$^nS_3 = \frac{^{pn}x_3}{^{pn}X_3} \tag{10.28}$$

If we make the assumption that there is no change in area ($\det {^n[\mathbf{F}]} = 1$), then

$$^{pn}X_1 {^{pn}}X_3 = {^{pn}}x_1 {^{pn}}x_3 \tag{10.29}$$

At this point, we can substitute the expressions in Equation 10.27 into Equation 10.29 and solve for the orientation of incremental extension for the last increment of strain, $^n\Theta$ (Fig. 10.14b):

$$\tan 2{^n\Theta} = \frac{2({^nx_1}{^nx_3} - {^nX_1}{^nX_3})}{\left({^nx_1}^2 - {^nx_3}^2 - {^nX_1}^2 + {^nX_3}^2\right)} \tag{10.30}$$

This equation describes the orientation of maximum extension relative to an arbitrary axis for a pure shear displacement from **X** to **x**. Now that we have calculated $^n\Theta$ for the nth increment, we can determine the coordinates for the start and end of the displacement in a reference frame parallel to the maximum stretch with Equation 10.27. We determine the principal strains using Equation 10.28. Then $^{pn}\mathbf{F}$ in the reference frame parallel to the principal stretches can be transformed into our arbitrary reference frame:

$$^nF_{ij} = R_{ik}R_{jl}{}^{pn}F_{kl}$$

We are now in a position to use the deformation gradient matrix for the last increment to "undeform" all the points along the fiber to their positions prior to the last increment of strain. Note that

$$^n\mathbf{X} = {}^n\mathbf{F}^{-1} \cdot {}^n\mathbf{x} \qquad (10.31)$$

Following the procedure described in Section 10.2.2, we calculate all the incremental deformation gradient tensors from the last increment to the first. The eigenvalues and eigenvectors of these tensors are used to characterize the cumulative incremental strain history (Fig. 10.15a). The finite strains are determined by multiplying the deformation gradient tensors in sequence and determining the eigenvalues and eigenvectors of the Green deformation tensor after each increment to construct the progressive finite strain history (Fig. 10.15b).

10.2.5 Geological applications of strain histories in cleavage-perpendicular planes

In a cleavage-perpendicular section that contains the maximum and minimum stretch orientations, the external rotation recorded by fibers can be interpreted as due to simple shear (i.e., rotational strain, Fig. 10.16b), or spin, where the rocks have rotated through a fixed orientation of pure shear stretching (Fig. 10.16a). The correct interpretation requires an assessment of the strain history in the context of geographic (vertical) and internal (bedding) reference frames. In the example here, the strain history is characterized by a gradual 25° clockwise rotation of the incremental extension direction (Fig. 10.15). A simple shear parallel to bedding is unlikely

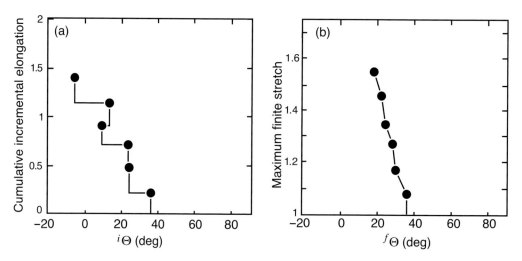

Figure 10.15 (a) Cumulative incremental strain history and (b) progressive finite strain history for the example in Figure 10.14.

10.2 Determination of a strain history

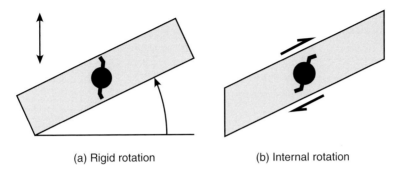

(a) Rigid rotation   (b) Internal rotation

Figure 10.16 Pressure shadow geometries for (a) rotation through a fixed extension direction and (b) simple shear parallel between bedding planes (internal rotation).

as the sense of internal rotation would be opposite to what is observed; the early fiber segments should rotate towards bedding (Fig. 10.16b), yet the early fiber segments are bedding-perpendicular (Figs. 10.15a and 10.16a). Both the magnitude and sense of rotation are consistent with the interpretation that early fiber growth occurred during layer-parallel shortening, but was followed by rigid rotation related to fold limb rotation through a fixed near-vertical stretching direction.

We finish with a MATLAB function, **Fibers**, that computes and plots the incremental and progressive strain histories of syntectonic fibers on cleavage-parallel (kk = 0) or cleavage-perpendicular (kk = 1) sections. The user should enter the reference plane, the center of the pyrite, and **x** and **y** points along the fiber from an image displayed in MATLAB. Try this function on the pressure shadows of Figures 10.11 and 10.14. Remember that, since these are antiaxial fiber growths, you should start digitizing the fiber at the margin of the pyrite grain (last increment of fiber growth).

```
function [cie,pfs] = Fibers(imageName,kk)
%Fibers determines the incremental and finite strain history of a fiber in
%a pressure shadow
%
%   USE: [cie,pfs] = Fibers(imageName,kk)
%
%   image: A character corresponding to the image filename, including
%          extension (eg. = 'fileName.jpg')
%   kk = An integer that indicates whether the fiber is on a cleavage
%        parallel (kk = 0), or cleavage perpendicular (kk = 1) section
%   cie = cumulative incremental elongation: column 1 = Incremental theta,
%         column 2 = cumulative incremental maximum elongation
%   pfs = progressive finite strain history: column 1 = Finite theta,
%         column 2 = maximum stretch magnitude
%
%   NOTE: Output theta angles are in radians

%Read and display image
IMG=imread(imageName);
imagesc(IMG);
```

```
%Prompt the user to define a reference plane. If the current reference
%plane is not satisfactory, the user can re-select the input points
a='n';
while a=='n'
    clf; %Clear figure
    imagesc(IMG); %Display image
    hold on;
    disp('Select two points along the reference plane, from left to right.');
    [refpx, refpy] = ginput(2);
    refpx = round (refpx); %Rounds imput x points to nearest integer
    refpy = round (refpy); %Rounds input y points to nearest integer
    plot(refpx,refpy,'--y','LineWidth',1.5);
    a=input('Would you like to keep the current reference plane? (y/n) ','s');
end

%Prompt the user to select the origin and fiber points from the image
%display. The origin is defined at the center of the pyrite sphere.
%The fiber points are selected sequentially along a single fiber path.
%If the current fiber path is not satisfactory, the user can re-select the
%input points
a='n';
while a=='n'
    clf; %Clear figure
    imagesc(IMG); %Display image
    hold on;
    plot(refpx,refpy, '--y', 'LineWidth',1.5)
    disp ('Select the origin point, center of pyrite sphere.');
    [xo, yo] = ginput(1); %Select center of grain as the origin
    xo=round(xo); yo=round(yo); %Rounds positions to nearest integer value
    plot (xo,yo,'ok','MarkerFaceColor','k', 'MarkerSize',8) %Plots origin
    %Digitize points along fiber
    disp ('Digitize points along the fiber');
    disp ('Left mouse button picks points');
    disp ('Right mouse button picks last point');
    x = []; y = []; n = 0; but = 1;
    while but == 1
        n = n + 1;
        [xi,yi,but] = ginput(1);
        xi=round(xi); %Rounds point coords to nearest integer
        yi=round(yi);
        plot (xi,yi,'-or','LineWidth',1.5); %Plots point
        x(n) = xi; y(n) = yi; %Add point to fiber path
    end
    a=input('Would you like to keep the current fiber path? (y/n) ', 's');
end
hold off;
```

10.2 Determination of a strain history

```
%Start calculation

%Switch y values from screen coordinates with (0,0) at the upper left
%corner to cartesian coordinates, with (0,0) at the lower left corner
nrow=size(IMG,1); %Number of rows in image
yo=nrow-yo;
y=nrow-y;
refpy = nrow-refpy;

%Set origin of coordinate system at center of pyrite sphere
x=x-xo;
y=y-yo;

%Rotate all points into a reference frame parallel to X1
phi=atan((refpy(2)-refpy(1))/(refpx(2)-refpx(1)));
Rot=[cos(phi) sin(phi);-sin(phi) cos(phi)];
vec=[x;y];
newvec=Rot*vec;
x=newvec(1,:);
y=newvec(2,:);

%Initialize some variables
cie = zeros(n-1,2);
rotmat = zeros(2,2,n-1);
finmat = zeros(2,2,n-1);
elong = zeros(1,n-1);
C = zeros(2,2,n-1);
pfs = zeros(n-1,2);

%Incremental, inverse modeling of pressure shadow (Backwards)
for i=1:n-1
    %If cleavage parallel section (Equation 10.21)
    if kk == 0
        cie(n-i,1)=atan((y(2)-y(1))/(x(2)-x(1)));
    %If cleavage perpendicular section (Equation 10.30)
    elseif kk == 1
        cie(n-i,1)=(atan((2*(x(2)*y(2)-x(1)*y(1)))/...
            (x(2)^2-y(2)^2-x(1)^2+y(1)^2)))/2;
    end
    Beta=[cos(cie(n-i,1)) sin(cie(n-i,1));-sin(cie (n-i,1))...
        cos(cie(n-i,1))];
    %If cleavage parallel face
    if kk == 0
        h=[x(1);y(1)];
        H=[x(2);y(2)];
        v0=H-h;
        v1=h/norm(h);
```

```
            v2=v0/norm(v0);
            Alpha=acos(dot(v1,v2));
            initlength=norm(h)*cos(Alpha);
            st1inc=(norm(v0)+initlength)/initlength;
            posmat=[st1inc 0;0 1];
        %If cleavage perpendicular section
        elseif kk == 1
            Bigx1=Beta*[x(1);y(1)];
            Bigx2=Beta*[x(2);y(2)];
            st1inc=(Bigx2(1)/Bigx1(1));
            st3inc=(Bigx2(2)/Bigx1(2));
            posmat=[st1inc 0;0 st3inc];
        end
        rotmat(:,:,n-i)=Beta'*posmat*Beta;
        elong(n-i)=st1inc-1;
        for j=1:n-i
            newposition = rotmat(:,:,n-i)\[x(j+1); y(j+1)];
            x(j)=newposition(1);
            y(j)=newposition(2);
        end
end

%Plot cummulative incremental maximum elongation
figure;
cie(:,2)=cumsum(elong); %Cummulative, incremental, maximum elongation
plot(cie(:,1)*180/pi,cie(:,2),'o');
xlabel('Theta incremental deg');
ylabel('Cumulative incremental elongation')
axis([-90 90 0 max(cie(:,2))+0.5]);

%Compute progressive finite strain (Forward)
finmat(:,:,1)=rotmat(:,:,1);
for i=2:n-1
    finmat(:,:,i)=rotmat(:,:,i)*finmat(:,:,i-1);
end
%Determine Cauchy deformation tensor
for i=1:n-1
    C(:,:,i)=finmat(:,:,i)'*finmat(:,:,i);
    %Stretch magnitude and orientation: Maximum eigenvalue and their
    %corresponding eigenvectors of Cauchy's tensor. Use MATLAB function eig
    [V,D]=eig(C(:,:,i));
    pfs(i,2)=sqrt(D(2,2));
    pfs(i,1)=atan(V(2,2)/V(1,2));
end

%Plot Progressive finite strain
figure
plot(pfs(:,1)*180/pi, pfs(:,2), 'o');
```

```
xlabel('Theta finite deg');
ylabel('Progressive Finite Strain');
axis([-90 90 1 max(pfs(:,2))+0.5]);
end
```

10.3 EXERCISES

1. Given pure shear deformation, derive an expression that relates the orientation of a line relative to the maximum stretching direction before deformation (θ) to the orientation of the line relative to the maximum stretching direction after deformation (θ') (Fig. 10.17).
2. Given simple shear deformation, derive an expression that relates the orientation of a line relative to the shear direction before deformation (ϕ) to the orientation of the line relative to the shear direction after deformation (ϕ') (Fig. 10.18).
3. The photo is an outcrop of a siltstone-shale sequence with bedding planes depicted by white lines (Figure 10.19). The dashed white line is parallel to the trace of quartz veins that were originally planar. (a) Assume that deformation is characterized by simple shear parallel to bedding in shale layers. If the veins in the undeformed state were parallel to the vein orientation in the more competent siltstone layers, what is the orientation and magnitude of the maximum stretch in the shale layers? (b) If the deformation was characterized by pure shear with extension parallel to bedding, what is the orientation and magnitude of the maximum stretch in the shale layers?
4. In the photomicrograph (Fig. 10.20), a circular, rigid siderite porphyroblast has an internal foliation (S_i) that is rotated relative to the penetrative foliation outside the porphyroblast (S_1). Either S_1 has rotated relative to a porphyroblast that does not rotate (pure shear), the porphyroblast has rotated relative to S_1 that does not rotate (simple shear parallel to S_1), or

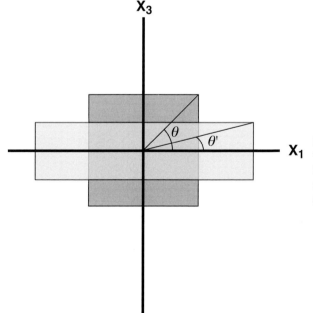

Figure 10.17 Pure shear deformation where a passive marker at an angle θ from the X_1 axis before deformation makes an angle θ' with the X_1 axis after deformation.

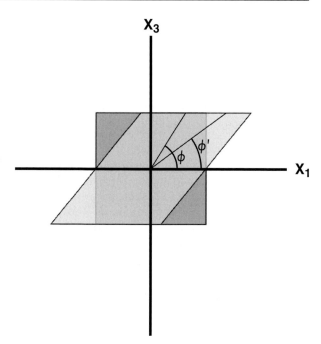

Figure 10.18 Simple shear deformation where a passive marker at an angle ϕ from the X_1 axis before deformation makes an angle Φ' with the X_1 axis after deformation.

Figure 10.19 Outcrop photo of pervasive quartz veins refracted at contacts between interbedded siltstone and shale layers from the Kodiak Formation of Afognak Island, Alaska.

10.3 Exercises

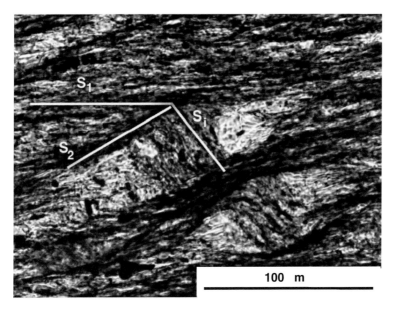

Figure 10.20 Siderite porphyroblast showing apparent rotation of an internal fabric (S_i) relative to external fabric (S_1) during development of S_2.

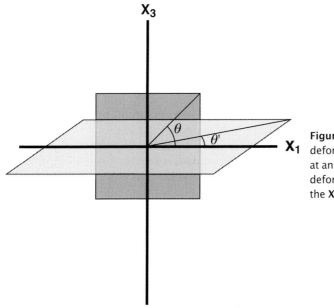

Figure 10.21 Sub-simple shear deformation where a passive marker at an angle θ from the X_1 axis before deformation makes an angle θ' with the X_1 axis after deformation.

both have rotated (general shear or simple shear oblique to S_1). (a) If we assume passive rotation of S_1 in response to pure shear (with coaxial stretching parallel to S_2), what is the magnitude of the maximum stretch? (b) If we assume simple shear parallel to S_1 and a rotation of an equant inclusion by $\gamma/2$ (Ghosh and Ramberg, 1976; Jeffrey, 1922), what is the magnitude of the maximum stretch? (c) If we assume general shear (with pure shear

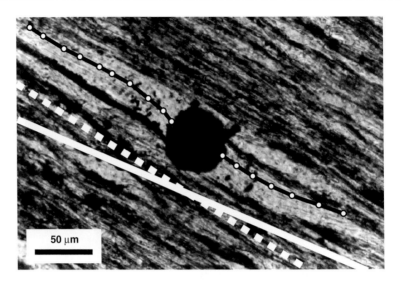

Figure 10.22 Pyrite pressure shadow with bedding shown by solid white line and cleavage by dashed line (Fisher *et al.*, 2002).

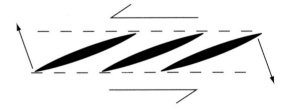

Figure 10.23 Sketch of en echelon veins. Shear zone boundaries shown by dashed lines.

   shortening perpendicular to S_1, simple shear parallel to S_1, and an incremental stretch parallel to S_2), what is the kinematic vorticity number W_k?
5. Given a general shear, derive an expression that relates the orientation of a line relative to the shear plane before deformation (θ) to the orientation of a line relative to the shear plane after deformation (θ') (Fig. 10.21).
6. Determine the cumulative incremental and progressive finite strain histories for the example depicted in Figure 10.22. (a) What is the magnitude of finite strain? (b) What is the total rotation of the maximum stretching direction? Is the external rotation likely due to rigid rotation or internal rotation? (c) If the bedding is parallel to the stable shear plane of sub-simple shear and the latest fiber segment is parallel to the incremental stretch, what is the kinematic vorticity number W_k? Hint: Use functions `Fibers` and `GeneralShear`.
7. For the en echelon vein set in Figure 10.23, assume that the veins are confined to the shear zone and that they open parallel to the incremental maximum stretch direction. What is the kinematic vorticity number W_k?

CHAPTER ELEVEN

Velocity description of deformation

11.1 INTRODUCTION

There are almost as many types of models as there are reasons for constructing them. At one extreme, a qualitative interpretation of the history of a region may be described as a "model." We have all seen titles like: "A Tectonic model for the Little Jackass Creek Quadrangle." At the other extreme, full-fledged mechanical models incorporate a complete set of constitutive relationships in a computational or analytical framework. In this chapter, we present one type of numerical model that falls between these two extremes. It is based on a limited set of largely kinematic and geometric assumptions, while ignoring forces, rock properties, equations of equilibrium, constitutive relationships, etc. The purpose of these *kinematic models* is to simulate structural geometries and visualize the evolution of structures through time. Because they can be executed quickly, kinematic models can be run thousands or millions of times to test large parameter spaces. Do not fall into the trap, however, of thinking that they "explain" the deformation!

Kinematic modeling uses ad-hoc velocity fields that satisfy known boundary conditions, and obey reasonable assumptions such as conservation of mass throughout deformation. Strictly speaking, the velocity fields used have no mechanical or dynamical significance. They are just convenient models to simulate observed structures from a descriptive (i.e., in terms of strain) rather than a genetic (i.e., in terms of stress) manner (Marrett and Peacock, 1999). A discussion of the advantages and disadvantages of kinematic with respect to mechanical modeling is beyond the scope of this book. The interested reader can consult Marrett and Peacock (1999), and Pollard (2000).

Besides being an excellent topic to illustrate the application of the concepts we have learned so far (e.g., coordinate transformations, vector operations), there are several advantages in using velocities to describe deformation (Waltham and Hardy, 1995): (1) The method is general and applicable to any kind of deformation, and (2) time-evolving parameters that are influenced by deformation – such as temperature, pressure, erosion, and sedimentation – can be easily modeled once the deformation velocities are specified. An additional advantage of kinematic models is often overlooked: If we can run a kinematic model forward, we can also run it backward. This proves to be an extremely useful property to solve a type of problems known as "inverse" problems.[1] In these problems, we are not so much interested in forward modeling deformation, but rather in finding the model that best replicates a structure as we observe it today. We will talk more about inverse problems in the next chapter.

11.2 THE CONTINUITY EQUATION

The starting assumption of all velocity models is that mass is conserved. Matter is neither created nor destroyed and little if any is converted into energy. This condition is specified by the *continuity equation*:

$$\frac{\partial \rho}{\partial t} + \rho \frac{\partial v_i}{\partial x_i} = 0$$

This equation basically says that the change in density (ρ) with respect to time (t) of a volume plus the flux of mass in and out of the volume (given by the second term of the equation, **v** is velocity) must be equal to zero. The continuity equation can also be written as

$$\frac{\partial \rho}{\partial t} + \rho \nabla \mathbf{v} = 0 \qquad (11.1)$$

where $\nabla \mathbf{v}$ is the *divergence of the velocity field*. If there are no changes in volume, such as compaction or thermal expansion during deformation, the density remains constant and Equation 11.1 reduces to

$$\nabla \mathbf{v} = \left[\frac{\partial v_1}{\partial x_1} + \frac{\partial v_2}{\partial x_2} + \frac{\partial v_3}{\partial x_3} \right] = 0 \qquad (11.2)$$

This condition is known as incompressibility. In some tectonic settings such as thrust belts, the wavelength of individual structures is short compared to their strike parallel dimension. In this case, we can assume *plane strain*. This means that there is no velocity parallel to strike (the $\mathbf{x_3}$ axis):

$$\frac{\partial v_3}{\partial x_3} = 0$$

Thus, cross-sectional area perpendicular to strike must be conserved and we can write the two-dimensional form of the incompressibility criterion:

$$\nabla \mathbf{v} = \left[\frac{\partial v_1}{\partial x_1} + \frac{\partial v_2}{\partial x_2} \right] = 0 \qquad (11.3)$$

How do we use this equation? In general our approach will be to assume a relationship for one of the two velocity components (usually v_1), and then use Equation 11.3 and the *boundary conditions* to calculate the other component (usually v_2). A few examples will make this clearer.

---

[1] Note that this is a different use of the term "inverse" than when we talked about inverting a matrix in a previous chapter.

11.3 Pure and simple shear in terms of velocities

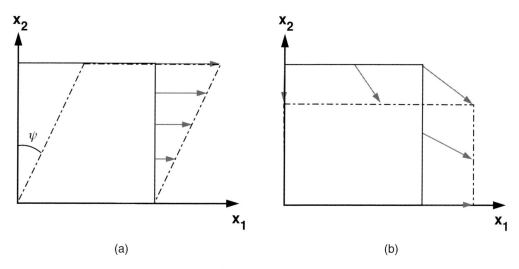

Figure 11.1 Velocity description of (a) simple shear, and (b) pure shear. Continuous rectangle shows the initial state, and dashed rectangle the deformed state. Gray arrows are velocity vectors.

11.3 PURE AND SIMPLE SHEAR IN TERMS OF VELOCITIES

Simple shear is a trivial example but a good place to start on understanding the velocity modeling approach. In simple shear (Fig. 11.1a), the velocity in the x_1 direction, v_1, is constant, and depends only on the x_2 coordinate:

$$v_1 = x_2 \tan \psi = x_2 \gamma \tag{11.4}$$

Therefore

$$\frac{\partial v_1}{\partial x_1} = 0$$

and from Equation 11.3

$$\frac{\partial v_2}{\partial x_2} = 0$$

Integrating to solve for v_2

$$v_2 = C \tag{11.5}$$

and since there is no movement in the x_2 direction ($v_2 = 0$, Fig. 11.1a), $C = 0$.

The case for pure shear deformation (Fig. 11.1b) is more interesting. The velocity in the x_1 direction, v_1, varies linearly with position in x_1, so we can write

$$v_1 = ax_1 \quad \text{and} \quad \frac{\partial v_1}{\partial x_1} = a \tag{11.6}$$

Using the incompressibility condition (Eq. 11.3):

$$\frac{\partial v_1}{\partial x_1} + \frac{\partial v_2}{\partial x_2} = a + \frac{\partial v_2}{\partial x_2} = 0 \quad \text{so} \quad \frac{\partial v_2}{\partial x_2} = -a$$

Integrating to solve for v_2:

$$v_2 = -ax_2 + C \tag{11.7}$$

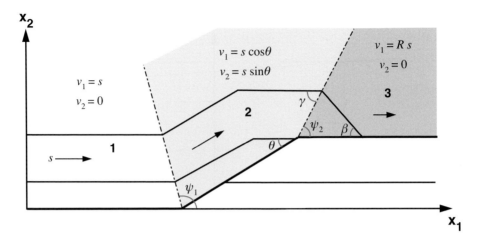

Figure 11.2 Geometric model of simple step, fault-bend folding with the controlling parameters together with the three velocity domains. Arrows show the velocity in each domain.

and using the boundary condition that there is no movement in the x_2 direction at the base of the block ($v_2 = 0$ at $x_2 = 0$, Fig. 11.1b), $C = 0$.

11.4 GEOLOGICAL APPLICATION: FAULT-RELATED FOLDING

In the upper brittle crust, folds are often associated with faults. In this section, we will use velocity models of deformation to describe two types of fault-related folds: those formed by movement of a rock sequence above a non-planar fault (i.e., fault-bend folds), and those formed by deformation at the tip of a propagating fault (i.e., fault-propagation folds).

11.4.1 Fault-bend folding

The basic equations for constant layer thickness fault-bend folding come from Suppe (1983). For a step off a horizontal decollement (Fig. 11.2):

$$\tan\theta = \frac{\sin 2\gamma}{(2\cos^2\gamma + 1)} \quad \beta = 180° - 2\gamma \quad R = \frac{\sin(\gamma - \theta)}{\sin\gamma} \tag{11.8}$$

where θ is the dip of the fault, γ is the axial angle of the fold, β is the dip of the forelimb, and R is the ratio that describes how slip diminishes across the hanging wall cutoff of the fold.

The model has three distinct regions or domains with different velocities, and in each domain the velocity is parallel to the local fault orientation (Fig. 11.2). Fault slip rate s is conserved between domains 1 and 2, but is consumed across domains 2 and 3 (Fig. 11.2). The horizontal (v_1) and vertical (v_2) velocities in the three domains are (Hardy, 1995)

$$\text{Domain 1: } v_1 = s, \quad v_2 = 0$$

$$\text{Domain 2: } v_1 = s\cos\theta, \quad v_2 = s\sin\theta \tag{11.9}$$

$$\text{Domain 3: } v_1 = Rs, \quad v_2 = 0$$

11.4 Geological application: Fault-related folding

You can see that since v_1 and v_2 within each domain are constant, the incompressibility criterion (area conservation, Eq. 11.3) is satisfied.

Besides incompressibility, another condition, known as *strain compatibility* (Chapter 9), should be fulfilled by the model. The three velocity domains must remain in contact, without overlaps or gaps. This condition requires that all points along a fault or a velocity boundary $f(x_1)$ obey the following equation (Waltham and Hardy, 1995):

$$^1v_2 - {}^1v_1 \frac{\partial f}{\partial x_1} = {}^2v_2 - {}^2v_1 \frac{\partial f}{\partial x_1} \tag{11.10}$$

where the superscripts 1 and 2 refer to the domains to be entered and exited, respectively. For the domains 1-2 boundary:

$$s\sin\theta - s\cos\theta \frac{\partial f}{\partial x_1} = -s\frac{\partial f}{\partial x_1}$$

which, rearranging, gives

$$\frac{\partial f}{\partial x_1} = \frac{\sin\theta}{(\cos\theta - 1)} \tag{11.11}$$

For $\theta = 30°$, $\tan^{-1}(\partial f/\partial x_1)$ or ψ_1 (Fig. 11.2) = 125°. This equation predicts that the velocity boundary between domains 1 and 2 is the bisector of the lower bend in the decollement. For the domains 2-3 boundary:

$$-Rs\frac{\partial f}{\partial x_1} = s\sin\theta - s\cos\theta \frac{\partial f}{\partial x_1}$$

and

$$R = \cos\theta - \frac{\sin\theta}{(\partial f/\partial x_1)} \tag{11.12}$$

This equation predicts the change of slip across the boundary between domains 2 and 3 for any inclination of the boundary (Hardy and Poblet, 1995; Hardy, 1995). When the boundary is that given by Suppe's equations (Eq. 11.8) for $\theta = 30°$, γ and $\psi_2 = 60°$ (Fig. 11.2). Using these values in Equation 11.12 gives a slip ratio R of 0.577, which is identical to that predicted by Equation 11.8.

Fault-bend folds grow as a result of kink band migration during two stages (Suppe, 1983). In the first stage (Fig. 11.3a), the displacement is less than the length of the ramp. The kink bands marked by AA' and BB' grow longer; the anticline increases in height and the crest decreases in width. The orientation of the boundary between domains 2 and 3 (kink A, Fig. 11.3a) is given by Equation 11.12. In the second stage (Fig. 11.3b), the displacement is greater than the length of the ramp; the crest of the anticline increases in width but stops growing in height. The orientation of the boundary between domains 2 and 3 (kink A, Fig. 11.3b) is given by Equation 11.11.

The function `FaultBendFold`, below, plots the evolution of a simple step, Mode 1 (Suppe, 1983) fault-bend fold. `FaultBendFold` uses the function `SuppeEquation` to compute γ from the input θ, and from these two, R (Eq. 11.8). The remaining part of the program deals with the identification of the velocity domains and the application of the velocities of Equation 11.9 to the bedding points, as the structure grows. To plot the evolution of a simple step fault-bend fold with a 25° dipping ramp, type in MATLAB®:

```
yp = [50,100,150,200,250]; %Beds datums
psect = [1000,500]; %Section parameters
pramp = [400,25*pi/180,100]; %Ramp parameters
```

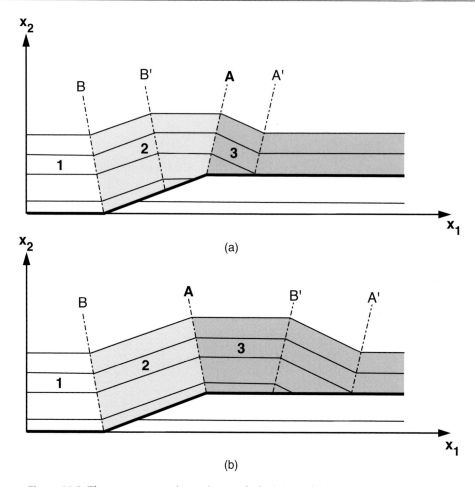

Figure 11.3 The two stages in the evolution of a fault-bend fold. (a) Fault displacement is less than the length of the ramp. (b) Fault displacement is greater than the length of the ramp. Numbers indicate the velocity domains.

```
pslip = [300,1]; %Slip parameters
frames = FaultBendFold(yp,psect,pramp,pslip); %Make fold
```

You will see a movie of the structure's evolution. To watch the movie again type:

```
movie(frames);
```

```
function frames = FaultBendFold(yp,psect,pramp,pslip)
%FaultBendFold plots the evolution of a simple step, Mode I fault bend fold
%
% USE: frames = FaultBendFold(yp,psect,pramp,pslip)
%
% yp = Datums or vertical coordinates of undeformed, horizontal beds
% psect = A 1 x 2 vector containing the extent of the section, and the
%         number of points in each bed
% pramp = A 1 x 3 vector containing the x coordinate of the lower bend in
```

11.4 Geological application: Fault-related folding

```
%         the decollement, the ramp angle, and the height of the ramp
% pslip = A 1 x 2 vector containing the total and incremental slip
% frames = An array structure containing the frames of the fold evolution
%         You can play the movie again just by typing movie(frames)
%
% NOTE: Input ramp angle should be in radians
%
% FaultBendFold uses function SuppeEquation

%Extent of section and number of points in each bed
extent = psect(1); npoint = psect(2);
%Make undeformed beds geometry: This is a grid of points along the beds
xp=0.0:extent/npoint:extent;
[XP,YP]=meshgrid(xp,yp);

%Fault geometry and slip
xramp = pramp(1); ramp = pramp(2); height = pramp (3);
slip = pslip(1); sinc = pslip(2);
%Number of slip increments
ninc=round(slip/sinc);

%Ramp angle cannot be greater than 30 degrees, and if it is 30 degrees,
%make it a little bit smaller to avoid convergence problems
if ramp > 30*pi/180
    error('ramp angle cannot be more than 30 degrees');
elseif ramp == 30*pi/180
    ramp=29.9*pi/180;
end

%Minimize Eq. 11.8 to obtain gamma from the input ramp angle (theta)
options=optimset('display','off');
gamma = fzero('SuppeEquation',1.5,options,ramp);
%Compute slip ratio R (Eq. 11.8)
R = sin(gamma - ramp)/sin(gamma);

%Make fault geometry
xf=[0 xramp xramp+height/tan(ramp) 1.5*extent];
yf=[0 0 height height];
%From the origin of each bed compute the number of points that are in the
%hanging wall. These points are the ones that will move
hwid = zeros(size(yp,2));
for i=1:size(yp,2)
    if yp(i) <= height
        hwid(i)=0;
        for j=1:size(xp,2)
            if xp(j) <= xramp + yp(i)/tan(ramp)
                hwid(i)= hwid(i)+1;
            end
```

```
            end
    else
        hwid(i)=size(xp,2);
    end
end

%Deform beds: Apply velocity fields of Eq. 11.9
%Loop over slip increments
for i=1:ninc
    %Loop over number of beds
    for j=1:size(XP,1)
        %Loop over number of hanging wall points in each bed
        for k=1:hwid(j)
            %If point is in domain 1
            if XP(j,k) < xramp - YP(j,k)*tan(ramp/2)
                XP(j,k) = XP(j,k) + sinc;
                YP(j,k) = YP(j,k);
            else
                %If point is in domain 2
                if YP(j,k) < height
                    XP(j,k) = XP(j,k) + sinc*cos(ramp);
                    YP(j,k) = YP(j,k) + sinc*sin(ramp);
                else
                    %If stage 1 of fault bend fold (Fig. 11.3a)
                    if i*sinc*sin(ramp) < height
                        %If point is in domain 2
                        if XP(j,k) < xramp + height/tan (ramp) +...
                                (YP(j,k)-height)*tan (pi/2-gamma)
                            XP(j,k) = XP(j,k) + sinc*cos (ramp);
                            YP(j,k) = YP(j,k) + sinc*sin (ramp);
                        %If point is in domain 3
                        else
                            XP(j,k)= XP(j,k) + sinc*R;
                            YP(j,k)= YP(j,k);
                        end
                    %If stage 2 of fault bend fold (Fig. 11.3b)
                    else
                        %If point is in domain 2
                        if XP(j,k) < xramp + height/tan (ramp)-...
                                (YP(j,k)-height)*tan (ramp/2)
                            XP(j,k)= XP(j,k) + sinc*cos (ramp);
                            YP(j,k)= YP(j,k) + sinc*sin (ramp);
                        %If point is in domain 3
                        else
                            XP(j,k) = XP(j,k) + sinc*R;
                            YP(j,k) = YP(j,k);
                        end
                    end
                end
```

11.4 Geological application: Fault-related folding

```
                    end
                end
            end
        end
        %Plot increment
        %Fault
        plot(xf,yf,'r-','LineWidth',2);
        hold on;
        %Beds
        for j=1:size(yp,2)
            %If below ramp
            if yp(j) <= height
                plot(XP(j,1:1:hwid(j)),YP(j,1:1:hwid (j)),'k-');
                plot(XP(j,hwid(j)+1:1:size(xp,2)),YP(j, hwid(j)+...
                    1:1:size(xp,2)),'k-');
            %If above ramp
            else
                plot(XP(j,:),YP(j,:),'k-');
            end
        end
        %Plot settings
        text(0.8*extent,1.75*max(yp),strcat('Slip = ', num2str(i*sinc)));
        axis equal;
        axis([0 extent 0 2.0*max(yp)]);
        hold off;
        %Get frame for movie
        frames(i) = getframe;
    end
end

function y = SuppeEquation(gamma,theta)
%SuppeEquation: First equation in Eq. 11.8 for fault bend folding

y = sin(2*gamma)/(2*(cos(gamma))^2+1) - tan(theta);
end
```

11.4.2 Similar folding over curving fault bends

Where bends in the faults are curved rather than straight, structural geologists commonly use inclined simple shear and the resulting similar folds to make kinematic models of folds over ramps. Hanging-wall, rollover anticlines above listric normal faults represent the most common application of this approach. Here, again, the velocity boundary condition of Equation 11.10 comes to our rescue. In this case, the velocity boundary condition is not a kink axial surface but the fault itself. As elsewhere in this chapter, we follow the development of Waltham and Hardy (1995).

To start, we define a coordinate system with x_1 horizontal and x_2 vertical and specify α, the angle between the vertical and the direction of inclined shear. The direction of inclined shear fixes the orientation of the x'_2 axis of a second coordinate system, with x'_1 perpendicular to it

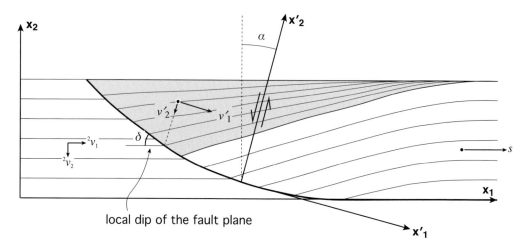

Figure 11.4 Geometric model of a similar fold produced by inclined shear with angle α. Two coordinate systems are used to solve this problem: one geographic (x_1-x_2) and another based on the direction of shear ($x'_1-x'_2$). Arrows show the velocities in the hanging wall and footwall.

(Fig. 11.4). The angle α then defines the two-dimensional coordinate transformation between the two coordinate systems:

$$a_{ij} = \begin{pmatrix} \cos\alpha & \cos(90+\alpha) \\ \cos(90-\alpha) & \cos\alpha \end{pmatrix} = \begin{pmatrix} \cos\alpha & -\sin\alpha \\ \sin\alpha & \cos\alpha \end{pmatrix}$$

Note that, normally, the coordinate transformation shown in Figure 11.4 would be considered a rotation by $-\alpha$, but we are using the common convention that antithetic shear (i.e., in the opposite sense from the main fault) is positive. Using the transformation of vectors equation, 3.6, we can write the relationship of the velocity of a point undergoing inclined shear in two coordinate systems as

$$\begin{aligned} v'_1 &= a_{11}v_1 + a_{12}v_2 = v_1\cos\alpha - v_2\sin\alpha \\ v'_2 &= a_{21}v_1 + a_{22}v_2 = v_1\sin\alpha + v_2\cos\alpha \end{aligned} \quad (11.13a)$$

and the reverse transformation:

$$\begin{aligned} v_1 &= a_{11}v'_1 + a_{21}v'_2 = v'_1\cos\alpha + v'_2\sin\alpha \\ v_2 &= a_{12}v'_1 + a_{22}v'_2 = -v'_1\sin\alpha + v'_2\cos\alpha \end{aligned} \quad (11.13b)$$

In the inclined simple shear model, the velocity perpendicular to the shearing, v'_1, is assumed to be constant and, furthermore, we have the boundary condition that where the fault is horizontal outside of the zone of inclined shear, the horizontal velocity is the fault slip rate s, and the vertical velocity is zero (Fig. 11.4). Using this condition and Equation 11.13a, we can thus write

$$v'_1 = s\cos\alpha$$

Substituting this expression into Equation 11.13b, we get

$$\begin{aligned} v_1 &= s\cos^2\alpha + v'_2\sin\alpha \\ v_2 &= -s\cos\alpha\sin\alpha + v'_2\cos\alpha \end{aligned} \quad (11.14)$$

We can now use the velocity boundary condition from Equation 11.10 for the listric normal fault shown in Figure 11.4, assuming that the footwall velocity is defined by $^2v_1$ and $^2v_2$ (Fig. 11.4). Substituting and rearranging Equations 11.14 to solve for v'_2:

11.4 Geological application: Fault-related folding

$$v'_2 = \frac{s\left[\cos\alpha\sin\alpha + \left(\frac{\partial f}{\partial x_1}\right)\cos^2\alpha\right] + {}^2v_2 - {}^2v_1\frac{\partial f}{\partial x_1}}{\cos\alpha - \frac{\partial f}{\partial x_1}\sin\alpha} \qquad (11.15)$$

where

$$\frac{\partial f}{\partial x_1} = \frac{\partial x_2}{\partial x_1} = \tan\delta$$

δ is the local dip of the fault at a position directly down the shear plane from the point of interest (Fig. 11.4). Equation 11.15 can be simplified for the case where the footwall is treated as stationary. The value for v'_2 can be substituted into Equations 11.14 to solve for v_1 and v_2 at any point. The function `SimilarFold`, below, plots the evolution of a similar fold with a fixed footwall. To produce a rollover with a shear angle α of 30° type:

```
yp = [50,100,150,200,250]; %Beds datums
psect = [1000,500]; %Section parameters
pslip = [200,1]; %Slip parameters
frames = SimilarFold(yp,psect,30*pi/180,pslip); %Make fold with 30 deg shear
```

Upon pressing return, you will be asked to digitize the geometry of the listric fault in a figure window. After that, you will see a movie of the fold's evolution.

```
function frames = SimilarFold(yp,psect,alpha,pslip)
%SimilarFold plots the evolution of a similar fold
%
% USE: frames = SimilarFold(yp,psect,alpha,pslip)
%
% yp = Datums or vertical coordinates of undeformed, horizontal beds
% psect = A 1 x 2 vector containing the extent of the section, and the
%         number of points in each bed
% alpha = Shear angle. Positive for shear antithetic to the fault and
%         negative for shear synthetic to the fault
% pslip = A 1 x 2 vector containing the total and incremental slip
% frames = An array structure containing the frames of the fold evolution
%          You can play the movie again just by typing movie(frames)
%
% NOTE: Use positive pslip for a normal fault

%Extent of section and number of points in each bed
extent = psect(1); npoint = psect(2);

%Make undeformed beds geometry: This is a grid of points along the beds
xp=0.0:extent/npoint:extent;
[XP,YP]=meshgrid(xp,yp);

%Slip and number of slip increments
slip = pslip(1); sinc = pslip(2);
ninc=round(slip/sinc);
```

```
%Prompt the user to select the geometry of the fault. If the current fault
%trajectory is not satisfactory, the user can re-select the input points
a='n';
while a=='n'
    %Plot beds
    for i=1:size(yp,2)
        plot(XP(i,:),YP(i,:),'k-');
        hold on;
    end
    axis equal;
    axis( [0 extent 0 2.0*max(yp)]);
    %Digitize fault
    disp ('Digitize a listric fault shallowing to the right');
    disp ('Left mouse button picks points');
    disp ('Right mouse button picks last point');
    fault = []; n = 0; but = 1;
    while but == 1
        n = n + 1;
        [xi,yi,but] = ginput(1);
        plot (xi,yi,'-or','LineWidth',1.5); %Plots point
        fault(n,1) = xi; fault(n,2) = yi; %Add point to fault
    end
    hold off;
    a=input('Would you like to keep the current fault? (y/n) ', 's');
end

%Sort fault points in x
fault = sortrows(fault,1);
xf = fault(:,1)';
yf = fault(:,2)';

%Find tangent of dip of fault segments: df/dx
dfx = zeros(1,n);
for i=1:n-1
    dfx(i) = (yf(i+1)-yf(i))/(xf(i+1)-xf(i));
end
dfx(n) = dfx(n-1);

%From the origin of each bed compute the number of points that are in the
%footwall. These points won't move
fwid = zeros(size(yp,2));
%Find y of fault below/above bed points
yfi = interp1(xf,yf,xp,'linear','extrap');
for i=1:size(yp,2)
    fwid(i)=0;
    for j=1:size(xp,2)
        if yp(i) < yfi(j)
            fwid(i) = fwid(i) + 1;
```

11.4 Geological application: Fault-related folding

```
            end
        end
end

%Coordinate transformation matrix between horizontal-vertical coordinate
%system and coordinate system parallel and perpendicular to shear direction
a11=cos(alpha);
a12=-sin(alpha);
a21=sin(alpha);
a22=a11;

%Transform fault and beds to coordinate system parallel and perpendicular
%to shear direction
xfS = xf*a11+yf*a12; %Fault
XPS = XP*a11+YP*a12; %Beds
YPS = XP*a21+YP*a22;

%Compute deformation
%Loop over slip increments
for i=1:ninc
    %Loop over number of beds
    for j=1:size(XPS,1)
        %Loop over number of bed points in hanging wall
        for k=fwid(j)+1:size(XPS,2)
            %Find local tangent of fault dip: df/dx
            if XPS(j,k) <= xfS(1)
                ldfx = dfx(1);
            elseif XPS(j,k) >= xfS(n)
                ldfx = dfx(n);
            else
                a = 'n'; L = 1;
                while a=='n'
                    if XPS(j,k) >= xfS(L) && XPS(j,k) < xfS(L+1)
                        ldfx = dfx(L);
                        a = 's';
                    else
                        L = L + 1;
                    end
                end
            end
            %Compute velocities perpendicular and along shear direction
            %Equations 11.13 and 11.15
            vxS = sinc*a11;
            vyS = (sinc*(a11*a21+ldfx*a11^2))/(a11-ldfx*a21);
            %Move point
            XPS(j,k) = XPS(j,k) + vxS;
            YPS(j,k) = YPS(j,k) + vyS;
        end
    end
```

```
%Transform beds back to geographic coordinate system
XP=XPS*a11+YPS*a21;
YP=XPS*a12+YPS*a22;

%Plot increment
%Fault
plot(xf,yf,'r-','LineWidth',2);
hold on;
%Beds
for j=1:size(yp,2)
    %Footwall
    plot(XP(j,1:1:fwid(j)), YP(j,1:1:fwid(j)),'k-');
    %Hanging wall
    plot(XP(j,fwid(j)+1:1:size(XP,2)),...
        YP(j,fwid(j)+1:1:size(XP,2)),'k-');
end
%Plot settings
text(0.8*extent,1.75*max(yp),strcat('Slip = ',num2str(i*sinc)));
axis equal;
axis( [0 extent 0 2.0*max(yp)]);
hold off;
%Get frame for movie
frames(i) = getframe;
end
end
```

11.4.3 Fault-propagation folding

Fault-propagation folds consume slip at the tip of a propagating fault. Here, we will concentrate on two types of fault-propagation folding velocity models: kink models, which include fixed axis and parallel models, and the trishear model.

Fixed axis kink model

The basic equations for fixed axis fault-propagation folding come from Suppe and Medwedeff (1990). For a step off a horizontal decollement (Fig. 11.5) and assuming no excess shear, the fold interlimb half angles $(\gamma_1, \gamma_i, \gamma_i^*, \gamma_e, \gamma_e^*)$ are related to the dip of the fault (θ) by the following equations:

$$\gamma_1 = \frac{180° - \theta}{2} \quad \gamma_e^* = \cot^{-1}\left[\frac{3 - 2\cos\theta}{2\sin\theta}\right] \quad \gamma_i^* = \gamma_1 - \gamma_e^*$$

$$\gamma_e = \cot^{-1}(\cot\gamma_e^* - 2\cot\gamma_1) \quad \gamma_i = \sin^{-1}\left[\frac{\sin\gamma_i^*\sin\gamma_e}{\sin\gamma_e^*}\right]$$

(11.16)

11.4 Geological application: Fault-related folding

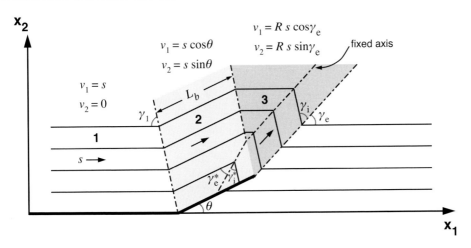

Figure 11.5 Geometric model of simple step, fixed axis fault-propagation folding with the controlling parameters together with the three velocity domains. Arrows show the velocity in each domain. Notice that the kink axis at the top of the forelimb is fixed (material does not flow through it).

The kink axis at the top of the forelimb is assumed to be fixed (no material flows through it), and the forelimb is allowed to thin or thicken (Fig. 11.5). In addition, the ratio of back limb (i.e., ramp) length, L_b, to fault slip, which is equivalent to the fault propagation to fault slip ratio (P/S) is (Suppe et al., 1992):

$$P/S = \frac{\cot \gamma_e^* - \cot \gamma_1}{\frac{1}{\sin \theta} - \frac{\sin \gamma_i / \sin \gamma_e}{\sin(\gamma_e + \gamma_i - \theta)}} + \frac{\sin(\gamma_1 + \theta)}{\sin \gamma_1} \qquad (11.17)$$

This ratio is constant in the model, and is equal to 2.0 (you can convince yourself by computing Eqs. 11.16 and 11.17 with different values of θ). The fixed axis model has three velocity domains (Fig. 11.5), and in each of these domains the velocities are (Hardy and Poblet, 1995)

$$\begin{aligned}
&\text{Domain 1: } v_1 = s, \quad v_2 = 0 \\
&\text{Domain 2: } v_1 = s \cos \theta, \quad v_2 = s \sin \theta \\
&\text{Domain 3: } v_1 = R s \cos \gamma_e, \quad v_2 = R s \sin \gamma_e
\end{aligned} \qquad (11.18)$$

where s is the fault slip rate, γ_e is the dip of the front axial surface and R is the change in slip across the boundary between domains 2 and 3 (Fig. 11.5). Notice again that since v_1 and v_2 within each domain are constant, the incompressibility criterion (Eq. 11.3) is satisfied. R is given by (Hardy and Poblet, 1995)

$$R = \frac{\sin(\gamma_1 + \theta)}{\sin(\gamma_1 + \gamma_e)} \qquad (11.19)$$

The following function, **FixedAxisFPF**, plots the evolution of a simple step, fixed axis fault-propagation fold. The structure of the program is similar to that of function **FaultBendFold**. From the input ramp angle θ, Equations 11.16, 11.17, and 11.19 are solved, and then the velocities of Equation 11.18 are applied to the bedding points as the structure grows. To make a fixed axis fault-propagation fold with a 20° dipping ramp, type:

```
yp = [50 100 150 200 250]; %Beds datums
psect = [1000 500]; %Section parameters
pramp = [400 20*pi/180]; %Fault parameters
pslip = [100 0.5]; %Slip parameters
frames = FixedAxisFPF(yp,psect,pramp,pslip); %Make fold
```

And to make one with a 40° dipping ramp type:

```
pramp = [400 40*pi/180]; %Fault parameters
frames = FixedAxisFPF(yp,psect,pramp,pslip); %Make fold
```

You will see thickening of the forelimb for the 20° dipping ramp, and thinning of the forelimb for the 40° dipping ramp. In the case of a step up from a decollement (Fig. 11.5), the forelimb thickens if the ramp angle $\theta < 29°$ and thins if $\theta > 29°$ (Suppe and Medwedeff, 1990). For $\theta = 29°$, bed length and thickness normal to bedding are preserved, and the fixed axis model is identical to the parallel model. Try running the program with other θ angles. You will find that it is not possible to produce an anticline with an overturned limb. That is a major limitation of the fixed axis model.

```
function frames = FixedAxisFPF(yp,psect,pramp,pslip)
%FixedAxisFPF plots the evolution of a simple step, fixed axis
%fault propagation fold
%
% USE: frames = FixedAxisFPF(yp,psect,pramp,pslip)
%
% yp = Datums or vertical coordinates of undeformed, horizontal beds
% psect = A 1 x 2 vector containing the extent of the section, and the
%         number of points in each bed
% pramp = A 1 x 2 vector containing the x coordinate of the lower bend in
%         the decollement, and the ramp angle
% pslip = A 1 x 2 vector containing the total and incremental slip
% frames = An array structure containing the frames of the fold evolution
%         You can play the movie again just by typing movie(frames)
%
% NOTE: Input ramp angle should be in radians

% Base of layers
base = yp(1);

%Extent of section and number of points in each bed
extent = psect(1); npoint = psect(2);
%Make undeformed beds geometry: This is a grid of points along the beds
xp=0.0:extent/npoint:extent;
[XP, YP]=meshgrid(xp,yp);

%Fault geometry and slip
xramp = pramp(1); ramp = pramp(2);
slip = pslip(1); sinc = pslip(2);
%Number of slip increments
ninc=round(slip/sinc);
```

11.4 Geological application: Fault-related folding

```
%Solve model parameters
%First equation of Eq. 11.16
gam1=(pi-ramp)/2.;
%Second equation of Eq. 11.16
gamestar = acot((3.-2.*cos(ramp))/(2.*sin(ramp)));
%Third equation of Eq. 11.16
gamistar=gam1-gamestar;
%Fourth equation of Eq. 11.16
game=acot(cot(gamestar)-2.*cot(gam1));
%Fifth equation of Eq. 11.16
gami = asin((sin(gamistar)*sin(game))/sin(gamestar));
%Ratio of backlimb length to total slip (P/S)(Eq. 11.17)
a1=cot(gamestar)-cot(gam1);
a2=1./sin(ramp)-(sin(gami)/sin(game))/sin(game+gami-ramp);
a3=sin(gam1+ramp)/sin(gam1);
lbrat=a1/a2 + a3;
%Change in slip between domains 2 and 3 (Eq. 11.19)
R=sin(gam1+ramp)/sin(gam1+game);
%From the origin of each bed compute the number of points that are in the
%hanging wall. These points are the ones that will move. Notice that this
%has to be done for each slip increment, since the fault propagates
hwid = zeros(ninc,size(yp,2));
for i=1:ninc
    uplift = lbrat*i*sinc*sin(ramp);
    for j=1:size(yp,2)
        if yp(j)-base<=uplift
            hwid(i,j)=0;
            for k=1:size(xp,2)
                if xp(k) <= xramp + (yp(j)-base)/tan(ramp)
                    hwid(i,j)=hwid(i,j)+1;
                end
            end
        else
            hwid(i,j)=size(xp,2);
        end
    end
end

%Deform beds. Apply velocity fields of Eq. 11.18
%Loop over slip increments
for i=1:ninc
    % Compute uplift
    lb = lbrat*i*sinc;
    uplift = lb*sin(ramp);
    lbh = lb*cos(ramp);
    % Compute point at fault tip
    xt = xramp + lbh;
    yt = base + uplift;
    %Loop over number of beds
```

```
        for j=1:size(XP,1)
            %Loop over number of hanging wall points in each bed
            for k=1:hwid(i,j)
                %If point is in domain 1
                if XP(j,k) < xramp - (YP(j,k)-base)/tan(gam1)
                    XP(j,k) = XP(j,k) + sinc;
                else
                    %If point is in domain 2
                    if XP(j,k) < xt - (YP(j,k)-yt)/tan(gam1)
                        XP(j,k) = XP(j,k) + sinc*cos(ramp);
                        YP(j,k) = YP(j,k) + sinc*sin(ramp);
                    else
                        %If point is in domain 3
                        if XP(j,k) < xt + (YP(j,k)-yt)/tan(game)
                            XP(j,k) = XP(j,k) + sinc*R*cos(game);
                            YP(j,k) = YP(j,k) + sinc*R*sin(game);
                        end
                    end
                end
            end
        end
    end
    %Plot increment
    %Fault
    xf=[0 xramp xramp+lbh];
    yf=[base base uplift+base];
    plot(xf,yf,'r-','LineWidth',2);
    hold on;
    %Beds
    for j=1:size(yp,2)
        %If beds cut by the fault
        if yp(j)-base <= uplift
            plot(XP(j,1:1:hwid(i,j)), YP(j,1:1:hwid(i,j)),'k-');
            plot(XP(j,hwid(i,j)+1:1:size(xp,2)), YP(j,hwid(i,j)+...
                1:1:size(xp,2)),'k-');
        %If beds not cut by the fault
        else
            plot(XP(j,:), YP(j,:),'k-');
        end
    end
    %Plot settings
    text(0.8*extent,1.75*max(yp),strcat('Slip = ',num2str(i*sinc)));
    axis equal;
    axis([0 extent 0 2.0*max(yp)]);
    hold off;
    %Get frame for movie
    frames(i) = getframe;
end
end
```

11.4 Geological application: Fault-related folding

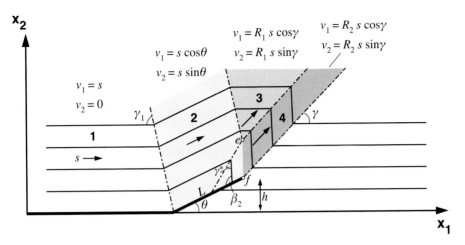

Figure 11.6 Geometric model of simple step, parallel fault-propagation folding with the controlling parameters together with the four velocity domains. Arrows show the velocity in each domain.

Parallel kink model

In the parallel model, bed length and layer thickness are preserved, and overturned limbs are allowed. The basic equations for a simple step, parallel fault-propagation fold (Fig. 11.6) are (Suppe and Medwedeff, 1990)

$$\frac{1 + 2\cos^2\gamma^*}{\sin 2\gamma^*} + \frac{\cos\theta - 2}{\sin\theta} = 0 \quad (11.20)$$

$$\gamma_1 = 90° - \theta/2; \quad \gamma = 90° + \gamma^* - \gamma_1; \quad \beta_2 = 180° - 2\gamma^*$$

where θ is the dip of the fault and $\gamma_1, \gamma, \gamma^*$ are axial angles as in Fig. 11.6.

The problem with the parallel model is that the backlimb length L_b is only equal to the fault length when $\theta = 29°$. Therefore, there is no direct way of deriving the fault propagation to fault slip ratio P/S. This problem was solved by Hardy (1997) using geometrical relations between fault slip and the height of the ramp (h), length of the ramp (L), and length of the forelimb (ef, Fig. 11.6). Here we just give the solution:

$$P/S = \frac{1}{\left[1 - \dfrac{\sin\theta}{\sin(2\gamma - \theta)}\right]} \quad (11.21)$$

Contrary to the fixed axis model, the P/S varies with θ in the parallel model. Figure 11.7 shows the variation of P/S for θ angles in the range 10–50°. For θ from 10 to 29°, the P/S increases approximately linearly from 1.55 to 2.0, whereas above this angle the P/S increases rapidly with θ reaching a value of 5.25 at $\theta = 50°$. At $\theta = 29°$ the P/S of the fixed axis and parallel models is identical.

The parallel model has four velocity domains (Fig. 11.6). The velocities in these domains are (Hardy, 1997)

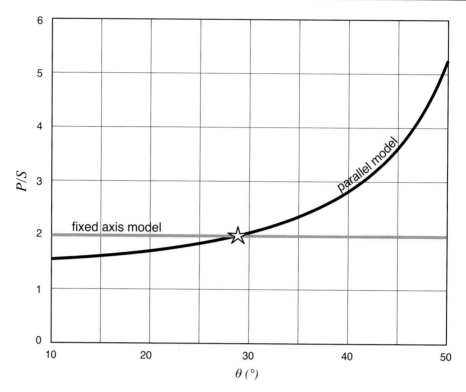

Figure 11.7 Variation of fault propagation to fault slip ratio (P/S) with step-up angle (θ), for simple step, fixed axis and parallel fault-propagation folds. Notice that the P/S is constant in fixed axis folds, while it is variable in parallel fault-propagation folds and increases with θ. Star shows the location where fixed axis and parallel models intersect and are identical.

$$\begin{aligned}
&\text{Domain 1: } v_1 = s, \quad v_2 = 0 \\
&\text{Domain 2: } v_1 = s\cos\theta, \quad v_2 = s\sin\theta \\
&\text{Domain 3: } v_1 = R_1 s\cos\gamma, \quad v_2 = R_1 s\sin\gamma \\
&\text{Domain 4: } v_1 = R_2 s\cos\gamma, \quad v_2 = R_2 s\sin\gamma
\end{aligned} \qquad (11.22)$$

where R_1 and R_2 are the slip ratios between the regions 2 and 3, and 2 and 4, respectively. R_1 and R_2 are given by (Hardy and Poblet, 2005)

$$R_1 = \frac{\sin(\gamma_1 + \theta)}{\sin(\gamma_1 + \gamma)}$$

$$R_2 = \frac{\sin\beta_2}{\sin(\beta_2 - \theta + \gamma)} \qquad (11.23)$$

The function `ParallelFPF`, below, plots the evolution of a simple step, parallel fault-propagation fold. The program's structure is similar to that of the program before: First the

11.4 Geological application: Fault-related folding

program computes the model parameters (Eqs. 11.20, 11.21 and 11.23), and then it applies the velocities of Equation 11.22 to the bedding points. **ParallelFPF** uses function **SuppeEquationTwo** to compute y^* from the input θ (Eq. 11.20). To make a parallel fault-propagation fold with a 20° dipping ramp, type:

```
yp = [50 100 150 200 250]; %Beds datums
psect = [1000 500]; %Section parameters
pramp = [400 20*pi/180]; %Fault parameters
pslip = [100 0.5]; %Slip parameters
frames = ParallelFPF(yp,psect,pramp,pslip); %Make fold
```

And to make one with a 40° dipping ramp type:

```
pramp = [400 40*pi/180]; %Fault parameters
frames = ParallelFPF(yp,psect,pramp,pslip); %Make fold
```

You will see that the 20° ramp produces a fold with an overturned forelimb, while the 40° ramp results in a fold with an upright forelimb. Overturned limbs are produced by ramp angles $\theta < 25°$ (Suppe and Medwedeff, 1990). Also, since the kink axis at the top of the forelimb is not fixed, there is flow of material through it. At $\theta < 29°$ material rolls from the forelimb onto the crest of the fold, whereas at $\theta > 29°$ material from the crest rolls onto the forelimb (Zapata and Allmendinger, 1996). At $\theta = 29°$ there is no flow of material through the kink axis (fixed axis and parallel models are identical). This has important implications for the geometry of growth strata as we will see in Section 11.5.

```
function frames = ParallelFPF(yp,psect,pramp,pslip)
%ParallelFPF plots the evolution of a simple step, parallel
%fault propagation fold
%
% USE: frames = ParallelFPF(yp,psect,pramp,pslip)
%
% yp = Datums or vertical coordinates of undeformed, horizontal beds
% psect = A 1 x 2 vector containing the extent of the section, and the
%         number of points in each bed
% pramp = A 1 x 2 vector containing the x coordinate of the lower bend in
%         the decollement, and the ramp angle
% pslip = A 1 x 2 vector containing the total and incremental slip
% frames = An array structure containing the frames of the fold evolution.
%          You can play the movie again just by typing movie(frames)
%
% NOTE: Input ramp angle should be in radians
%
% ParallelFPF uses function SuppeEquationTwo

% Base of layers
base = yp(1);

%Extent of section and number of points in each bed
extent = psect(1); npoint = psect(2);
%Make undeformed beds geometry: This is a grid of points along the beds
xp=0.0:extent/npoint:extent;
```

```
[XP, YP]=meshgrid(xp,yp);

%Fault geometry and slip
xramp = pramp(1); ramp = pramp(2);
slip = pslip(1); sinc = pslip(2);
%Number of slip increments
ninc=round(slip/sinc);

%Solve model parameters
%Solve first equation in Eq. 11.20 by minimizing SuppeEquationTwo
options=optimset('display','off');
gamstar = fzero('SuppeEquationTwo',0.5,options,ramp);
%Solve second equation in Eq. 11.20
gam1 = pi/2. - ramp/2.;
%Solve third equation in Eq. 11.20
gam = pi/2.+gamstar-gam1;
%Solve fourth equation in Eq. 11.20
bet2 = pi - 2.*gamstar;
%Other angle for computation
kap = pi - bet2 + ramp;
%Eq. 11.21
lbrat = 1./(1.-sin(ramp)/sin(2.*gam-ramp));
%Eq. 11.23
R1=sin(gam1+ramp)/sin(gam1+gam);
R2=sin(bet2)/sin(bet2-ramp+gam);

%From the origin of each bed compute the number of points that are in the
%hanging wall. These points are the ones that will move. Notice that this
%has to be done for each slip increment, since the fault propagates
hwid = zeros(ninc,size(yp,2));
for i=1:ninc
    uplift = lbrat*i*sinc*sin(ramp);
    for j=1:size(yp,2)
        if yp(j)-base<=uplift
            hwid(i,j)=0;
            for k=1:size(xp,2)
                if xp(k) <= xramp + (yp(j)-base)/tan(ramp)
                    hwid(i,j)=hwid(i,j)+1;
                end
            end
        else
            hwid(i,j)=size(xp,2);
        end
    end
end

%Deform beds: Apply velocity fields of Eq. 11.22
%Loop over slip increments
```

11.4 Geological application: Fault-related folding

```
for i=1:ninc
    % Compute uplift
    lb = lbrat*i*sinc;
    uplift = lb*sin(ramp);
    lbh = lb*cos(ramp);
    % Compute distance ef in Figure 11.6
    ef=uplift/sin(2.*gamstar);
    % Compute fault tip
    xt=xramp+lbh;
    yt=base+uplift;
    % Compute location e in Figure 11.6
    xe=xt+ef*cos(kap);
    ye=yt+ef*sin(kap);
    %Loop over number of beds
    for j=1:size(XP,1)
        %Loop over number of hanging wall points in each bed
        for k=1:hwid(i,j)
            %If point is in domain 1
            if XP(j,k) < xramp - (YP(j,k)-base)/tan(gam1)
                XP(j,k) = XP(j,k) + sinc;
            else
                % if y lower than y at e
                if YP(j,k) < ye
                    %If point is in domain 2
                    if XP(j,k) < xt + (YP(j,k)-yt)/tan(kap)
                        XP(j,k) = XP(j,k) + sinc*cos(ramp);
                        YP(j,k) = YP(j,k) + sinc*sin(ramp);
                    else
                        %If point is in domain 4
                        if XP(j,k) < xt + (YP(j,k)-yt)/ tan(gam)
                            XP(j,k) = XP(j,k) + sinc*R2*cos (gam);
                            YP(j,k) = YP(j,k) + sinc*R2*sin (gam);
                        end
                    end
                % if y higher than y at e
                else
                    %If point is in domain 2
                    if XP(j,k) < xe - (YP(j,k)-ye)/tan(gam1)
                        XP(j,k) = XP(j,k) + sinc*cos(ramp);
                        YP(j,k) = YP(j,k) + sinc*sin(ramp);
                    else
                        %If point is in domain 3
                        if XP(j,k) < xe + (YP(j,k)-ye)/tan(gam)
                            XP(j,k) = XP(j,k) + sinc*R1*cos(gam);
                            YP(j,k) = YP(j,k) + sinc*R1*sin(gam);
                        else
                            %If point is in domain 4
                            if XP(j,k) < xt + (YP(j,k)-yt)/tan (gam)
```

```
                                        XP(j,k) = XP(j,k) + sinc*R2*cos (gam);
                                        YP(j,k) = YP(j,k) + sinc*R2*sin (gam);
                                    end
                                end
                            end
                        end
                    end
                end
            end
        end
        %Plot increment
        %Fault
        xf=[0 xramp xramp+lbh];
        yf=[base base uplift+base];
        plot(xf,yf,'r-','LineWidth',2);
        hold on;
        %Beds
        for j=1:size(yp,2)
            %If beds cut by the fault
            if yp(j)-base <= uplift
                plot(XP(j,1:1:hwid(i,j)), YP(j,1:1:hwid(i,j)),'k-');
                plot(XP(j,hwid(i,j)+1:1:size(xp,2)),...
                    YP(j,hwid(i,j)+1:1:size(xp,2)),'k-');
            %If beds not cut by the fault
            else
                plot(XP(j,:), YP(j,:),'k-');
            end
        end
        %Plot settings
        text(0.8*extent,1.75*max(yp),strcat('Slip = ',num2str(i*sinc)));
        axis equal;
        axis([0 extent 0 2.0*max(yp)]);
        hold off;
        %Get frame for movie
        frames(i) = getframe;
    end
end

function y = SuppeEquationTwo(gamstar,ramp)
%SuppeEquationTwo: First equation in Eq. 11.20 for parallel fault
%propagation folding

y = (1.+2.*cos(gamstar)*cos(gamstar))/sin(2.*gamstar) +...
    (cos(ramp)-2.)/sin(ramp);
end
```

Trishear model

In the kink models above, the relations between fault parameters (e.g., ramp angle) and fold parameters (e.g., interlimb angles) are established at the start of deformation, and the fold

11.4 Geological application: Fault-related folding

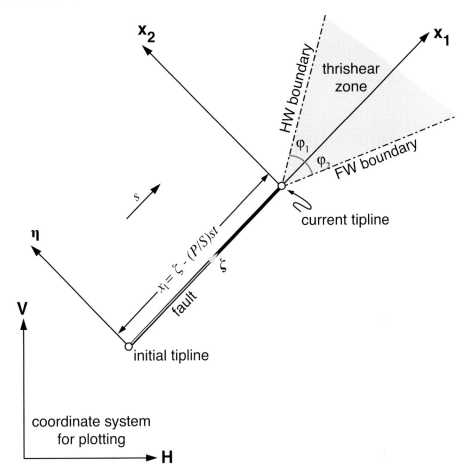

Figure 11.8 Geometry of the trishear model showing the coordinate systems used in the model.

grows in a similar fashion by increasing in size but not changing in shape. This allows the derivation of mathematical rules to predict fold geometry from fault geometry and vice versa (e.g., Eqs. 11.16 and 11.20). Kink models, however, cannot explain some of the features commonly observed in fault propagation folds, such as footwall synclines and changes in stratigraphic thickness and dip on forelimbs. These features are better explained by the *trishear kinematic model* (Allmendinger, 1998; Erslev, 1991).

The velocity field for the trishear model was derived by Zehnder and Allmendinger (2000); here we follow the same line of reasoning. In trishear, the displacement along the fault is accommodated by deformation in a triangular shear zone radiating from the fault tip (Fig. 11.8). The footwall is held fixed and the hanging wall moves rigidly along the fault at the fault slip rate s. The movement of the fault tip is determined by the P/S, which in trishear, unlike kink models, is an input parameter independent of ramp angle θ or other parameters. The velocity field is defined in a coordinate system attached to the fault tip and with axes parallel and perpendicular to the fault line ($\mathbf{x}_1 - \mathbf{x}_2$ in Fig. 11.8). We seek to construct a velocity field in the trishear zone that conserves area, is continuous, and matches the velocities at the hanging wall and footwall boundaries of the zone. The boundary conditions are

$$v_1 = s \quad v_2 = 0 \quad \text{on} \quad x_2 = x_1 \tan\varphi_1$$
$$v_1 = 0 \quad v_2 = 0 \quad \text{on} \quad x_2 = -x_1 \tan\varphi_2 \tag{11.24}$$

To find the velocities, we will choose a v_1 field consistent with Equation 11.24, and then determine v_2 from Equations 11.24 and 11.3 (incompressibility). Let us assume that the trishear zone is symmetric, $\varphi_1 = \varphi_2 = \varphi$. The simplest v_1 field that you can think of is where v_1 varies linearly from one side of the triangular shear zone to the other:

$$v_1 = \frac{s}{2}\left(\frac{x_2}{x_1 \tan\varphi} + 1\right)$$

We can make this equation more general by specifying a "concentration factor" c, which allows for non-linear variation in v_1 as a function of the power $1/c$. To simplify writing the equations, let $m = \tan\varphi$:

$$v_1 = \frac{s}{2}\left[\text{sgn}(x_2)\left(\frac{|x_2|}{x_1 m}\right)^{1/c} + 1\right] \quad x_1 > 0 \quad -x_1 m \leq x_2 \leq x_1 m \quad c \geq 1 \tag{11.25}$$

where $\text{sgn}(x_2)$ denotes the sign of x_2. It can easily be seen that the above field satisfies the v_1 boundary conditions in Equation 11.24. To find v_2, we differentiate Equation 11.25 with respect to x_1, invoke incompressibility (Eq. 11.3),

$$\frac{\partial v_2}{\partial x_2} = -\frac{\partial v_1}{\partial x_1}$$

and integrate with respect to x_2 giving

$$v_2 = \frac{sm}{2(1+c)}\left(\frac{|x_2|}{x_1 m}\right)^{(1+c)/c} + C$$

The constant of integration, C, is found by using the v_2 boundary conditions in Equation 11.24. The resulting v_2 field in the trishear zone is

$$v_2 = \frac{sm}{2(1+c)}\left[\left(\frac{|x_2|}{x_1 m}\right)^{(1+c)/c} - 1\right] \tag{11.26}$$

For $c = 1$, the v_1 velocity distribution (Eq. 11.25) is linear in x_2, producing a strain rate that is nearly uniform with respect to x_2. We call this field the "linear field." Velocity vectors and v_1, v_2 variations across the trishear zone for this case are plotted in Figure 11.9a for $\varphi = 30°$. As c increases, the deformation concentrates towards the center of the trishear zone, producing non-uniform strain rates. Figure 11.9b shows the velocity vectors and v_1, v_2 variations for $c = 3$ and $\varphi = 30°$. Note that as $x_2 \to -x_1 \tan\varphi$, $v_2/v_1 \to -\tan\varphi$, i.e., as the footwall trishear boundary is approached, the velocity vectors are parallel to the boundary.

Given the velocity field of Equations 11.25 and 11.26, the resultant deformation is easily computed. We introduce three coordinate systems (Fig. 11.8): the horizontal-vertical system (**H** – **V**), which is used for inputting and plotting the data; the $\zeta - \eta$ system, which is attached to the initial fault tip and is stationary; and the $\mathbf{x_1}-\mathbf{x_2}$ system, which is attached to the fault tip and moves at a speed $(P/S)s$. At the start of the deformation, the $\zeta - \eta$ and $\mathbf{x_1}-\mathbf{x_2}$ systems overlap, and at a later time t they are related by

$$x_1 = \zeta - (P/S)st \quad x_2 = \eta \tag{11.27}$$

The normal flow of a trishear program is then as follows: (1) Bedding data are entered in the **H** – **V** coordinate system; (2) the data are transformed to the $\zeta - \eta$ system; (3) at each increment

11.4 Geological application: Fault-related folding

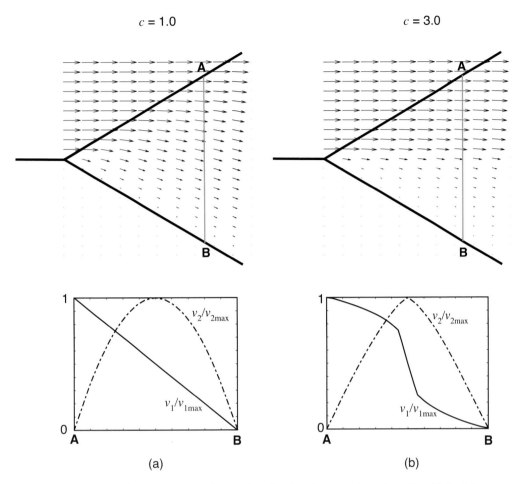

Figure 11.9 Velocity vectors and variation of velocities v_1 and v_2 along line AB for (a) symmetric, linear v_1 field ($c = 1.0$), and (b) symmetric, non-linear v_1 field with $c = 3.0$.

of deformation x_1, x_2 (Eq. 11.27) and v_1, v_2 (Eqs. 11.25 and 11.26) are computed, and ζ, η are updated accordingly; and (4) the data are transformed back to the $\mathbf{H} - \mathbf{V}$ coordinate system and plotted. The following MATLAB function, `Trishear`, carries out all these steps. `Trishear` uses function `VelTrishear` to compute the velocity field. To make a contractional, trishear fault-propagation fold with initial fault tip ($x_1 = 300$, $x_2 = 50$), ramp angle $= 30°$, $P/S = 1.5$, trishear angle $= 60°$, fault slip $= 100$ units, and concentration factor $= 1.0$, type:

```
yp = [50 80 110 140 170]; %Beds datums
psect = [1000 500]; %Section parameters
tparam = [300 50 30*pi/180 1.5 60*pi/180 100 1.0]; %Trishear parameters
sinc = 1.0; %Slip parameters
frames = Trishear(yp,psect,tparam,sinc); %Make trishear fold
```

You will see that the geometry of the fold is not similar, and that geometry and finite strain vary along and across the stratigraphy with proximity to the fault tip. Trishear fold geometries are richer than those of kink models. The drawback, however, is that because trishear folds are not similar, there are no mathematical or geometric rules to relate fold geometry to fault geometry. We will talk about the solution to this problem in the next chapter.

```
function frames = Trishear(yp,psect,tparam,sinc)
%Trishear plots the evolution of a 2D trishear fault propagation fold
%
% USE: frames = Trishear(yp,psect,tparam,sinc)
%
% yp = Datums or vertical coordinates of undeformed, horizontal beds
% psect = A 1 x 2 vector containing the extent of the section, and the
%           number of points in each bed
% tparam = A 1 x 7 vector containing: the x coordinate of the fault tip
%           (entry 1), the y coordinate of the fault tip (entry 2), the
%           ramp angle (entry 3), the P/S (entry 4), the trishear angle
%           (entry 5), the fault slip (entry 6), and the concentration
%           factor (entry 7)
% sinc = slip increment
% frames = An array structure containing the frames of the fold evolution
%           You can play the movie again just by typing movie(frames)
%
% NOTE: Input ramp and trishear angles should be in radians.
%       For reverse faults use positive slip and slip increment.
%       For normal faults use negative slip and slip increment
%
% Trishear uses function VelTrishear

%Extent of section and number of points in each bed
extent = psect(1); npoint = psect(2);
%Make undeformed beds geometry: This is a grid of points along the beds
xp=0.0:extent/npoint:extent;
[XP, YP]=meshgrid(xp,yp);

% Model parameters
xt = tparam(1); %x fault tip
yt = tparam(2); %y fault tip
ramp = tparam(3);%Ramp angle
ps = tparam(4); %P/S
tra = tparam(5); %Trishear angle
m = tan(tra/2); %Tangent of half trishear angle
slip = tparam(6); %Fault slip
c = tparam(7); %Concentration factor
%Number of slip increments
ninc=round(slip/sinc);

%Transformation matrix from geographic to fault coordinates
a11=cos(ramp);
a12=cos(pi/2-ramp);
a21=cos(pi/2+ramp);
a22=a11;
```

11.4 Geological application: Fault-related folding

```
% Transform to coordinates parallel and perpendicular to the fault, and
% with origin at initial fault tip
FX=(XP-xt)*a11+(YP-yt)*a12;
FY=(XP-xt)*a21+(YP-yt)*a22;

%Run trishear model
%Loop over slip increments
for i=1:ninc
    %Loop over number of beds
    for j=1:size(FX,1)
        %Loop over number of points in each bed
        for k=1:size(FX,2)
            %Solve trishear in a coordinate system attached to current
            %fault tip (Eq. 11.27)
            xx=FX(j,k)-(ps*i*abs(sinc));
            yy=FY(j,k);
            %Compute velocity (Eqs. 11.25 and 11.26)
            [vx,vy]=VelTrishear(xx,yy,sinc,m,c);
            %Update FX, FY coordinates
            FX(j,k)=FX(j,k)+vx;
            FY(j,k)=FY(j,k)+vy;
        end
    end
    %Transform back to horizontal-vertical XP, YP coordinates for plotting
    XP=(FX*a11+FY*a21)+xt;
    YP=(FX*a12+FY*a22)+yt;
    %Make fault geometry
    xtf=xt+(ps*i*abs(sinc))*cos(ramp);
    ytf=yt+(ps*i*abs(sinc))*sin(ramp);
    XF=[xt xtf];
    YF=[yt ytf];
    %Make trishear boundaries
    axlo=0:10:300;
    htz=axlo*m;
    ftz=-axlo*m;
    XHTZ=(axlo*a11+htz*a21)+xtf;
    YHTZ=(axlo*a12+htz*a22)+ytf;
    XFTZ=(axlo*a11+ftz*a21)+xtf;
    YFTZ=(axlo*a12+ftz*a22)+ytf;
    %Plot increment. Fault
    plot(XF, YF,'r-','LineWidth',2);
    hold on;
    % Hanging wall trishear boundary
    plot(XHTZ, YHTZ,'b-');
    % Footwall trishear boundary
    plot(XFTZ, YFTZ,'b-');
    % Beds: Split hanging wall and footwall points
    hw = zeros(1,size(XP,2));
```

```
        fw = zeros(1,size(XP,2));
        xhb = zeros(size(XP,1),size(XP,2));
        yhb = zeros(size(XP,1),size(XP,2));
        xfb = zeros(size(XP,1),size(XP,2));
        yfb = zeros(size(XP,1),size(XP,2));
        for j=1:size(XP,1)
            hw(j)=0.0;
            fw(j)=0.0;
            for k=1:size(XP,2)
                %If hanging wall points
                if XP(j,k)<=xt+(YP(j,k)-yt)/tan(ramp),
                    hw(j)=hw(j)+1;
                    xhb(j,hw(j))=XP(j,k);
                    yhb(j,hw(j))=YP(j,k);
                %if footwall points
                else
                    fw(j)=fw(j)+1;
                    xfb(j,fw(j))=XP(j,k);
                    yfb(j,fw(j))=YP(j,k);
                end
            end
            plot(xhb(j,1:1:hw(j)),yhb(j,1:1:hw(j)),'k-');
            plot(xfb(j,1:1:fw(j)),yfb(j,1:1:fw(j)),'k-');
        end
        %Plot settings
        text(0.8*extent,1.75*max(yp),strcat('Slip = ',num2str(i*sinc)));
        axis equal;
        axis( [0 extent 0 2.0*max(yp)]);
        hold off;
        %Get frame for movie
        frames(i) = getframe;
end
end

function [vx, vy] = VelTrishear(xx,yy,sinc,m,c)
%VelTrishear: Symmetric, linear vx trishear velocity field
%Equation 6 of Zehnder and Allmendinger 2000

%If behind the fault tip
if xx <0.
    %If hanging wall
    if yy >=0.
        vx = sinc;
        vy = 0.;
    %If footwall
    elseif yy<0.
        vx=0.;
        vy=0.;
```

11.4 Geological application: Fault-related folding

```
          end
  %If ahead the fault tip
  elseif xx>=0.
      %If hanging wall
      if yy>=xx*m
          vx=sinc;
          vy=0.;
      %If footwall
      elseif yy<=-xx*m
          vx=0.;
          vy=0.;
      %If inside the trishear zone
      else
          %Some variables to speed up the computation
          a=1+c; b=1/c; d=a/c; ayy=abs(yy); syy = yy/ayy;
          %Eq. 11.25
          vx=(sinc/2.)*(syy*realpow(ayy/(xx*m),b)+1);
          %Eq. 11.26
          vy=(sinc/2.)*(m/a)*(realpow(ayy/(xx*m),d)-1);
      end
  end
  end
```

11.4.4 Modeling sedimentation: Growth strata

Once the velocity fields of the fault-related fold models are specified, it is easy to model time-dependent processes such as sedimentation during growth of the structure (syntectonic sedimentation). To illustrate this, we will follow a simple approach. We introduce a ratio G that describes the relation between regional subsidence and local uplift of the anticlinal crest. When $G = 1.0$ (subsidence = uplift), the anticlinal crest is always at the surface and sedimentation takes place only on the flanks of the structure. When $G > 1.0$ (subsidence > uplift), sedimentation takes place on the crest as well as the flanks, but strata on the crest are thinner than those on the flanks. The crestal uplift rate for the fault-bend fold and trishear models is just $\sin\theta s$ and for the simple step, fixed axis, and parallel fold models is twice that ($2\sin\theta s$; Hardy and Poblet, 2005). In addition, we will assume that the basin always fills to the top with strata and that the growing fold has no effect on the local sedimentation. Essentially, a background regional sedimentation fills the basin to the top, and concepts such as base level, erosion, etc. are not considered. These assumptions are somewhat naive, but they are sufficient to investigate the pattern of syntectonic growth strata in the different fault-related fold models.

The functions **FaultBendFoldGrowth**, **FixedAxisFPFGrowth**, **ParallelFPFGrowth**, and **TrishearGrowth** plot the evolution of syntectonic sedimentation in the fault-bend fold, fixed axis, parallel, and trishear models, respectively. These functions use the assumptions introduced before. For the purpose of illustration, we just include here function **TrishearGrowth**. To plot the evolution of syntectonic sedimentation in a trishear fold with subsidence versus uplift rate $G = 2.0$, type:

```
yp = [50 80 110 140 170]; %Beds datums
psect = [1000 500]; %Section parameters
```

```
tparam = [300 50 30*pi/180 1.5 60*pi/180 100 1.0]; %Trishear parameters
sinc = 1.0; %Slip parameters
G = 2.0; %Subsidence versus uplift rate
frames = TrishearGrowth(yp,psect,tparam,sinc, G); %Make trishear fold

function frames = TrishearGrowth(yp,psect,tparam,sinc, G)
%Trishear plots the evolution of a 2D trishear fault propagation fold and
%adds growth strata for a given subsidence versus uplift rate
%
% USE: frames = TrishearGrowth(yp,psect,tparam,sinc, G)
%
% yp = Datums or vertical coordinates of undeformed, horizontal beds
% psect = A 1 x 2 vector containing the extent of the section, and the
%         number of points in each bed
% tparam = A 1 x 7 vector containing: the x coordinate of the fault tip
%          (entry 1), the y coordinate of the fault tip (entry 2), the
%          ramp angle (entry 3), the P/S (entry 4), the trishear angle
%          (entry 5), the fault slip (entry 6), and the concentration
%          factor (entry 7)
% sinc = slip increment
% G = Subsidence versus uplift rate
% frames = An array structure containing the frames of the fold evolution.
%          You can play the movie again just by typing movie(frames)
%
% NOTE: Input ramp and trishear angles should be in radians.
%       For reverse faults use positive slip and slip increment.
%       For normal faults use negative slip and slip increment
%
% TrishearGrowth uses function VelTrishear

% Top of layers
top = yp(size(yp,2));

%Extent of section and number of points in each bed
extent = psect(1); npoint = psect(2);
%Make undeformed beds geometry: This is a grid of points along the beds
xp=0.0:extent/npoint:extent;
[XP, YP]=meshgrid(xp,yp);

% Model parameters
xt = tparam(1); %x fault tip
yt = tparam(2); %y fault tip
ramp = tparam(3);%Ramp angle
ps = tparam(4); %P/S
tra = tparam(5); %Trishear angle
m = tan(tra/2); %Tangent of half trishear angle
slip = tparam(6); %Fault slip
c = tparam(7); %Concentration factor
```

11.4 Geological application: Fault-related folding

```
%Number of slip increments
ninc=round(slip/sinc);

%Transformation matrix from geographic to fault coordinates
a11=cos(ramp);
a12=cos(pi/2-ramp);
a21=cos(pi/2+ramp);
a22=a11;

% Make ten growth strata
nincG=round(ninc/10);
% Initialize count of growth strata to 1
countG = 1;

% Transform to coordinates parallel and perpendicular to the fault, and
% with origin at initial fault tip
FX=(XP-xt)*a11+(YP-yt)*a12;
FY=(XP-xt)*a21+(YP-yt)*a22;

%Run trishear model
%Loop over slip increments
for i=1:ninc
    %Loop over number of beds
    for j=1:size(FX,1)
        %Loop over number of points in each bed
        for k=1:size(FX,2)
            %Solve trishear in a coordinate system attached to current
            %fault tip (Eq. 11.27)
            xx=FX(j,k)-(ps*i*abs(sinc));
            yy=FY(j,k);
            %Compute velocity (Eqs. 11.25 and 11.26)
            [vx,vy]=VelTrishear(xx,yy,sinc,m,c);
            %Update FX, FY coordinates
            FX(j,k)=FX(j,k)+vx;
            FY(j,k)=FY(j,k)+vy;
        end
    end
    %Transform back to horizontal-vertical XP, YP coordinates for plotting
    XP=(FX*a11+FY*a21)+xt;
    YP=(FX*a12+FY*a22)+yt;
    %Make fault geometry
    xtf=xt+(ps*i*abs(sinc))*cos(ramp);
    ytf=yt+(ps*i*abs(sinc))*sin(ramp);
    XF=[xt xtf];
    YF=[yt ytf];
    %Make trishear boundaries
    axlo=0:10:300;
    htz=axlo*m;
```

```
ftz=-axlo*m;
XHTZ=(axlo*a11+htz*a21)+xtf;
YHTZ=(axlo*a12+htz*a22)+ytf;
XFTZ=(axlo*a11+ftz*a21)+xtf;
YFTZ=(axlo*a12+ftz*a22)+ytf;
%Plot increment. Fault
plot(XF,YF,'r-','LineWidth',2);
hold on;
% Hanging wall trishear boundary
plot(XHTZ, YHTZ,'b-');
% Footwall trishear boundary
plot(XFTZ, YFTZ,'b-');
% Beds: Split hanging wall and footwall points
hw = zeros(1,size(XP,2));
fw = zeros(1,size(XP,2));
xhb = zeros(size(XP,1),size(XP,2));
yhb = zeros(size(XP,1),size(XP,2));
xfb = zeros(size(XP,1),size(XP,2));
yfb = zeros(size(XP,1),size(XP,2));
for j=1:size(XP,1)
    hw(j)=0.0;
    fw(j)=0.0;
    for k=1:size(XP,2)
        %If hanging wall points
        if XP(j,k)<= xt+(YP(j,k)-yt)/tan(ramp),
            hw(j)=hw(j)+1;
            xhb(j,hw(j))=XP(j,k);
            yhb(j,hw(j))=YP(j,k);
        %If footwall points
        else
            fw(j)=fw(j)+1;
            xfb(j,fw(j))=XP(j,k);
            yfb(j,fw(j))=YP(j,k);
        end
    end
    %If Pregrowth strata
    if (j <= size(yp,2))
        plot(xhb(j,1:1:hw(j)),yhb(j,1:1:hw(j)),'k-');
        plot(xfb(j,1:1:fw(j)),yfb(j,1:1:fw(j)),'k-');
    %If Growth strata
    else
        plot(xhb(j,1:1:hw(j)),yhb(j,1:1:hw(j)),'g-');
        plot(xfb(j,1:1:fw(j)),yfb(j,1:1:fw(j)),'g-');
    end
end
%Plot settings
text(0.8*extent,1.75*max(yp),strcat('Slip = ',num2str(i*sinc)));
axis equal;
```

11.4 Geological application: Fault-related folding

```
        axis([0 extent 0 2.0*max(yp)]);
        hold off;
        %Get frame for movie
        frames(i) = getframe;
        %Add growth strata. Careful: Intersections pregrowth-growth strata are
        %not calculated. Growth strata will not look right for subsidence rate
        %lower than uplift rate G < 1.0
        if (i == countG*nincG)
            %Make growth strata
            %Update top
            top = top + nincG*sinc*sin(ramp)*G;
            % Make bed geometry
            xm=i*sinc:extent/npoint:extent+i*sinc;
            [GXP, GYP]=meshgrid(xm,top);
            %Transform to coordinate axes parallel and perpendicular to the
            %fault, and with origin at initial fault tip location
            GFX=(GXP-xt)*a11+(GYP-yt)*a12;
            GFY=(GXP-xt)*a21+(GYP-yt)*a22;
            %Add to beds
            FX = [FX; GFX];
            FY = [FY; GFY];
            % update count of growth strata
            countG = countG + 1;
        end
    end
end
```

Figure 11.10 shows the growth strata geometries for the fault-bend fold (Fig. 11.10a), fixed axis (Fig. 11.10b), parallel (Fig. 11.10c), and trishear (Fig. 11.10d) models, for $G = 2.0$. The kink models (Fig. 11.10a–c) require instantaneous rotation of strata as they pass through a kink axial surface. The complexity derives from understanding how particles behave with respect to the kink axes. Fundamentally, there are two possibilities: (1) The kink axes move with the material. This type of kink axis is referred to as a *fixed* or *passive* kink axis (dashed line axes in Fig. 11.10a–c). (2) The material flows through the kink axis, which is then called an *active* kink axis (continuous line axes in Fig. 11.10a–c). Within the growth strata, the fixed and the active kink axes merge at the depositional surface, forming an upward-narrowing kink band or growth triangle (Fig. 11.10a–c). In the trishear model (Fig. 11.10d) on the other hand, changes in dip and thickness of the growth strata are indicative of progressive rotation of the material as the fold limb is gradually rotated into its final orientation. Understanding growth strata geometries is not simple. The functions above facilitate this task.

As we said before, our model of syntectonic sedimentation is quite simplistic. More insight can be gleaned from an approach that considers the effect of both background sedimentation and local erosion, transport, and deposition as a result of fold growth (Hardy et al., 1996). The velocity analysis facilitates this approach, but requires that one be very clear about reference frames (Waltham and Hardy, 1995). In Chapter 7, we introduced the concept of Lagrangian and Eulerian reference frames. Modeling simultaneous tectonics and sedimentation generally requires an Eulerian reference frame so that the two can be treated as simultaneous, rather than sequential processes (Hardy and Poblet, 1995). In an Eulerian reference frame, different particles with different velocities may happen to be at a fixed coordinate at different times.

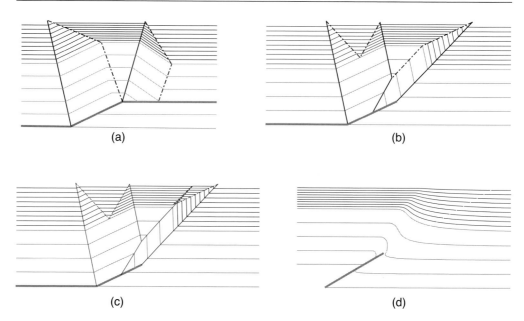

Figure 11.10 Growth strata geometries for (a) fault-bend fold, (b) fixed axis, (c) parallel, and (d) trishear models. In all cases subsidence to uplift rate $G = 2.0$. In (a) to (c) active kink axes are indicated by continuous lines, and passive kink axes by dashed lines.

One's point of view is of particles flowing through our fixed coordinate system with time due to different processes, rather than the coordinate system being fixed to moving particles in the Lagrangian case. We will leave this topic for now though the interested reader can find excellent summaries in the literature (Hardy and Poblet, 1995; Hardy *et al.*, 1996; Waltham and Hardy, 1995).

11.5 EXERCISES

1. Compare the fold geometries of rollovers produced by inclined shear with angles 0, 15, 30, −15, and −30°. In all cases try to use the same geometry for the listric normal fault. Hint: Use function **SimilarFold**.
2. Compare the fold geometries of fixed axis and parallel fault-propagation folds for $\theta = 10, 20, 30, 40$, and $50°$. Make the comparison in terms of dip and thickness of the forelimb, and vergence of the fold. Hint: Use functions **FixedAxisFPF** and **ParallelFPF**.
3. Compare the geometries of trishear folds for $P/S = 1.0, 1.5, 2.0$, and 2.5. In all cases use ramp angle = 30°, trishear angle = 60°, fault slip = 100 units, and concentration factor = 1.0. Hint: Use function **Trishear**.
4. Compare the geometries of trishear folds for trishear angles 40, 50, 60, and 70°. In all cases use ramp angle = 30°, $P/S = 1.5$, fault slip = 100 units, and concentration factor = 1.0. Hint: Use function **Trishear**.
5. Compare the geometries of trishear folds for concentration factor $c = 1.0, 1.5, 2.0$, and 3.0. In all cases use ramp angle = 30°, $P/S = 1.5$, trishear angle = 60°, and fault slip = 100 units. Hint: Use function **Trishear**.

11.5 Exercises

6. There are several velocity fields that can fulfill the incompressibility condition in a symmetric trishear zone. One of these fields is the "sine velocity field" corrected from Zehnder and Allmendinger (2000):

$$v_1 = \frac{s(\sin\beta + 1)}{2} \quad v_2 = \frac{sm}{\pi}(\cos\beta + \beta\sin\beta - \pi/2)$$

where $\beta = (x_2\pi)/(2x_1 m)$ and $m = \tan\varphi$. Write a MATLAB function that implements this velocity field and use it in the function **Trishear**. Compare the fold geometries produced by the linear v_1 field and the sine field. Hint: Use function **VelTrishear** as the base of your new "sine velocity" function.

7. Compare the growth strata geometries of fixed axis and parallel fault propagation folds for θ = 10, 20, 30, 40, and 50° and G = 2.0. Make the comparison in terms of the location of fixed and active kink axes. Hint: Use functions **FixedAxisFPFGrowth** and **ParallelFPFGrowth**.

8. Compare the growth strata geometries of trishear folds for P/S = 1.0, 1.5, 2.0, and 2.5. In all cases use a ramp angle = 30°, trishear angle = 60°, fault slip = 100 units, concentration factor = 1.0 and G = 2.0. What happens when the P/S = 1.0? Hint: Use function **TrishearGrowth**.

9. Modify the function **SimilarFold** to simulate growth strata.

10. Discuss a methodology to compute finite strain in velocity models of deformation. Modify one of the fault-related fold velocity models to plot strain ellipses.

CHAPTER
TWELVE

Error analysis

12.1 INTRODUCTION

Structural geologists have a love–hate relationship with uncertainty. Ask any one of us how much uncertainty is associated with a single strike and dip measurement and we will readily admit that natural surfaces are highly irregular at various scales and so there are probably "a few degrees" slop in our measurements. Because one can collect only a relatively small number of measurements per day in the field, we certainly aren't going to repeat a single bedding measurement 20 or more times just to get "good statistics" at a single location! Nonetheless, when we calculate a mean vector of a bunch of, say, paleocurrent directions (Chapter 2) or a best-fit fold axis (Chapter 5), we routinely calculate and report the confidence intervals, or rather more likely a computer program written by someone else calculates them for us. If you have read this far in the book, however, you now know how to calculate them for yourself! Two recent trends brought on by the digital age are forcing structural geologists to reexamine their relationship with uncertainty.[1]

First, the availability of large digital data sets and their incorporation into routine geological studies have exploded in the last couple of decades. We have access to digital elevation models sampled on a 30 m grid, GoogleEarth imagery with a resolution of less than 5 m, GPS data sets with thousands of individual stations, and nearly instant access to hundreds or thousands of aftershocks that follow a large earthquake. These data sets allow us to analyze vastly more data than previously; the era of collecting just 30 or 40 measurements in the field per day is long gone. The digital data can be analyzed quantitatively and that means taking into account uncertainties.

Second, we now routinely use models to interpret the data that we collect and therefore are forced to confront the question, "How well does our model fit the data that we are trying to

---

[1] In this chapter and, indeed, throughout the book, we use the terms "error" and "uncertainty" interchangeably.

explain?" A best-fit fold axis represents the approximation of a natural curvilinear surface with a cylindrical fold model. The mean vector representing an average paleocurrent direction implies a model of unidirectional current flow. The concept of least squares fitting of a model to data has been touched on at several points in this book (Chapters 5, 7, and 8) in the context of specific geological problems. A full treatment of statistics and error analysis is beyond the scope of this book and there are many fine texts where these topics are exhaustively explored, both more authoritatively and more entertainingly than we could ever do (among many others, the ones we have found to be particularly helpful include: Bevington and Robinson, 2003; Fisher *et al.*, 1987; Press *et al.*, 1986; Taylor, 1997). There is one topic, however, that deserves special mention, particularly as we go from uncertainties in data sets that we've measured directly to models that are calculated from data with inherent uncertainties; that topic is error propagation.

12.2 ERROR PROPAGATION

Suppose that, through repeated measurements, we had determined the uncertainties on two parameters, a and b; we'll call the uncertainties δa and δb, respectively. Now we want to calculate c, the sum of a and b. What is the uncertainty on the calculated value, c? The highest and lowest likely values of c are

$$c_{max} = a + b + (\delta a + \delta b)$$
$$c_{min} = a + b - (\delta a + \delta b)$$

In general, we can say that the maximum probable error on the calculated parameter, c, is

$$\delta c_{max} = \delta a + \delta b \qquad (12.1)$$

We have just propagated the errors on a and b to determine the maximum error on c. We will come back to the question of whether or not δc_{max} is the most likely error or not, below.

To go beyond this somewhat trivial example, one might develop similar equations for progressively more complicated cases, but it turns out that all of the specific cases can be encapsulated in one general rule:

$$\delta q_{max} = \left|\frac{\partial q}{\partial a}\right|\delta a + \left|\frac{\partial q}{\partial b}\right|\delta b + \cdots + \left|\frac{\partial q}{\partial z}\right|\delta z \qquad (12.2)$$

where q is any function of $(a, b, ..., z)$. You can see that Equation 12.1 can be derived from Equation 12.2 as the partial derivative of c with respect to a and that with respect to b are both equal to 1.

Although Equation 12.2 gives the maximum error in q, it is not the most probable error if uncertainties in the measured parameters are independent and random and thus the errors follow a Gaussian distribution (i.e., the typical "bell-shaped" curve). Under these conditions, the calculated error is essentially the square root of the sum of the squares of the measured errors:

$$\delta q = \sqrt{\left(\frac{\partial q}{\partial a}\delta a\right)^2 + \left(\frac{\partial q}{\partial b}\delta b\right)^2 + \cdots + \left(\frac{\partial q}{\partial z}\delta z\right)^2} \qquad (12.3)$$

Let's formalize this a bit and look at a somewhat more complicated case where the two variables involved may have some degree of correlation. In Equation 7.16, the standard deviation of a series of measurements was described:

$$\sigma = \sqrt{\frac{1}{N-1}\sum_{k=1}^{N}(u_k - \bar{u})^2}$$

and the variance was defined as the square of the standard deviation. The *covariance* of two parameters a and b is

$$\sigma_{ab} = \frac{1}{n}\sum_{i=1}^{n}(a_i - \bar{a})(b_i - \bar{b}) \quad (12.4)$$

where $\bar{a}$ is the average of the a measurements and $\bar{b}$ is the average of the b measurements. With this definition, we can now come up with an expression for the variance of a calculated parameter as a function of the variances and covariances of the measured parameters:

$$\sigma_c^2 = \left(\frac{\partial c}{\partial a}\right)^2 \sigma_a^2 + \left(\frac{\partial c}{\partial b}\right)^2 \sigma_b^2 + 2\frac{\partial c}{\partial a}\frac{\partial c}{\partial b}\sigma_{ab} \quad (12.5)$$

If a and b are truly independent and random, the covariance, σ_{ab}, will go to zero and Equation 12.5 simply becomes the square of Equation 12.3. If a and b are highly correlated, your error will still never be greater than Equation 12.2.

12.3 GEOLOGICAL APPLICATION: CROSS-SECTION BALANCING

Balanced cross sections have been a staple of the structural geologist's toolbox for more than half a century, thanks largely to pioneering work in the Canadian Rocky Mountains (e.g., Bally et al., 1966; Dahlstrom, 1969, 1970; Price and Mountjoy, 1970). Many reasons exist for constructing these sections: They may be used to project data to depth and interpret structural geometry for exploration or scientific purposes, or the magnitudes of shortening calculated from balanced cross sections may become input for palinspastic restorations or geodynamic models. Their utility, and limitations, were perhaps best described by Clint Dahlstrom (1969) when he wrote: "If a cross section passes the geometric tests [i.e., is balanced] it could be correct… On the other hand, a cross section that does not pass the geometric tests could not possibly be correct." We would now like to go the extra step and – rather than the binary decision: might be correct or definitely not correct – ask the question: What is the uncertainty associated with balanced cross sections?

Cross sections are models fit to data that come from a variety of sources: outcrop measurements, stratigraphic sections, and subsurface data such as seismic reflection surveys and well data. Each of these input data sources has uncertainty associated with it. How representative are the outcrop measurements, what are the stratigraphic thickness variations, how well do we know subsurface velocities in order to convert from time to depth? There are other uncertainties; the most important being the choice of a specific kinematic fold-fault model (Chapter 11). Although structural geologists who construct balanced cross sections are painfully aware of these uncertainties, despite half a century of use, there has been no formal way of incorporating them into the final model in a meaningful way. Before investigating one promising approach to this problem, we need a brief review.

12.3.1 Line-length and area balancing

In Chapter 11, we showed that all cross-section balancing stems, fundamentally, from the continuity and incompressibility equations (Eqs. 11.1 and 11.2). If we make a further

12.3 Geological application: Cross-section balancing

Type of balance	Dimension	Assumptions	Fold kinematic model
Volume	3D	Density of rocks constant during deformation, no compaction, no addition or subtraction of material	Non-cylindrical
Area	2D	Plane strain, cross-sectional area preserved	Cylindrical folding (parallel, similar, or trishear)
Length	1D	Linear strain, no bedding thickness changes, shear parallel to layers so the layers are lines of no finite elongation	Parallel

Table 12.1 Types of cross section balancing and assumptions

assumption of plane strain, the two-dimensional version of the incompressibility equation (Eq. 11.3) is what structural geologists call area balancing (Mitra and Namson, 1989). Finally, we can reduce the problem to one dimension by assuming a parallel fold model in which bedding thickness does not change and the stratigraphic horizons are lines of no finite longitudinal strain (Chapter 10). The relationship of these different dimensions to the kinematics of folding and the general assumptions in each case are shown in Table 12.1.

The majority of published balanced cross sections are line-length balanced sections. Line-length balanced cross sections are a subset of area balanced sections because they include all of the assumptions of area and volume balancing. In line-length balancing, the additional assumption is that "parallel" folding occurs via shear parallel to bedding (i.e., "flexural slip" folding); thus, the stratigraphic layers are lines of no finite longitudinal extension. This allows us to calculate the shortening simply by measuring the bed length in the deformed state and drawing the same bed length as a straight line in the undeformed state. The majority of shortening in this model occurs where faults shear across layers. In addition to a restrictive folding model, the previous sentence highlights an additional hidden weakness of line-length balanced cross sections: only faults that produce obvious offset at the scale of the section are included in the shortening estimate. The implicit assumption is that faults with displacements smaller than the scale of the cross section do not contribute to shortening, a concept that is seldom tested (Marrett and Allmendinger, 1990, 1992).

People who construct line-length balanced sections (Fig. 12.1) will often say that the magnitude of shortening is a "minimum estimate." This error arises in cases where the thrust fault is emergent and hanging wall cutoffs have been eroded (Fig. 12.1b). The resulting saw-toothed gap on the restored section represents the minimum eroded bed length needed to line up the stratigraphic horizons of hanging wall and footwall. Because the structural geologist does not know how much bed length has been eroded, s/he simply makes the gap as small as possible.

Though significant, the error resulting from eroded hanging wall cutoffs is hardly the only error, in fact it is an error largely dependent on a single specific model. There may be many different individual line-length models that can fill the requisite cross-sectional area. Likewise, critical unknowns, such as the exact depth to the decollement across a deformed section, contribute to the apparent paradox that two (or more) line-length balanced sections drawn along the same profile line commonly have different "minimum" estimates.

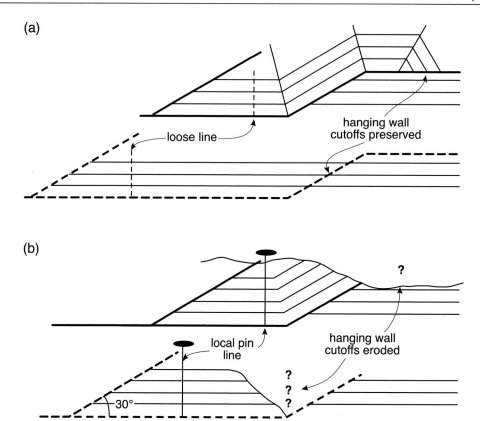

Figure 12.1 Line-length balancing using a parallel, kink fold geometry. (a) Shows the case where the hanging wall cutoffs are preserved so that the restoration of the upper plate, and the undeformed trajectory of the trailing thrust trace, is unambiguous. (b) Shows the more typical case where hanging wall cutoffs are eroded. In this case, a local pin line is needed to restore the thrust plate and the footwall trace of the trailing thrust. The gap marked by question marks is the source of the statement that such reconstructions are minimum estimates of shortening.

One approach to a more accurate and rigorous error estimation would be for the geologist to construct numerous different cross sections along the same profile line and look at the distribution of shortening estimate; essentially a hand-crafted Monte Carlo simulation. To develop reliable statistics, it would take tens to hundreds of sections in the same exact area, each having a slightly different starting geometry. Therein lies the fundamental reason why virtually all line-length balanced sections are presented without error estimates: because a single section, even one constructed using commercial software packages, can take weeks to construct, the task of drawing hundreds in the same place is both too daunting and too tedious to contemplate!

To quantify the uncertainties on shortening magnitudes, we will do the error analysis via area balancing. This has numerous advantages: (1) Because line-length sections are a subset of area balancing, a single area balance encompasses all possible line-length solutions. (2) Area balance is independent of specific, two-dimensional fold-fault kinematic model (e.g., parallel kink folds, similar folds, trishear folds, disharmonic folding, etc.; Chapter 11). (3) Unlike line-length sections, area balancing also accounts for shortening due to deformation on structures

12.3 Geological application: Cross-section balancing

too small to be depicted on the cross section. Finally, (4) because one can describe the areas of both the deformed and restored sections as explicit equations, the errors can be formally (analytically) propagated through those equations, obviating the need to draw multiple sections.

The origin of area balancing dates to the turn of the twentieth century and the work of Chamberlin (1910; 1919; 1923) and Buxtorf (1916). Those two pioneering authors first used the condition that the deformed cross-sectional area should equal the initial area to calculate the depth to the decollement; a technique known as excess area balancing (Mitra and Namson, 1989). Chamberlin first used the term "thin-shelled" to refer to the Appalachian Valley and Ridge province and "thick-shelled" to describe the Colorado Rocky Mountains because he concluded from area balancing that the decollement of the former was shallower, and that of the latter deeper, in the crust. Chamberlin's analysis was flawed but his insight significant. Rodgers (1949) introduced slightly modified versions of these terms, "thin-skinned" and "thick-skinned," which are widely used today.

12.3.2 Error propagation in a simple area balance

We'll start with a simple "crustal" area balance, which can be thought of as nothing more than a long, skinny box being deformed into a short, fat box (Fig. 12.2). The areas of the two boxes must be equal, so we can calculate the shortening as a function of the measurable dimensions of the boxes:

$$X_1 X_2 = x_1 x_2 \quad \text{and} \quad S = X_1 - x_1$$

Rearranging and solving for S:

$$S = \frac{x_1 x_2 - x_1 X_2}{X_2} = x_1 x_2 X_2^{-1} - x_1 \quad (12.6)$$

To calculate the shortening, we only need to know the modern-day dimensions (the width and thickness of the plateau, x_1, x_2) and make some estimate of the initial crustal thickness (X_2). This calculation is nothing more than the classic area balance–depth to decollement equation, where the unknown is the shortening, S.

Now, let's see how the errors propagate through Equation 12.6. To do so, we need to calculate the three partial differential equations:

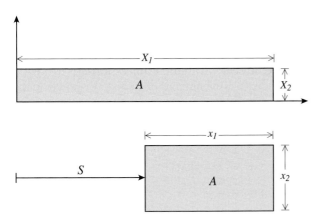

Figure 12.2 Simple area balance from an initial undeformed state at the top to a horizontally shortened and vertically thickened deformed state at the bottom. The two areas, A, are equal.

Parameter	Value (km)	Uncertainty (km)	Error in S due to parameter (km)
x_1	200	1	1
x_2	70	5	28.6
X_2	35	10	114.3

Table 12.2 Simple crustal area balance with errors

$$\frac{\partial S}{\partial x_2} = x_1 X_2^{-1} \quad \frac{\partial S}{\partial X_2} = -x_1 x_2 X_2^{-2} \quad \text{and} \quad \frac{\partial S}{\partial x_1} = x_2 X_2^{-1} - 1$$

We can now write the complete equation for the error in calculation of S as a function of the errors in the measured or estimated parameters:

$$\delta S = \sqrt{\left(\frac{\partial S}{\partial x_1}\delta x_1\right)^2 + \left(\frac{\partial S}{\partial x_2}\delta x_2\right)^2 + \left(\frac{\partial S}{\partial X_2}\delta X_2\right)^2}$$
$$= \sqrt{[(x_2 X_2^{-1} - 1)\delta x_1]^2 + [(x_1 X_2^{-1})\delta x_2]^2 + [(-x_1 x_2 X_2^{-2})\delta X_2]^2}$$
(12.7)

With Equation 12.7 in hand, we can enter some realistic values, shown in Table 12.2. Using Equation 12.6, you can see that these values yield a shortening, $S = 200$ km (logically enough since we have doubled the crustal thickness). We can now calculate the uncertainty of this shortening value from Equation 12.7:

$$\delta S = \sqrt{(1)^2 + (28.6)^2 + (114.3)^2} = 118 \, \text{km}$$

So, our shortening estimate would be 200 ± 118 km! This calculation tells us two very important things: First, there is a high degree of uncertainty – in this case greater than 50% – in crustal scale balancing. Second, almost all of the uncertainty comes from the error in estimate of initial crustal thickness, X_2. Even if we assumed that the initial crustal thickness error was only ± 5 km (instead of the 10 km used in the above calculation), the final shortening estimate would still be ± 64 km! For a more realistic crustal scale balance, we could include some additional fluxes: (1) material lost from the system by erosion, (2) material added to the system by magmatism, and (3) material lost by tectonic erosion.

12.3.3 Error propagation using a more general area balance

One can calculate analytically the area of a polygon of any shape and number of vertices and this enables us to capture the area of a polygon that envelops the deformed region of pre-growth strata (Fig. 12.3) (Judge and Allmendinger, 2011). To start, we define a matrix to hold all the vertices of our deformed polygon. In this matrix, the number of rows corresponds to the number of vertices in the polygon and, within an individual row, the first column contains the x_1 value and the second the x_2 value of a single vertex in two dimensions:

$$x(n,2) = \begin{bmatrix} x_{11} & x_{12} \\ x_{21} & x_{22} \\ \vdots & \vdots \\ x_{n1} & x_{n2} \end{bmatrix} \begin{matrix} \text{first vertex} \\ \\ \\ n^{\text{th}} \text{ vertex} \end{matrix}$$

12.3 Geological application: Cross-section balancing

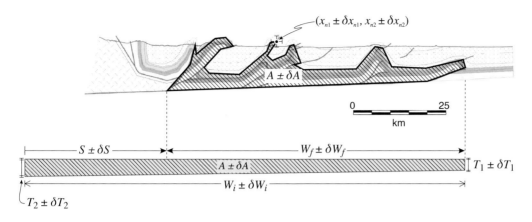

Figure 12.3 General area balance for the southernmost Subandean belt in northwestern Argentina (Echavarría et al., 2003), showing the parameters that go into the equations described in Section 12.3.3.

The area of any polygon of any shape (except one that crosses itself) can be written as (e.g., Harris and Stocker, 1998)

$$A = \frac{1}{2}\sum_{i=1}^{n}\left(x_{i1}x_{(i+1)2} - x_{(i+1)1}x_{i2}\right) \quad \text{if } i+1>n,\ i+1=1 \tag{12.8}$$

where n is the number of vertices in the polygon and (x_{i1}, x_{i2}) are the locations of the vertices. The uncertainty on the deformed area (δA) is a function of the uncertainty $(\delta x_{i1}, \delta x_{i2})$ on each specific vertex. To get things in the form of Equations 12.2 and 12.3, we need to calculate the partial differentials of A with respect to each of the components of each vertex. To do this, it will help to calculate Equation 12.8 for a simple four-vertex polygon:

$$A = \frac{1}{2}(x_{11}x_{22} - x_{21}x_{12} + x_{21}x_{32} - x_{31}x_{22} + x_{31}x_{42} - x_{41}x_{32} + x_{41}x_{12} - x_{11}x_{42})$$

Gathering terms, and recalling that the first index indicates the vertex number:

$$A = \frac{1}{2}(x_{11}(x_{22} - x_{42}) + x_{21}(x_{32} - x_{12}) + x_{31}(x_{42} - x_{22}) + x_{41}(x_{12} - x_{32})) \tag{12.9a}$$

and the same equation in terms of the x_2 components:

$$A = \frac{1}{2}(x_{12}(x_{41} - x_{21}) + x_{22}(x_{11} - x_{31}) + x_{32}(x_{21} - x_{41}) + x_{42}(x_{31} - x_{11})). \tag{12.9b}$$

Now, we can calculate the partial derivatives of each of the four vertices:

$$\frac{\partial A}{\partial x_{11}} = \frac{(x_{22} - x_{42})}{2} \quad \text{and} \quad \frac{\partial A}{\partial x_{12}} = \frac{(x_{41} - x_{21})}{2} \tag{12.10a}$$

$$\frac{\partial A}{\partial x_{21}} = \frac{(x_{32} - x_{12})}{2} \quad \text{and} \quad \frac{\partial A}{\partial x_{22}} = \frac{(x_{11} - x_{31})}{2} \tag{12.10b}$$

$$\frac{\partial A}{\partial x_{31}} = \frac{(x_{42} - x_{22})}{2} \quad \text{and} \quad \frac{\partial A}{\partial x_{32}} = \frac{(x_{21} - x_{41})}{2} \tag{12.10c}$$

$$\frac{\partial A}{\partial x_{41}} = \frac{(x_{12} - x_{32})}{2} \quad \text{and} \quad \frac{\partial A}{\partial x_{42}} = \frac{(x_{31} - x_{11})}{2} \tag{12.10d}$$

In general, if you inspect Equations 12.8 and 12.10 carefully, you see that we can write

$$A = \frac{1}{2}\sum_{i}^{n}(x_{i1}x_{k2} - x_{k1}x_{i2}) \quad \text{and} \quad \frac{\partial A}{\partial x_{i1}} = \frac{(x_{k2} - x_{m2})}{2} \quad \text{and} \quad \frac{\partial A}{\partial x_{i2}} = \frac{(x_{m1} - x_{k1})}{2}$$

where $k = i + 1$ and $m = i - 1$ (12.11)
if $k > n \Rightarrow k = k - n$
if $m < 1 \Rightarrow m = m + n$

The maximum error on the deformed area, $\delta A_{\max}$, propagated from the errors on the input vertices, is

$$\delta A_{\max} = \sum_{i=1}^{n}\left(\left|\frac{\partial A}{\partial x_{i1}}\right|\delta x_{i1} + \left|\frac{\partial A}{\partial x_{i2}}\right|\delta x_{i2}\right) \quad (12.12a)$$

and the Gaussian error, if all the components were independent and random would be

$$\delta A_{\text{Gaussian}} = \sqrt{\sum_{i=1}^{n}\left(\left(\frac{\partial A}{\partial x_{i1}}\delta x_{i1}\right)^2 + \left(\frac{\partial A}{\partial x_{i2}}\delta x_{i2}\right)^2\right)} \quad (12.12b)$$

To calculate the error on the deformed area, you would substitute the partial differentials of the deformed area with respect to each vertex in Equations 12.11 into Equations 12.12.

The error on the deformed area is only the start of this problem because we now have to calculate the shortening, S, which is a function of the initial width minus the final width (Fig. 12.3). First, the appropriate equation for the initial width, W_i, is

$$W_i = \frac{2A}{(T_1 + T_2)} = 2A(T_1 + T_2)^{-1} \quad (12.13)$$

where A is given by Equation 12.8, because the deformed and undeformed areas must be the same, and T_1 and T_2 are the stratigraphic thicknesses on the right and left sides of the cross section (Fig. 12.3). The uncertainty on the initial width is

$$\delta W_{i\max} = \left|\frac{\partial W_i}{\partial A}\right|\delta A + \left|\frac{\partial W_i}{\partial T_1}\right|\delta T_1 + \left|\frac{\partial W_i}{\partial T_2}\right|\delta T_2 \quad (12.14a)$$

$$\delta W_i = \sqrt{\left(\frac{\partial W_i}{\partial A}\delta A\right)^2 + \left(\frac{\partial W_i}{\partial T_1}\delta T_1\right)^2 + \left(\frac{\partial W_i}{\partial T_2}\delta T_2\right)^2} \quad (12.14b)$$

where

$$\frac{\partial W_i}{\partial A} = \frac{2}{(T_1 + T_2)} \quad \frac{\partial W_i}{\partial T_1} = \frac{-2A}{(T_1 + T_2)^2} \quad \text{and} \quad \frac{\partial W_i}{\partial T_2} = \frac{-2A}{(T_1 + T_2)^2}$$

and δA comes from Equation 12.12a if maximum error, or 12.12b if Gaussian error. The uncertainties in stratigraphic thickness would be determined from the actual stratigraphic variations in the field.

Finally, the shortening and shortening error are given by

$$S = W_f - W_i \quad (12.15)$$

$$\delta S_{\max} = \left|\frac{\partial S}{\partial W_i}\right|\delta W_i + \left|\frac{\partial S}{\partial W_f}\right|\delta W_f = |-\delta W_i| + |\delta W_f| \quad (12.16a)$$

$$\delta S = \sqrt{\left(\frac{\partial S}{\partial W_i}\delta W_i\right)^2 + \left(\frac{\partial S}{\partial W_f}\delta W_f\right)^2} = \sqrt{(-\delta W_i)^2 + (\delta W_f)^2} \quad (12.16b)$$

12.3 Geological application: Cross-section balancing

In this section, we have derived just the basic equations for error propagation in a general area balance. In the process, several important questions and assumptions have been glossed over. Perhaps the most pressing question is: How does one define the enveloping polygon? The approach of Judge and Allmendinger (2011) is to increase the number of vertices in the enveloping polygon (Fig. 12.3a) until the solutions for the shortening magnitude and the error stabilize. In their analyses, the solutions stabilize at 20 to 25 vertices, but this number will vary depending on the complexity and length of starting cross section. Additional questions might include: Should one use the maximum or the Gaussian error? What is the relative contribution of depth to decollement, stratigraphic thickness, and eroded hanging wall cutoffs in the overall shortening uncertainty? You will get a chance to explore some of these questions in the exercises at the end of the chapter.

The following MATLAB® function, **BalCrossErr**, computes the magnitude and error of shortening, deformed area, and initial width in an area balance calculation. The user needs to input stratigraphic thicknesses and their uncertainties on both sides of the section, as well as the pre-growth strata polygon vertices, their uncertainties and locations (decollement, surface, subsurface, or eroded). Total (kk = 0), stratigraphic thickness (kk = 1), decollement (kk = 2), eroded (kk = 3), surface (kk = 4), or subsurface (kk = 5) vertices related errors can be computed by the program.

```
function [short,shortp,defa,inw] = BalCrossErr(strat,vert,kk)
%BalCrossErr computes the shortening error in area balanced cross sections.
%The algorithm was originally written by Phoebe A. Judge
%
% USE: [short,shortp,defa,inw] = BalCrossErr(strat,vert,kk)
%
% strat= 1 x 5 vector with east stratigraphic thickness (entry 1),
%        west strat. thickness (entry 2), error on east strat
%        thickness (entry 3), error on west strat thickness
%        (entry 4), and error on final width (entry 5)
% vert = number of vertices x 5 vector with x coordinates of vertices
%        (column 1), y coords of vertices (column 2), errors in x
%        coords of vertices (column 3), errors in y coords of vertices
%        (column 4), and vertices tags (column 5). The vertices tags are
%        as follows: 1 = Vertex at decollement, 2 = Vertex at surface,
%        3 = Vertex at subsurface, 4 = Vertex at eroded hanging-wall
%        cutoff
% kk = A flag to indicate whether the program computes total errors
%        (kk = 0), errors due to stratigraphy only (kk = 1), errors due to
%        vertices at decollement only (kk = 2), errors due to vertices in
%        eroded hanging walls only (kk = 3), errors due to surface
%        vertices
%        (kk = 4), or errors due to subsurface vertices (kk = 5)
% short = Shortening magnitude and its gaussian and maximum errors
% shortp = Shortening percentage and its gaussian and maximum errors
% defa = Deformed area and its gaussian and maximum errors
% inw = Initial width and its gaussian and maximum errors
%
% NOTE: The user selects the length units of the problem. Typical length
% units are kilometers
```

```
%Stratigraphic thicknesses
E1 = strat(1);      %E strat thickness
W1 = strat(2);      %W strat thickness
dE1 = strat(3);     %Uncertainty on E strat thickness
dW1 = strat(4);     %Uncertainty on W strat thickness
dx2 = strat(5);     %Uncertainty on the final width

%Vertices
X = vert(:,1);      %x coordinate
Y = vert(:,2);      %y coordinate
dX = vert(:,3);     %Uncertainty in x
dY = vert(:,4);     %Uncertainty in y
Loc = vert(:,5);    %Vertex location
n = size(vert,1);   %Number of vertices

%If only errors due to stratigraphy
if kk == 1
   dx2 = 0.0; %Make uncertainty on the final width zero
   dX = dX * 0.0; %Make errors in vertices locations zero
   dY = dY * 0.0;
%If only errors due to vertices
elseif kk > 1
   dE1 = 0.0; %Make errors in stratigraphy zero
   dW1 = 0.0;
   dx2 = 0.0; %Make error in final width zero
   for i=1:n
       %If only errors due to decollement vertices
       if kk == 2
           if Loc(i) ~= 1
               dX(i) = 0.0; %Make errors in other vertices zero
               dY(i) = 0.0;
           end
       %If only errors due to eroded hanging walls
       elseif kk == 3
           if Loc(i) ~= 4
               dX(i) = 0.0; %Make errors in other vertices zero
               dY(i) = 0.0;
           end
       %If only errors due to surface vertices
       elseif kk == 4
           if Loc(i) ~= 2
               dX(i) = 0.0; %Make errors in other vertices zero
               dY(i) = 0.0;
           end
       %If only errors due to subsurface vertices
       elseif kk == 5
           if Loc(i) ~= 3
```

12.3 Geological application: Cross-section balancing

```
                    dX(i) = 0.0; %Make errors in other vertices zero
                    dY(i) = 0.0;
                end
            end
        end
end

%Initialize output variables
short = zeros(1,3); shortp = zeros(1,3);
defa = zeros(1,3); inw = zeros(1,3);

%Deformed area
%Calculate area of deformed state
aX = [X; X(1)];
aY = [Y; Y(1)];
XArea = 0.5*(aX(1:n).*aY(2:n+1) - aX(2:n+1).*aY(1:n));
defa(1) = (abs(sum(XArea)));
%Calculate gaussian uncertainty of deformed area
aX = [X(n); aX];
aY = [Y(n); aY];
dAx = 0.5*(aY(3:n+2) - aY(1:n));
dAy = 0.5*(aX(3:n+2) - aX(1:n));
delAx = (dAx.*dX).^2;
delAy = (dAy.*dY).^2;
%Sum the X and Y components
SdelAx = sum(delAx);
SdelAy = sum(delAy);
%take the square root of the sum of individual components
defa(2) = sqrt(SdelAx+SdelAy);
%Calculate maximum uncertainty of deformed area
dAxM = abs(dAx); dAyM = abs(dAy);
delAxM = dAxM.*dX;
delAyM = dAyM.*dY;
%sum the X and Y components
SdelAxM = sum(delAxM);
SdelAyM = sum(delAyM);
%Add everything together to get the maximum uncertainty in Area
defa(3) = SdelAxM+SdelAyM;

%Original width
%Calculate the original width assuming constant Area
inw(1) = defa(1)/(((E1)/2)+((W1)/2));
%Calculate final width from the imported polygon
x21 = max(X) - min(X);
%Calculate gaussian uncertainty of the original width
ddA = 1/(((E1)/2)+((W1)/2)); %partial of x1 wrt Area
ddE1 = -(2*defa(1))/((E1)^2+(2*E1*W1)+(W1)^2); %partial of x1 wrt E1
ddW1 = -(2*defa(1))/((E1)^2+(2*E1*W1)+(W1)^2); %partial of x1 wrt W1
```

```
inw(2) = sqrt(((ddA*defa(2))^2)+((ddE1*dE1)^2)+((ddW1*dW1)^2));
%Calculate maximum uncertainty of the original width
inw(3) = ((abs(ddA*defa(3)))+(abs(ddE1*dE1))+(abs(ddW1*dW1)));

%Shortening
%Calculate shortening
short(1) = inw(1)-x21;
%Calculate gaussian uncertainty in shortening
short(2) = sqrt((inw(2))^2+(dx2)^2);
%Calculate maximum uncertainty in shortening
short(3) = (inw(3)+dx2);
%Calculate percent shortening
shortp(1) = (1-(x21/inw(1)))*100;
%Calculate gaussian uncertainty of percent shortening
ddx1 = x21/((inw(1))^2); %partial of S wrt x1
ddx2 = -1/inw(1); %partial of S wrt x2
shortp(2) = sqrt(((ddx1*inw(2))^2)+((ddx2*dx2)^2))*100;
%Calculate maximum uncertainty in shortening
shortp(3) = (abs(ddx1*inw(3))+(abs(ddx2*dx2)))*100;
end
```

12.4 UNCERTAINTIES IN STRUCTURAL DATA AND THEIR REPRESENTATION

Generally speaking, structural data result from a process that involves three sequential steps: data acquisition, processing, and interpretation. All these three steps generate uncertainties of different magnitude and nature. Take, for example, seismic reflection surveys on which so many exploration targets and balanced cross sections depend: uncertainties in acquisition are prominent in land surveys (especially if operating over rough terrains), but offshore acquisition generates almost no uncertainties. In processing, uncertainties are mainly due to migration (relocation of reflectors to their true positions). During the interpretation phase of a project, detection of seismic markers, picking horizons, and interpreting faults are all potential sources of error. Finally, interpreted horizons and faults should be converted from time to depth, generating potentially huge errors that can account for as much as 50% or more of the total uncertainties (Thore et al., 2002). Migration and time-to-depth generated uncertainties can be quantified based on their associated velocity fields, although this is not straightforward (Thore et al., 2002). Errors due to interpretation are more difficult to estimate, and their determination may require the analysis of several interpretations from different people (Bond et al., 2007).

There is, however, a unifying feature about uncertainties: in order to be implemented, uncertainties need to be described in terms of magnitude, direction, and correlation length (a measure of how uncertainties in one region are correlated with those in another region; Thore et al., 2002). To illustrate these concepts, we refer to the simple, two-dimensional example of Figure 12.4. The folded bed (black line in Figure 12.4) consists of n points. Each of these points has uncertainties in location in x_1 and x_2 (i.e., uncertainty direction). These uncertainties follow a normal probability distribution, with a mean μ equal to the observed or measured x_1 and x_2, and a standard deviation σ (i.e., uncertainty magnitude). The uncertainties in location are correlated along the bed up to a maximum distance or correlation length l_c. A typical way to describe the spatial dependence of uncertainties along the profile is through a spherical

12.4 Uncertainties in structural data and their representation

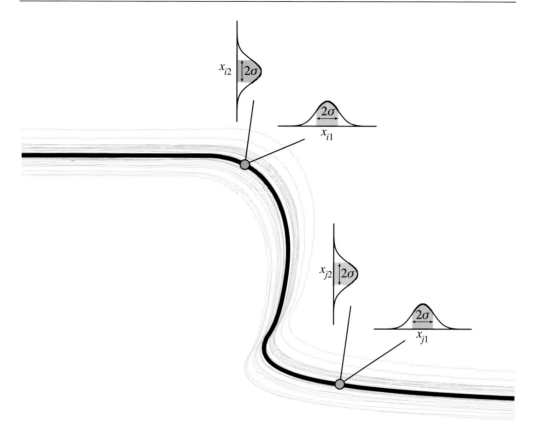

Figure 12.4 Strategy to generate realizations from a folded bed (dark line). Each point on the bed has uncertainties in x_1 and x_2 that follow a normal probability distribution, with a mean equal to the measured location and a standard deviation σ. Uncertainties within a distance along the bed lower than the correlation length are correlated. Gray lines are realizations.

variogram model (Davis, 2002). For this model, the $n \times n$ matrix $\mathbf{R}$ that describes the correlation of the uncertainties in location is

$$R_{ij} = \begin{cases} 1 + 0.5(h^3 - 3h), & h < 1 \\ 0, & h \geq 1 \end{cases} \quad (12.17)$$

where h is l_{ij}/l_c, and l_{ij} is the distance along the bed between points i and j (Figure 12.4). The $n \times n$ covariance matrix $\mathbf{C}$ (a matrix whose ij element is a measure of how uncertainties in points i and j change together) is

$$\mathbf{C} = \mathbf{SRS} \quad (12.18)$$

where $\mathbf{S}$ is an $n \times n$ diagonal matrix with the diagonal elements equal to the standard deviation σ.

Given the covariance matrix $\mathbf{C}$, we can generate different realizations (i.e., synthetic data sets that obey the observations and their uncertainties) of the bed following a procedure known as the Cholesky or square root method (Oliver et al., 2008). Basically, we decompose the covariance matrix $\mathbf{C}$ into the product of a matrix and its transpose using Cholesky factorization:[2]

---

[2] Cholesky factorization is a form of triangular decomposition that can be applied to positive definite matrices (those matrices that have all eigenvalues greater than zero; Lindfield and Penny, 1999).

$$\mathbf{C} = \mathbf{L}\mathbf{L}^T \tag{12.19}$$

We then generate a column vector **z** of $n \times 1$ independent, random numbers from 0 to 1. Finally, we can compute the $\mathbf{x}_1$ and $\mathbf{x}_2$ locations of points in the realization as (Aster *et al.*, 2005)

$$\begin{aligned} \mathbf{x}_{1(\text{realization})} &= \mathbf{x}_{1(\text{observed})} + \mathbf{L}\mathbf{z} \\ \mathbf{x}_{2(\text{realization})} &= \mathbf{x}_{2(\text{observed})} + \mathbf{L}\mathbf{z} \end{aligned} \tag{12.20}$$

The MATLAB function `BedRealizations`, below, generates realizations of a bed in two dimensions using the Cholesky method. The function relies on two MATLAB functions: `chol` which performs the Cholesky factorization, and `randn` which generates random numbers. In addition to the bed data and number of realizations, the user needs to input the standard deviation σ and correlation length l_c of the uncertainties. `BedRealizations` calls function `CorrSpher` (also below) which computes the correlation matrix **R** for the spherical variogram model.

```
function rlzt = BedRealizations(xp,yp,N,sigma,corrl)
%BedRealizations generates and plots realizations of a bed using a
%spherical variogram and the Cholesky method
%
%   USE: rlzt = BedRealizations(xp,yp,N,sigma,corrl)
%
%   xp = column vector with x locations of points along bed
%   yp = column vector with y locations of points along bed
%   N = number of realizations
%   sigma = Variance
%   corrl = Correlation length
%   rlzt = npoints x 2 x N+1 matrix with bed realizations. The first
%          realization in this matrix is the input xp, yp bed
%
%   BedRealizations uses function CorrSpher

%Number of points along bed
nj = max(size(xp));

%Variance matrix
Sf = zeros(nj,nj);
for i=1:nj
    for j=1:nj
        if i==j
            Sf(i,j)=sigma;
        end
    end
end

%Calculate correlation matrix using spherical variogram model
Rf=CorrSpher(xp,yp,corrl);
%Calculate covariance matrix (Cf)
Cf=Sf*Rf*Sf;
%Cholesky decomposition of covariance matrix. Here we use the MATLAB
%function chol
```

12.4 Uncertainties in structural data and their representation

```
[L,p] = chol(Cf,'lower');
if p > 0
    error ('Cf not positive definite');
end

%Initialize realizations
rlzt = zeros(nj,2,N+1);

%Start figure
figure;
hold on;
gray = [0.75 0.75 0.75];

%Generate realizations
for i=1:N+1
    %First realization is the bed itself
    if i == 1
        rlzt(:,1,i) = xp;
        rlzt(:,2,i) = yp;
    %Other Realizations
    else
        %Compute uncertainty in horizontal and vertical
        z = randn(nj,1);
        lz = L*z;
        %Add to observed data to generate realization
        rlzt(:,1,i) = xp + lz;
        rlzt(:,2,i) = yp + lz;
    end
    % Plot realization
    plot(rlzt(:,1,i),rlzt(:,2,i),'.','MarkerEdgeColor',gray);
end

%plot bed in black
plot(rlzt(:,1,1),rlzt(:,2,1),'k.');
hold off;
axis equal;
end

function r = CorrSpher(xp,yp,laj)
%CorrSpher calculates the correlation matrix for a spherical variogram
%
%   USE: r=CorrSpher(xp,yp,laj)
%
%   xp = vector with x locations of points along bed
%   yp = vector with y locations of points along bed
%   laj = correlation length
%   r = correlation matrix
```

```
%Number of points along bed
nj = max(size(xp));

%Initialize correlation matrix
r = zeros(nj,nj);

%Compute correlation matrix
for i=1:nj
    for j=1:nj
        %Find distance v between points i and j along bed
        v = 0.0;
        minind = min(i,j); %minimum index
        maxind = max(i,j); %maximum index
        for k = minind:1:maxind-1
            v = v + sqrt((xp(k)-xp(k+1))^2 + (yp(k)-yp(k+1))^2);
        end
        %Compute variogram entry
        h = v/laj;
        %If within correlation length
        if h < 1.0
            r(i,j)=1.0+0.5*(-3*h + h^3);
        end
    end
end
```

12.5 GEOLOGICAL APPLICATION: TRISHEAR INVERSE MODELING

In Chapter 11, we introduced the trishear kinematic model as a way to produce richer, more complex fault-propagation fold geometries and strain fields than those of kink-based models. However, when modeling natural fault-propagation folds, the main limitation of trishear is that, contrary to the kink models, trishear is incremental and there are no mathematical or geometrical rules to derive the model parameters from the observed fold geometry. There are two solutions to this problem: One can run trishear models forward to see how well they deform the beds to reproduce their final geometry. Implicit in this modeling is the assumption that one knows the initial geometry of the beds. Alternatively, one can run trishear models backward to see how well they unfold the beds to their original, approximately planar orientations (Allmendinger, 1998). In practice, this second strategy is easier because the initial state (planar beds) is much simpler than the final state (complexly folded beds), and there are simple statistical descriptions of the initial state. In two dimensions, for example, the goodness of fit of a model can be evaluated by how well the model restores a bed to a straight line. A *merit* or *objective* function f_{obj} is used to measure the fit. By convention this function is low when the fit is good. f_{obj} can be easily estimated by a simple least-squares linear regression of the restored bed profile. The MATLAB function **BackTrishear**, below, restores a folded bed in two dimensions using an input combination of trishear parameters (a trishear model), and returns an estimate of f_{obj}. **BackTrishear** uses function **regress** (MATLAB Statistics Toolbox) to perform the linear regression of the restored bed profile.

12.5 Geological application: Trishear inverse modeling

```
function chisq = BackTrishear(xp,yp,tparam,sinc)
%BackTrishear retrodeforms bed for the given trishear parameters and return
%sum of square of residuals (chisq)
%
%   USE: chisq = BackTrishear(xp,yp,tparam,sinc)
%
%   xp = column vector with x locations of points along bed
%   yp = column vector with y locations of points along bed
%   tparam = A 1 x 7 vector with the x and y coordinates of the fault tip
%            (entries 1 and 2), the ramp angle (entry 3), the  P/S (entry 4),
%            the trishear angle (entry 5), the fault slip (entry 6),  and the
%            concentration factor (entry 7)
%   sinc = slip increment
%   chisq = sum of square of residuals (objective function)
%
%   NOTE: Input ramp and trishear angles should be in radians
%         For reverse faults use positive slip and slip increment
%         For normal faults use negative slip and slip increment
%         The MATLAB Statistics Toolbox is needed to run this function
%
%   BackTrishear uses function VelTrishear

% Model parameters

xtf = tparam(1); %x current fault tip
ytf = tparam(2); %y current fault tip
ramp = tparam(3);%Ramp angle
psr = tparam(4)*-1.0; %P/S: Multiply by -1 because we are restoring bed
tra = tparam(5); %Trishear angle
m = tan(tra/2); %Tangent of half trishear angle
slip = tparam(6); %Fault slip
c = tparam(7); %Concentration factor
ninc=round(slip/sinc); %Number of slip increments
sincr = slip/ninc*-1.0; %Slip increment: Multiply by -1 (restoring bed)

%Transformation matrix from geographic to fault coordinates
a11=cos(ramp);
a12=cos(pi/2-ramp);
a21=cos(pi/2+ramp);
a22=a11;

% Transform to coordinates parallel and perpendicular to the fault, and
% with origin at current fault tip
fx=(xp-xtf)*a11+(yp-ytf)*a12;
fy=(xp-xtf)*a21+(yp-ytf)*a22;

% Restore
for i=1:ninc
```

```
    for j=1:size(fx,1)
        % Solve trishear in a coordinate system attached to current
        % fault tip. Note: First retrodeform and then move tip back
        xx=fx(j)-(psr*(i-1)*abs(sincr));
        yy=fy(j);
        % compute velocity
        [vx,vy]=VelTrishear(xx,yy,sincr,m,c);
        % UPDATE fx, fy coordinates
        fx(j)=fx(j)+vx;
        fy(j)=fy(j)+vy;
    end
end

%Fit straight line to restored bed. Use MATLAB function regress (MATLAB
%Statistics Toolbox) to compute linear regression. b(1) is the intercept
%and b(2) the slope of the line
XX = [ones(size(fx)) fx];
YY = fy;
b = regress(YY,XX);

%Compute chisq (objective function) = Sum of square of residuals between
%straight line and restored bed
chisq = sum((fy-b(1)-b(2)*fx).^2.);
end
```

We now have a way to assess the goodness of fit of a trishear model. But now the question is: Within all possible trishear models, what is the model with the best fit or lowest f_{obj}? This is in essence an inverse (minimization) problem (Aster et al., 2005). There are several ways to solve this problem. The easiest is to establish a grid of possible trishear models (defined by minimum and maximum limits, and step sizes of the parameters), and to systematically test each one of these models to find out the one with the lowest f_{obj} (the best-fit model). This grid-search method (Allmendinger, 1998) is robust (you are guaranteed to find the model with the lowest f_{obj} in the grid), but it is quite inefficient. A grid search of all parameters of a two-dimensional trishear model (Chapter 11) may involve testing hundreds of thousands of models, and even at the speed of today's personal computers this can take hours.

The other strategy involves the use of optimization methods. Optimization algorithms do not systematically explore the parameter space as the grid-search method does, but rather traverse the space in search of the best-fit model. A good analogy is to imagine the parameter space to be a rough terrain with valleys and hills. The grid-search method would explore the entire terrain systematically to find the lowest point. The optimization algorithms on the other hand would be like a ball moving down the terrain under the force of gravity. This of course is much more efficient than the grid-search method. The problem with optimization algorithms is that they can be caught in local minima. Depending on the energy and size of the ball, it might get stuck in a local valley before getting to the lowest point (Cardozo and Aanonsen, 2009). A detailed description of optimization methods is beyond the scope of this book and there are several fine texts that introduce this topic in a more authoritative manner (e.g., Nocedal and Wright, 1999; and Aster et al., 2005).

A depth-converted seismic section of the Santa Fe Springs anticline and the underlying Puente Hills thrust fault in the Los Angeles Basin (Shaw and Shearer, 1999) illustrates the concept of

12.5 Geological application: Trishear inverse modeling

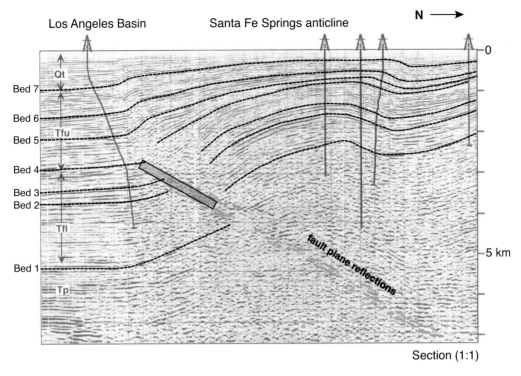

Figure 12.5 Depth-converted seismic section of the Santa Fe Springs anticline in the Los Angeles Basin. Dashed lines are interpreted beds and the rectangle is the area where the fault tip can be located. Seismic and well data from Shaw and Shearer (1999).

optimization and trishear inverse modeling. The thrust fault is well defined in the seismic data (Fig. 12.5). The ramp angle is $29°$. The location of the fault tip is not exactly known and is assumed to be along the fault within the gray rectangle in Figure 12.5. The distance along the fault between the lower end of the rectangle and the possible location of the fault tip is defined here as *lft*. We can run a trishear inversion to search for the *lft*, P/S, trishear angle, and fault slip that best fit the structure in Figure 12.5. The MATLAB function `InvTrishear`, below, is designed for this purpose. The function takes the coordinates of a bed in two dimensions and a guess of the four parameters above, and estimates the model (i.e., the combination of trishear parameters) that best restores the bed. `InvTrishear` uses our previous function `BackTrishear` to obtain a value of f_{obj} for the currently tested model. The inversion (f_{obj} minimization) is done through the MATLAB function `fmincon` (MATLAB Optimization Toolbox), which performs a constrained (limits on the searched parameters), gradient-based optimization.

```
function [xbest,fval,flag] = InvTrishear(xp,yp,tparams,sinc,maxit)
%InvTrishear performs inverse trishear modeling using a constrained,
%gradient based optimization method
%
%   [xbest,fval,flag] = InvTrishear(xp,yp,tparams,sinc,maxit)
%
%   xp = column vector with x locations of points along bed
%   yp = column vector with y locations of points along bed
%   tparams = A 1 x 8 vector with the x and y coordinates of the
```

```
%              lowest possible location of the fault tip (entries 1 and 2),
%              the distance along the fault line from the lowest to the
%              highest possible locations of the fault tip (lft, entry 3),
%              the ramp angle (entry 4), the P/S (entry 5), the trishear angle
%              (entry 6), the fault slip (entry 7), and the concentration
%              factor (entry 8)
% sinc = slip increment
% maxit = maximum number of iterations in the optimized search
% xbest = Best-fit model
% fval = Objective function value of best-fit model
% flag = Integer that indicates if the model converged (flag > 0)
%
% NOTE: Input ramp and trishear angles should be in radians
%       The search is for the best-fit slip, trishear angle, P/S, and lft
%       The MATLAB Optimization Toolbox is needed to run this function
%
% InvTrishear uses function BackTrishear

%Trishear parameters for BackTrishear
tparam = zeros(1,7);

%Known values
xtt = tparams(1); %Coordinates of lowest possible location of fault tip
ytt = tparams(2);
tparam(3) = tparams(4); %Ramp angle
tparam(7) = tparams(8); %Concentration factor

%Set initial guess (x0), minimum (lb), and maximum  (ub) parameters limits
%Entries in these vectors are: [slip trishear angle P/S lft]
%These entries should be in the same order of magnitude
%The values and scaling below only work for the Santa Fe Springs anticline
%Change lb and ub if you want to search over a larger or smaller parameter
%space
sf = 1.0e-3; %scaling for slip and lft
x0= [tparams(7)*sf tparams(6) tparams(5) tparams(3)*sf/2.]; %initial guess
lb = [0. 40.*pi/180. 1.5 0.0]; %lower limit
ub = [15. 80.*pi/180. 3.5 tparams(3)*sf]; %upper limit

%Optimization settings: Display off, maximum number of iterations, and type
%of algorithm. Use MATLAB function optimset (MATLAB  Optimization Toolbox)
options = optimset('Display','off','MaxIter',maxit,...
    'Algorithm','active-set');

%Compute best-fit model using constrained, gradient  based optimization
%method. Use MATLAB function fmincon (MATLAB Optimization Toolbox)
[xbest,fval,flag] = fmincon(@objfun,x0, [], [], [], [],lb,ub,@confun,...
                    options);
```

12.5 Geological application: Trishear inverse modeling

```
%Supporting functions

%Function to compute the objective function for a given combination of
%parameters x
function f = objfun(x)
tparam(6) = x(1)/sf; %Slip: Return to its non-scaled value
tparam(5) = x(2); %Trishear angle
tparam(4) = x(3); %P/S
lft = x(4)/sf; %lft: Return to its non-scaled value
tparam(2) = ytt + lft*sin(tparam(3)); %x fault tip
tparam(1) = xtt + lft*cos(tparam(3)); %y fault tip
f = BackTrishear(xp,yp,tparam,sinc); %Compute objective function
end

%Function for constrained optimization method fmincon
function [c, ceq] = confun(x)
% Nonlinear inequality constraints
c = [];
% Nonlinear equality constraints
ceq = [];
end
end
```

Running **InvTrishear** for bed 4 of Figure 12.5, with an initial guess a_0 of [1.0 km, 2.5, 60°, 7.5 km] (*lft, P/S*, trishear angle, and fault slip), minimum limits a_{min} of [0, 1.5, 40°, 0], and maximum limits a_{max} of [2 km, 3.5, 80°, 15 km] produces a best-fit estimate a_f of [1.45 km, 2.52, 71°, 6.7 km] (for bed 4 with about 500 points, the computation takes 20 seconds!). A forward model of a smoothed version of the restored beds using the best-fit parameters a_f is shown in Figure 12.6. This model fits well beds 4 and 7 and not so well beds 1 to 3.

One can try to refine this analysis by using different beds in the section, changing parameter limits, etc., but here we are interested in another, perhaps more profound issue. The fold data in Figure 12.5 have errors of various kinds, including imaging and interpretation errors (Section 12.4). These errors introduce some uncertainty in the estimated best-fit parameters a_f. How can we estimate the uncertainties of a_f? Figure 12.7 shows a strategy to do this. Basically, from the observed data set we generate several synthetic data sets (i.e., realizations) as outlined in Section 12.4. Inverse modeling (f_{obj} minimization) of these realizations gives a set of simulated best-fit parameters ($a_{f1}, a_{f2}, ...$) that are distributed around the best-fit model for the observed data a_{f0}. From these, we can determine the probability distribution and uncertainties of a_f. Since the synthetic data sets are generated randomly from the observed data and there is no conditioning between synthetics, this technique is known as a randomized maximum likelihood method (RML, Oliver *et al.*, 2008). The MATLAB function **RMLMethod** below performs this type of analysis.

```
function [xbesti,fvali] = RMLMethod(xp,yp,tparams,sinc,maxit,N,sigma,corrl)
%RMLMethod runs a Monte Carlo type, trishear inversion analysis for a
%folded bed
%
% USE: [xbest,fval] = RMLMethod(xp,yp,tparams,sinc,maxit,N,sigma,corrl)
```

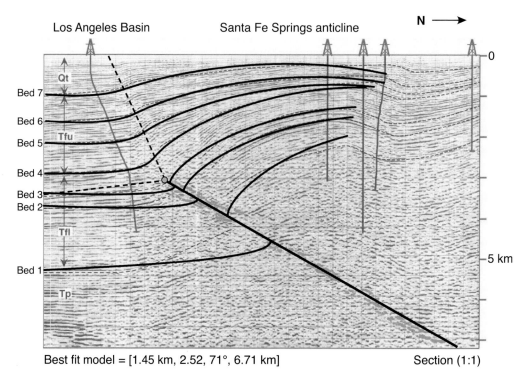

Figure 12.6 Best-fit trishear model (black tick lines) for the Santa Fe Springs anticline. The best-fit model was obtained by inversion of bed 4. The entries in the best-fit vector correspond to location of fault tip along fault line *lft*, *P/S*, trishear angle, and fault slip.

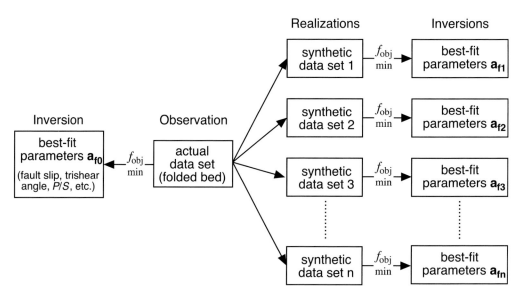

Figure 12.7 Strategy to estimate the uncertainty of the best-fit parameters a_f. From a measured data set (observations) several synthetic data sets are generated (realizations). Inverse modeling of these realizations gives a set of simulated best-fit parameters (a_{f1}, a_{f2} ...) from which we can determine the uncertainties of a_f.

12.5 Geological application: Trishear inverse modeling

```
%
% xp = column vector with x locations of points along bed
% yp = column vector with y locations of points along bed
% tparams = A vector of guess trishear parameters as in function
%         InvTrishear
% sinc = slip increment
% maxit = maximum number of iterations in the optimized search
% N = number of realizations
% sigma = Variance
% corrl = Correlation length
% xbest = Best-fit models for realizations
% fval = Objective function values of best-fit models
%
% NOTE: Input ramp and trishear angles should be in radians
%
% RMLMethod uses function BedRealizations and InvTrishear

%Generate realizations
rlzt = BedRealizations(xp,yp,N,sigma,corrl);

%Initialize xbesti and fvali
xbesti=zeros(N+1,4);
fvali=zeros(N+1,1);

%Find best-fit model for each realization
count = 1;
for i=1:N+1
    [xbest,fval,flag] = InvTrishear(rlzt(:,1,i),rlzt(:,2,i),tparams,...
                    sinc,maxit);
    % if the function converges to a solution
    if flag > 0
        xbesti(count,:)=xbest;
        fvali(count,:)=fval;
        %Output realization number and fval
        disp(['Realization ',num2str(i),' fval = ',num2str(fval)]);
        %Increase count
        count = count + 1;
    end
end

%Remove not used elements of xbesti and fvali
xbesti(count:N+1,:)=[];
fvali(count:N+1,:)=[];
end
```

Figures 12.8 and 12.9 show the result of applying the RML method to the Santa Fe Springs anticline. Figure 12.8 shows bed 4 (Fig. 12.5) and its realizations (gray lines). One thousand realizations were created using a standard deviation σ (uncertainty in location) of 50 m and a correlation length l_c of 100 km (l_c must be large to obtain smooth fold profiles). This makes the

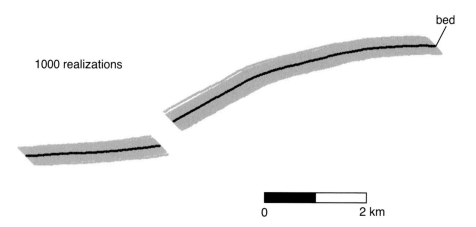

Figure 12.8 Realizations of bed 4 in Santa Fe Springs anticline (Figure 12.5). Realizations are based on a standard deviation of 50 m and a correlation length of 100 km.

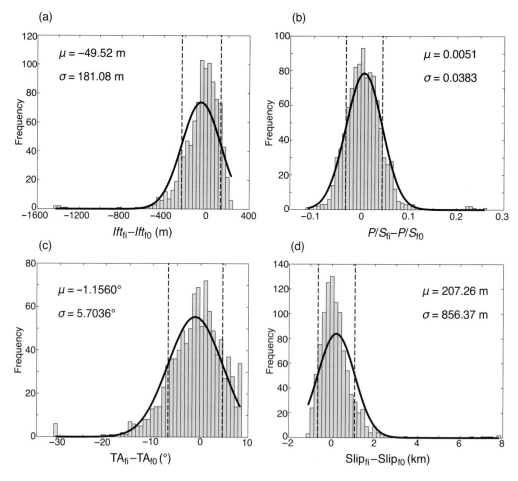

Figure 12.9 Statistics of the trishear inversions of bed 4 realizations in Figure 12.8. (a) to (d) are histograms for uncertainties in (a) location of fault tip along fault projection lft, (b) P/S, (c) trishear angle, and (d) fault slip. In (a) to (d), the black thick line is a normal distribution fit to the histogram. Dashed lines delimit the 68% confidence interval.

realizations fall within a distance of ±200 m from the interpreted bed. This uncertainty is comparable with the possible uncertainty due to time-to-depth conversion of the seismic data, where a 10% error in a seismic velocity of 4 km/s (a reasonable value for the Tertiary sediments in the Los Angeles Basin) at 1 s two-way travel time would yield an uncertainty in depth of ±200 m.

Figure 12.9 shows the statistics of the inversions of the 1000-bed realizations of Figure 12.8. Thanks to the fast optimization methods, the 1000 inversions took a couple of hours. The statistics is shown as deviations of the synthetic best-fit parameters $\mathbf{a}_{fi}$ with respect to the observed ones $\mathbf{a}_{fo}$. Normal fits to the probability distribution of $\mathbf{a}_{fi}-\mathbf{a}_{fo}$ indicate that the σ errors in *lft*, P/S, trishear angle, and fault slip are 181 m, 0.04, 5.7°, and 856 m (Figure 12.9). In other words, there is 68% chance that the true best-fit parameter values fall within intervals in *lft*, P/S, trishear angle, and fault slip of 1216–1578 m, 2.48–2.56, 65–76°, and 6.06–7.8 km (regions limited by dashed lines in Figure 12.9). Notice that the major uncertainties are in trishear angle and fault slip. The structure can be fit with relatively high trishear angle and low fault slip, or vice versa (Figure 12.9 c, d). Based on trishear inverse modeling of the section in Figure 12.5, Allmendinger and Shaw (2000) found that the Puente Hills thrust initiated at the same location as the 1987 M6.0 Whittier Narrows earthquake. This observation, which has important implications for earthquake hazard assessment, is within the 68% confidence intervals of our analysis. You will get the chance to try the RML method in the exercises section.

12.6 EXERCISES

1. Use Equations 12.9 through 12.16 to calculate the uncertainty on shortening in the simple crustal area balance in Section 12.3.2. Do you get the same answer? If not why not? Discuss your results.
2. Derive a set of equations, similar to Equations 12.15 and 12.16, that gives the uncertainty on percentage shortening rather than just the magnitude of shortening.
3. The following table of numbers represents a 30-vertex polygon for the southernmost Subandean belt in northwestern Argentina (Echavarría et al., 2003), the same one shown in Figure 12.3. Calculate the uncertainty in shortening magnitude and percentage. Then, redo your analysis to investigate the relative importance of uncertainties in eroded hanging wall cutoffs, decollement, and stratigraphic thickness in contributing to the overall uncertainty. Hint: Use function `BalCrossErr`.

T_1 (km)	T_2 (km)	Error in T_1 (km)	Error in T_2 (km)
2.9	4.6	0.29	0.46

Stratigraphy

x_1 (km)	x_2 (km)	Error in x_1 (km)	Error in x_2 (km)	Tag*
−23.075 114 01	−8.587 204 59	0.75	0.75	1
−93.403 839 88	−11.033 838 30	0.75	0.75	1
−87.167 322 58	−4.173 669 27	0.8	0.8	3
−85.824 072 70	0.911 490 99	0.1	0.1	2
−85.536 233 44	−1.439 196 30	0.8	0.8	3
−80.451 073 18	2.686 499 76	3.0	3.0	4
−77.620 653 78	2.974 339 02	3.0	3.0	4

(cont.)

x_1 (km)	x_2 (km)	Error in x_1 (km)	Error in x_2 (km)	Tag*
−75.60577896	0.67162494	0.1	0.1	2
−78.91593045	−0.09594642	0.8	0.8	3
−83.42541220	−4.94124063	0.8	0.8	3
−82.03418911	−6.42841014	0.8	0.8	3
−73.97468982	−6.76422261	0.8	0.8	3
−66.53884227	−0.67162494	0.8	0.8	3
−65.62735128	1.63108914	3.0	3.0	4
−63.90031572	1.87095519	3.0	3.0	4
−66.44289585	−1.82298198	0.8	0.8	3
−65.00369955	−3.07028544	0.8	0.8	3
−61.45368201	−1.58311593	0.8	0.8	3
−60.35029818	0.71959815	0.1	0.1	2
−59.29488756	1.10338383	3.0	3.0	4
−57.37595916	−3.74191038	0.8	0.8	3
−59.19894114	−4.17366927	0.8	0.8	3
−56.60838780	−5.51691915	0.8	0.8	3
−42.40831764	−5.32502631	0.8	0.8	3
−37.27518417	0.52770531	0.8	0.8	3
−35.16436293	0.14391963	0.8	0.8	3
−32.76570243	−4.74934779	0.8	0.8	3
−23.45889969	−5.13313347	0.8	0.8	3
−13.43249880	−2.87839260	0.8	0.8	3
−12.80884707	−4.94124063	0.8	0.8	3

*Tag key indicates the setting of the vertex: 1 – decollement; 2 – point on the land surface; 3 – normal point in the subsurface; 4 – eroded (i.e., point above the erosional surface).

Enveloping polygon for deformed area

4. Supplementary data file "Problem 12.4" contains the digitized contacts of beds 3 and 4 in Figure 12.5. It also has a fault file with the lowest and highest possible locations of the Puente Hills thrust tip. Run an RML analysis for beds 3 and 4. In each case use 1000-bed realizations and uncertainties, limits, and guess parameters similar to the ones used in Section 12.5.

References

Albee, H. F. & H. L. Cullins (1975). *Geologic Map of the Poker Peak Quadrangle, Bonneville County, Idaho*, U.S. Geological Survey, Geologic Quadrangle Map GQ 1260.

Allison, I. (1984). The pole of the Mohr diagram. *Journal of Structural Geology*, **6**, 331-333.

Allmendinger, R. W. (1998). Inverse and forward numerical modeling of trishear fault-propagation folds. *Tectonics*, **17**, 640-656.

Allmendinger, R. W. & J. H. Shaw (2000). Estimation of fault propagation distance from fold shape: Implications for earthquake seismicity. *Geology*, **28**, 1099-1102.

Allmendinger, R. W., J. P. Loveless, M. E. Pritchard & B. Meade (2009). From decades to epochs: Spanning the gap between geodesy and structural geology of active mountain belts. *Journal of Structural Geology*, **31**, 1409-1422, doi: 10.1016/j.jsg.2009.08.008.

Angelier, J. (1984). Tectonic analysis of fault slip data sets. *Journal of Geophysical Research*, **89**, 5835-5848.

Aster, R. C., B. Borchers & C. H. Thurber (2005). *Parameter Estimation and Inverse Problems*. Amsterdam: Elsevier Academic Press.

Bally, A. W., P. L. Gordy & G. A. Stewart (1966). Structure, seismic data, and orogenic evolution of southern Canadian Rocky Mountains. *Bulletin Canadian Petroleum Geology*, **14**, 337-381.

Bevington, P. R. & D. K. Robinson (2003). *Data Reduction and Error Analysis for the Physical Sciences*. New York: McGraw-Hill.

Bobyarchick, A. R. (1986). The eigenvalues of steady flow in Mohr space. *Tectonophysics*, **122**, 35-51.

Bond, C. E., A. D. Gibbs, Z. K. Shipton & S. Jones (2007). What do you think this is? "Conceptual uncertainty" in geoscience interpretation. *GSA Today*, **17**, 4.

Bott, M. H. P. (1959). The mechanics of oblique slip faulting. *Geological Magazine*, **96**, 109-117.

Buxtorf, A. (1916). Prognosen und Befunden beim Hauensteinbasis und Grencherbergtunnel und die Bedeutung der letzteren für die Geologie des Juragebirges. *Verhandlungen der Naturforschenden Gesellschaft in Basel*, **27**, 184-205.

Cardozo, N. & S. Aanonsen (2009). Optimized trishear inverse modeling. *Journal of Structural Geology*, **31**, 546-560.

Cardozo, N. & R. W. Allmendinger (2009). SSPX: A program to compute strain from displacement/velocity data. *Computers and Geosciences*, **35**, 1343-1357, doi: 10.1016/j.cageo.2008.05.008.

Chamberlin, R. T. (1910). The Appalachian folds of central Pennsylvania. *Journal of Geology*, **18**, 228-251.

Chamberlin, R. T. (1919). The building of the Colorado Rockies. *Journal of Geology*, **27**, 145-164.

Chamberlin, R. T. (1923). On the crustal shortening of the Colorado Rockies. *American Journal of Science*, **6**, 215-221.

Charlesworth, H. A. K., C. W. Langenberg & J. Ramsden (1976). Determining axes, axial planes, and sections of macroscopic folds using computer-based methods. *Canadian Journal of Earth Science*, **13**, 54-65.

Cladouhos, T. T. & R. W. Allmendinger (1993). Finite strain and rotation from fault slip data. *Journal of Structural Geology*, **15**, 771-784.

Cutler, J. & D. Elliott (1983). The compatibility equations and the pole of the Mohr circle. *Journal of Structural Geology*, **10**, 287-297.

Dahlstrom, C. D. A. (1969). Balanced cross sections. *Canadian Journal of Earth Sciences*, **6**, 743-757.

Dahlstrom, C. D. A. (1970). Structural geology in the eastern margin of the Canadian Rocky Mountains. *Bulletin of Canadian Petroleum Geology*, **18**, 332-406.

Davis, J. C. (2002). *Statistics and Data Analysis in Geology*. Hoboken, NJ: John Wiley.

DePaor, D. G. (1983). Orthographic analysis of geological structures – I. Deformation theory. *Journal of Structural Geology*, **5**, 255-277.

Durney, D. W. & J. G. Ramsay (1973). Incremental strains measured by syntectonic crystal growths. In K. A. De Jong & R. Scholten, eds., *Gravity and Tectonics*. New York: John Wiley & Sons, pp. 67-96.

Echavarría, L., R. Hernández, R. W. Allmendinger & J. Reynolds (2003). Subandean thrust and fold belt of northwestern Argentina: Geometry and timing of the Andean evolution. *AAPG Bulletin*, **87**, 965-985.

Elliott, D. (1972). Deformation paths in structural geology. *Bulletin of the Geological Society of America*, **83**, 2621-2635.

Erslev, E. A. (1991). Trishear fault-propagation folding. *Geology*, **19**, 617-620.

Fisher, D., C. Y. Lu & H. T. Chu (2002). Taiwan Slate Belt: Insights into the ductile interior of an arc-continent collision. In T. Byrne and C. S. Liu, eds., *Geology and Geophysics of an Arc-Continent Collision*. Boulder, CO: Geological Society of America, Special Paper 358, pp. 93-106.

Fisher, N. I., T. L. Lewis & B. J. Embleton (1987). *Statistical Analysis of Spherical Data*. Cambridge: Cambridge University Press.

Fjeldskaar, W., M. Ter Voorde, H. Johansen *et al.* (2004). Numerical simulation of rifting in the northern Viking Graben: The mutual effect of modelling parameters. *Tectonophysics*, **382**, 189-212, doi: 10.1016/j.tecto.2004.01.002.

Fossen, H. (2010). *Structural Geology*. Cambridge: Cambridge University Press.

Fossen, H. & B. Tikoff (1993). The deformation matrix for simultaneous simple shearing, pure shearing, and volume change, and its application to transpression-transtension tectonics. *Journal of Structural Geology*, **15**, 413-422.

Gephart, J. W. (1990). Stress and the direction of slip on fault planes. *Tectonics*, **9**, 845-858.

Ghosh, S. K. & H. Ramberg (1976). Reorientation of inclusions by combination of pure shear and simple shear. *Tectonophysics*, **34**, 1-70.

References

Hardy, S. (1995). A method for quantifying the kinematics of fault-bend folding. *Journal of Structural Geology*, **17**, 1785-1788, doi: 10.1016/0191-8141(95)00077-Q.

Hardy, S. (1997). A velocity description of constant-thickness fault-propagation folding. *Journal of Structural Geology*, **19**, 893-896.

Hardy, S. & J. Poblet (1995). The velocity description of deformation, Paper 2: Sediment geometries associated with fault-bend and fault-propagation folds. *Marine and Petroleum Geology*, **12**, 165-176.

Hardy, S. & J. Poblet (2005). A method for relating fault geometry, slip rate and uplift data above fault-propagation folds. *Basin Research*, **17**, 417-424.

Hardy, S., J. Poblet, K. McClay & D. Waltham (1996). Mathematical modelling of growth strata associated with fault-related fold structures. In P. G. Buchanan & D. A. Nieuwland, eds., *Modern Developments in Structural Interpretation, Validation and Modelling*. London: The Geological Society, pp. 265-282.

Harris, J. W. & H. Stocker (1998). *Handbook of Mathematics and Computational Science*. New York: Springer-Verlag.

Holt, W. E., B. Shen-Tu, J. Haines & J. Jackson (2000). On the determination of self-consistent strain rate fields within zones of distributed continental deformation. In M. A. Richards et al., eds., *The History and Dynamics of Global Plate Motions*. Washington, DC: American Geophysical Union, Geophysical Monograph, pp. 113-141.

Jackson, J. A. & D. P. McKenzie (1988). The relationship between plate motions and seismic moment tensors, and the rates of active deformation in the Mediterranean and Middle East. *Geophysical Journal*, **93**, 45-73.

Jaeger, J. C. & N. G. W. Cook (1979). *Fundamentals of Rock Mechanics*. London: Chapman and Hall.

Jeffery, G. B. (1922). The motion of ellipsoidal particles immersed in a viscous fluid. *Proceedings of the Royal Society of London, Series A*, **102**, 161-179, doi: doi:10.1098/rspa.1922.0078.

Judge, P. A. & R. W. Allmendinger (2011). Assessing uncertainties in balanced cross sections. *Journal of Structural Geology*, **33**, 458-467, doi: 10.1016/j.jsg.2011.01.006.

Klotz, J., D. Angermann, G. Michel *et al.* (1999). GPS-derived deformation of the Central Andes including the 1995 Antofagasta Mw = 8.0 Earthquake. *Pure and Applied Geophysics*, **154**, 709-730.

Kostrov, V. V. (1974). Seismic moment and energy of earthquakes, and seismic flow of rock. *Izvestiya, Academy of Sciences, USSR, Physics of the Solid Earth*, **1**, 23-44.

Kreemer, C., W. E. Holt & A. J. Haines (2003). An integrated global model of present-day plate motions and plate boundary deformation. *Geophysical Journal International*, **154**, 8-34.

Lindfield, G. R. & J. E. T. Penny (1999). *Numerical Methods Using MATLAB*. Englewood Cliffs, NJ: Prentice Hall.

Lisle, R. J. & D. M. Ragan (1988). Brevia: Strain from three stretches: a simple Mohr circle solution. *Journal of Structural Geology*, **10**, 905-906.

Malvern, L. E. (1969). *Introduction to the Mechanics of a Continuous Medium*. Englewood Cliffs, NJ: Prentice-Hall.

Mandl, G. & G. K. Shippam (1981). Mechanical model of thrust sheet gliding and imbrication. In K. R. McClay & N. J. Price, eds., *Thrust and Nappe Tectonics*. London: Geological Society of London, Special Publications 9, pp. 79-98.

Marrett, R. A. & R. W. Allmendinger (1990). Kinematic analysis of fault-slip data. *Journal of Structural Geology*, **12**, 973-986.

Marrett, R. A. & R. W. Allmendinger (1992). The amount of extension on "small" faults: An example from the Viking Graben. *Geology*, **20**, 47-50.

Marrett, R. A. & D. C. P. Peacock (1999). Strain and stress. *Journal of Structural Geology*, **21**, 1057–1063.

Marshak, S. & G. Mitra (1988). *Basic Methods of Structural Geology*. Englewood Cliffs, NJ: Prentice Hall.

McKenzie, D. (1978). Some remarks on the development of sedimentary basins. *Earth and Planetary Science Letters*, **40**, 25–32.

Means, W. D. (1976). *Stress and Strain: Basic Concepts of Continuum Mechanics for Geologists*. New York: Springer-Verlag.

Means, W. D., B. E. Hobbs, G. S. Lister & P. F. Williams (1980). Vorticity and non-coaxiality in progressive deformations. *Journal of Structural Geology*, **2**, 371–378.

Menke, W. (1984). *Geophysical Data Analysis: Discrete Inverse Theory*. Orlando, FL: Academic Press.

Merle, O. (1986). Patterns of stretch trajectories and strain rates within spreading-gliding nappes. *Tectonophysics*, **124**, 211–222.

Mitra, S. (1978). Microscopic deformation mechanisms and flow laws in quartzites within the South Mountain anticline. *Journal of Geology*, **86**, 129–152.

Mitra, S. & J. Namson (1989). Equal-area balancing. *American Journal of Science*, **289**, 563–599.

Molnar, P. (1983). Average regional strain due to slip on numerous faults of different orientations. *Journal of Geophysical Research*, **88**, 6430–6432.

Nickelsen, R. P. (1979). Sequence of structural stages of the Alleghany Orogeny, at the Bear Valley strip mine, Shamokin, Pennsylvania. *American Journal of Science*, **279**, 225–271.

Nocedal, J. & S. J. Wright (1999). *Numerical Optimization*. New York: Springer-Verlag, Springer Series in Operations Research.

Nye, J. F. (1985). *Physical Properties of Crystals: Their Representation by Tensors and Matrices*. Oxford: Oxford University Press.

Oliver, D. S., A. C. Reynolds & N. Liu (2008). *Inverse Theory for Petroleum Reservoir Characterization and History Matching*. Cambridge: Cambridge University Press.

Pollard, D. D. (2000). Strain and stress: Discussion. *Journal of Structural Geology*, **22**, 1359–1368.

Pollard, D. D. & R. C. Fletcher (2005). *Fundamentals of Structural Geology*. Cambridge: Cambridge University Press.

Press, W. H., B. P. Flannery, S. A. Teukolsky & W. T. Vetterling (1986). *Numerical Recipes: The Art of Scientific Computing*. Cambridge: Cambridge University Press.

Price, R. A. & E. W. Mountjoy (1970). Geologic structure of the Canadian Rocky Mountains between Bow and Athabasca rivers: A progress report. In J. O. Wheeler ed., *Structure of the Southern Canadian Cordillera*. Geological Association of Canada, pp. 7–25.

Ragan, D. M. (2009). *Structural Geology: An Introduction to Geometrical Techniques*. Cambridge: Cambridge University Press.

Ramberg, H. (1975). Particle paths, displacement and progressive strain applicable to rocks. *Tectonophysics*, **28**, 1–37.

Ramsay, J. G. (1967). *Folding and Fracturing of Rocks*. New York: McGraw-Hill.

Ramsay, J. G. & M. I. Huber (1983). *The Techniques of Modern Structural Geology. Volume 1: Strain Analysis*. London: Academic Press.

Rodgers, J. (1949). Evolution of thought on structure of middle and southern Appalachians. *American Association of Petroleum Geologists Bulletin*, **33**, 1643–1654.

Shaw, J. H. & P. M. Shearer (1999). An elusive blind-thrust fault beneath metropolitan Los Angeles. *Science*, **283**, 1516–1518.

References

Shen, Z., D. D. Jackson & B. X. Ge (1996). Crustal deformation across and beyond the Los Angeles Basin from geodetic measurements. *Journal of Geophysical Research,* **101**, 27,957–27,980.

Snyder, J. P. (1987). *Map Projections: A Working Manual.* Washington, DC: U.S. Geological Survey, Professional Paper 1395.

Suppe, J. (1983). Geometry and kinematics of fault-bend folding. *American Journal of Science,* **283**, 684–721.

Suppe, J. & D. Medwedeff (1990). Geometry and kinematics of fault-propagation folding. *Eclogae Geologicae Helvetiae,* **83**, 409–454.

Suppe, J., G. T. Chou & S. C. Hook (1992). Rates of folding and faulting determined from growth strata. In K. R. McClay, ed., *Thrust Tectonics.* London: Chapman & Hall, pp. 105–121.

Taylor, J. R. (1997). *An Introduction to Error Analysis: The Study of Uncertainties in Physical Measurements.* Sausalito, CA: University Science Books.

Thore, P., A. Shtuka, M. Lecour, T. Ait-Ettajer & R. Cognot (2002). Structural uncertainties: Determination, management, and applications. *Geophysics,* **67**, 840.

Truesdell, C. (1953). Two measures of vorticity. *Journal of Rational Mechanics and Analysis,* **2**, 173–217.

USGS Eastern Region (2000). Map Projections. http://egsc.usgs.gov/isb/pubs/MapProjections/projections.html (January 27, 2011).

Waltham, D. & S. Hardy (1995). The velocity description of deformation, Paper 1: Theory. *Marine and Petroleum Geology,* **12**, 153–163.

Willett, S. D., C. Beaumont & P. Fullsack (1993). Mechanical model for the tectonics of doubly vergent compression orogens. *Geology,* **21**, 371–374.

Zapata, T. R. & R. W. Allmendinger (1996). Growth strata record of instantaneous and progressive limb rotation, Precordillera thrust belt and Bermejo Basin, Argentina. *Tectonics,* **15**, 1065–1083.

Zehnder, A. T. & R. W. Allmendinger (2000). Velocity field for the trishear model. *Journal of Structural Geology,* **22**, 1009–1014.

Index

arrays 7
axes 26

balanced cross sections 256
   area balancing 257, 258, 259
   depth to decollement 259
   error propagation 259, 260
   Gaussian error on area 262
   line-length balancing 257
   maximum error on area 262
   minimum shortening estimate 257
   shortening error 262
   thick-skinned 259
   thin-skinned 259

Cholesky factorization 267
continuity equation 218
coordinate systems 23–25
   Cartesian 24
   east-north-up 25
   left-handed 24, 78
   north-east-down (NED) 25, 28, 114
   right-handed 24, 78
   spherical 23

deformation 1, 120
   elongation 120, 124, 125, 169
   gradient 123, 126
   quadratic elongation 120
   stretch 120, 123, 125
   translation 123, 125
Delaunay triangulation 157
direction cosines 28, 31, 185
displacement 165
   Euler 124
   field 121, 185, 188, 190, 194
   gradient 121, 124, 173
   Lagrange 124
   path 185, 193, 196, 203
   vector 123

error propagation 2, 255
   error in quadrature 255
   maximum error 255
Eulerian frame 251
external rotation 200

fault
   decollement 220, 230
   inversion for stress 116
   listric 225
   movement plane 115, 151
   principal stress ratio 116
   propagation to slip (P/S) 231
   Puente Hills thrust 279
   stress on arbitrary plane 113–116
fibers
   antitaxial 200, 201, 209
   syntaxial 200
fold
   best-fit axis 91
   cylindrical 26, 92
   down-plunge projection 51–53
   fault-bend fold 220, 221
   fault-propagation fold 220, 230, 231, 235, 236
   footwall synclines 241
   kink bands 221
   orientation matrix 93
   parallel folding 257
   profile view 51
   rollover anticline 225, 227
   Santa Fe Springs anticline 273
   similar fold 227
functions 8

geometric moment 150
growth strata 237, 247
   active kink axis 251
   fixed or passive kink axis 251
   growth triangle 251
   instantaneous rotation 251

progressive rotation 251
subsidence vs. uplift (G) 247
syntectonic sedimentation 247

incompressibility 218, 221, 231, 242
inverse problem 218, 272
  grid search method 272
  objective function 270
  optimization 272
  trishear inverse modeling 270

kinematic models 217
  concentration factor 242
  fault-bend folding 220
  fault-propagation folding 230
  fault-related folding 220
  fixed axis kink model 230, 231
  inclined simple shear 226
  parallel kink model 235
  similar folding 225
  trishear model 240
kinematic vorticity number 195, 196
kinematics 183, 199

Lagrangian frame 252
linear algebra 6
lineation 26
loops 7

magnitudes 1
map projections 18-22
  azimuthal 19
  conformal 20
  datum 19
  developable surface 19
  eastings 20
  equidistant 20
  false northings 22
  geoid 18
  latitude 18
  longitude 18
  NAD83 19
  northings 20
  UTM 20-22
  WGS84 19
MATLAB 6
matrix
  addition 70
  antisymmetric (skew) 72, 83
  asymmetric 83
  cofactors 74
  conformable 71
  design 156
  determinant 74, 79
  diagonal 70
  dyad product 71, 72
  identity 70, 78
  inverse 76-77, 78
  Kronecker delta 70, 105
  multiplication 71
  orientation 93

orthogonal 72, 78
principal diagonal 70
square 70
symmetric 72, 83
transpose 49, 50, 72, 78
Mohr circle 168
  3D stress 98
  finite strain 177
  infinitesimal strain 143
  pole 108, 177, 178
  stress 108-111
  tensor transformation 88

notation
  Gibbs dyadic 169
  indicial 27, 66-67, 82
  matrix 66, 69
  summation convention 67, 68

orientations 1, 8, 31
  azimuth format 12
  bipolar distribution 91
  dip direction 1, 12
  dip, apparent 1, 39
  dip, true 1, 39
  girdle distribution 91
  orientation matrix 93
  pitch 1, 12
  plunge 1
  quadrant format 12
  rake 1, 12, 39
  right-hand rule 12
  strike 1
  trend 1
orthographic projection 3-4, 51
  folding line 3

partial derivatives 123
pressure shadow 199, 200, 201, 205, 209

radians 6
rotation 2, 55-56
  axis, antisymmetric tensor 91
  internal 190, 200
  of axes 46-48

scalars 25, 81
seismic moment 150
seismic reflection 266
shear
  angular 140, 170
  antithetic 226
  engineering shear strain 141, 188
    rate 193
  general 192, 196
  inclined 225
  parallel 257
  pure 183, 186, 192, 196, 200
  shear strain 140, 171
  simple 188, 190, 192, 196, 200, 208
  tensor shear strain 141

spherical projection 8-18
spin 200, 208
statistics
  Bingham 93
  correlation coefficient 131
  covariance 131, 256, 267
  Fisher 37
  least squares 92
  standard deviation 129, 255
  variance 129, 256
stereonet 12-15, 62
  equal angle projection 16
  equal area projection 17, 20
  great circle 12, 58
  lower hemisphere 12, 26
  primitive 12
  rotations 14-15
  small circle 15, 58
  upper hemisphere 12
strain 2
  compatibility 179
  dilatation 142
  finite 165, 188
    elongation 170
    quadratic elongation 170
    stretch 170, 190
    volume ratio 171
  history
    coaxial 200
    cumulative incremental 184, 205, 208
    non-coaxial 200, 205
    progressive finite 186, 190, 196, 204, 205, 208
  infinitesimal 135, 136, 151
    axes 153
    ellipse 143, 190
    principal strains 146
    principal stretches 143, 184
    tensor 138, 143, 151, 178, 179
  invariants 141
  irrotational 186, 200
  plane strain 218
  principal stretches 184, 185
  rate 131, 193
  rotational 190
  volume ratio 171
  volume strain 142
stress 120
  biaxial 111
  Cauchy's Law 101, 114
  compression 104
  conjugate shear 100
  cylindrical 111
  deviatoric 112-113
  force 98
  hydrostatic 111, 112
  mean stress 112
  Mohr circle 108-111
    3D stress 109
  pole to 108
  normal 99, 105
  on arbitrary plane 113-116

  principal axes 104, 114
  principal stress ratio 116
  pure shear 111
  shear 99, 104
  spherical 111
  tension 104
  tensor 101
  traction 98, 114
  triaxial 111
  uniaxial 111
summation convention, Einstein 67, 68
  dummy suffix 68
  free suffix 68

tensor 45, 81
  antisymmetric 138, 158
  asymmetric moment 152
  Cauchy deformation 168, 176, 177, 185, 186
    gradient 128
  characteristic (secular) equation 90
  deformation gradient 125, 155, 176, 184, 185, 188, 190, 193, 195, 203, 205, 208
    rate 193
  displacement gradient 135, 146, 147, 150, 151, 155, 157, 168
  dyad (tensor) product 84
  eigenvalue 90, 91, 105, 185, 189, 205, 208
  eigenvector 90, 91, 105, 152, 185, 186, 188, 189, 200, 205, 208
  Eulerian finite strain 167
  Eulerian displacement gradient 128
  field tensor 104
  Green deformation 168, 189, 195, 205, 208
    gradient 128
  infinitesimal strain 138, 146, 151, 158
  invariants 90
  Lagrangian displacement gradient 128, 166, 167
  Lagrangian finite strain 167
  linear vector operator 84
  magnitude ellipsoid 89
  Mohr circle 88, 108-111
  principal axes 83
  representation quadric 89
  rotation 138, 146, 151, 168, 190
  rotation axis 91
  second order (rank) 82
  seismic moment 152
  stress tensor 101
  stretch, left 174, 175
  stretch, right 174, 190
  symmetric 138
  transformations 85-87, 113, 115
transformation
  Cauchy 123
  coordinate 44, 121, 165
  Green 123
  orthogonality relations 48, 77
  position vector 50
  Pythagorean Theorem 47

rotation of axes 46-48
tensor transformations 85-87, 113, 115
transformation matrix 48, 69, 72, 78, 84, 85, 108, 143
translation of axes 45
vector transformations 48-50, 67, 115
trigonometry
  plane 5
  spherical 5

uncertainty 2, 129, 254, 256, 266
  Cholesky or square root method 267
  correlation coefficient 131
  correlation length 266
  covariance 131, 256, 267
  Gaussian distribution 129
  in best-fit parameters 275
  Monte Carlo simulation 258
  randomized maximum likelihood 275
  realizations 267
  spherical variogram 266
  standard deviation 129, 255
  variance 129, 256

variables 7
vector 25-40, 81
  addition 33, 35
  axial 138
  base 29
  cross product 34, 39, 114
  direction cosines 28, 31
  displacement 123
  dot product 34, 39, 47, 71, 72
  dyad product 71, 72, 84
  Fisher statistics 37
  magnitude 27
  mean 34-36
  resultant 35
  scalar multiplication 33
  transformations 48-50, 67, 77, 115
  unit 27, 30
velocity
  divergence 218
  domains 220
  linear trishear field 242
  pure shear 219
  simple shear 219